Recent Progress in Plant Biochemistry and Molecular Biology

Recent Progress in Plant Biochemistry and Molecular Biology

Edited by **Harvey Parker**

SYRAWOOD
PUBLISHING HOUSE
New York

Published by Syrawood Publishing House,
750 Third Avenue, 9th Floor,
New York, NY 10017, USA
www.syrawoodpublishinghouse.com

Recent Progress in Plant Biochemistry and Molecular Biology
Edited by Harvey Parker

International Standard Book Number: 978-1-68286-287-2 (Hardback)

Printed in China.

Contents

Preface

This book has been a concerted effort by a group of academicians, researchers and scientists, who have contributed their research works for the realization of the book. This book has materialized in the wake of emerging advancements and innovations in this field. Therefore, the need of the hour was to compile all the required researches and disseminate the knowledge to a broad spectrum of people comprising of students, researchers and specialists of the field.

Plant biochemistry and molecular biology are significant sub-disciplines of botany. They are closely linked and help in understanding plant mechanisms. This book aims to delve into the advanced topics in these fields such as plant genetics, plant biotechnology, plant diversity, conservation biology, interactions between plants and other organisms, etc. This book brings together case studies from different parts of the world to keep the readers updated with the latest concepts in these fields. This book will prove beneficial to students and professionals alike.

At the end of the preface, I would like to thank the authors for their brilliant chapters and the publisher for guiding us all-through the making of the book till its final stage. Also, I would like to thank my family for providing the support and encouragement throughout my academic career and research projects.

Editor

1

Identification of a Bioactive Compound against Adult T-cell Leukaemia from Bitter Gourd Seeds

Hisahiro Kai [1,2,*], Ena Akamatsu [2], Eri Torii [2], Hiroko Kodama [2,3], Chizuko Yukizaki [3], Isao Akagi [2,4], Hisatoshi Ino [5], Yoichi Sakakibara [6], Masahito Suiko [6], Ikuo Yamamoto [7], Akihiko Okayama [7], Kazuhiro Morishita [8], Hiroaki Kataoka [9] and Koji Matsuno [1]

[1] Department of Pharmaceutical Health Sciences, School of Pharmaceutical Sciences, Kyushu University of Health and Welfare, 1714-1 Yoshino-machi, Nobeoka, Miyazaki 882-8508, Japan; E-Mail: kjmtsn@phoenix.ac.jp

[2] Research Promotion Bureau for Collaboration of Regional Entities, Miyazaki Prefectural Industrial Support Foundation, 16500-2 Higashi-Kaminaka, Sadowara-cho, Miyazaki, Miyazaki 880-0303, Japan; E-Mails: rojopino@mac.com (E.A.); ushio16@fc.miyazaki-u.ac.jp (E.T.); hiroyoko33@hotmail.com (H.K.); akagi046@chem.agri.kagoshima-u.ac.jp (I.A.)

[3] Miyazaki Prefectural Food Research and Development Center, 16500-2 Higashi-Kaminaka, Sadowara-cho, Miyazaki, Miyazaki 880-0303, Japan; E-Mail: yukizaki@iri.pref.miyazaki.jp

[4] Department of Biochemical Science and Technology, Faculty of Agriculture, Kagoshima University, 1-21-24 Korimoto, Kagoshima, Kagoshima 890-0065, Japan

[5] Miyazaki Agricultural Experiment Station, 5805 Shimonaka, Sadowara-cho, Miyazaki, Miyazaki 880-0212, Japan; E-Mail: ino-hisatoshi@pref.miyazaki.lg.jp

[6] Department of Biochemistry and Applied Biosciences, Faculty of Agriculture, University of Miyazaki, 1-1 Gakuenkibanadai-nishi, Miyazaki, Miyazaki 889-2192, Japan; E-Mails: ysakaki@cc.miyazaki-u.ac.jp (Y.S.); msuiko@cc.miyazaki-u.ac.jp (M.S.)

[7] Department of Rheumatology, Infectious Diseases and Laboratory Medicine, Faculty of Medicine, University of Miyazaki, 5200 Kihara Kiyotake, Miyazaki, Miyazaki 889-1692, Japan; E-Mails: yamamoto@fc.miyazaki-u.ac.jp (I.Y.); okayama@med.miyazaki-u.ac.jp (A.O.)

[8] Division of Tumor and Cellular Biochemistry, Department of Medical Sciences, Faculty of Medicine, University of Miyazaki, 5200 Kihara Kiyotake, Miyazaki, Miyazaki 889-1692, Japan; E-Mail: kmorishi@fc.miyazaki-u.ac.jp

[9] Section of Oncopathology and Regenerative Biology, Department of Pathology, Faculty of Medicine, University of Miyazaki, 5200 Kihara Kiyotake, Miyazaki, Miyazaki 889-1692, Japan; E-Mail: mejina@med.miyazaki-u.ac.jp

* Author to whom correspondence should be addressed; E-Mail: hkai@phoenix.ac.jp

Abstract: In our previous report, an 80% ethanol bitter gourd seed extract (BGSE) was found to suppress proliferation of adult T-cell leukemia (ATL) cell lines. The present study aimed to identify the bioactive compounds from BGSE specific against ATL. From the result of an HPLC-MS analysis, α-eleostearic acid (α-ESA) was present in BGSE at 0.68% ± 0.0022% (±SD, n = 5). In the cell proliferation test, α-ESA potently suppressed proliferation of two ATL cell lines (ED and Su9T01; IC_{50} = 8.9 and 29.3 μM, respectively) more than several other octadecanoic acids. However, α-ESA moderately inhibited phytohemagglutinin-activated human peripheral blood mononuclear cells (PBMC; IC_{50} = 31.0 μM). These results suggest that BGSE-derived α-ESA has potential as a functional food constituent because of its activity against ATL, particularly against ED cells. Moreover, α-ESA might be effective for the prevention of moderate adverse effects of ATL on normal T cells.

Keywords: adult T-cell leukemia; bitter gourd seed extract; α-eleostearic acid; phytohemagglutinin-activated human peripheral blood mononuclear cell

1. Introduction

Adult T-cell leukemia (ATL) occurs in a small population of human T-cell leukemia virus type I (HTLV-I) infected individuals. After transmission of HTLV-I, 2%–5% of carriers are likely to develop ATL after a long latency period (30–50 years) [1]. These patients have been frequently identified as being from a restricted area of tropical regions [2]. It is currently very difficult to effectively treat patients with ATL using existing therapeutic methods, and most clinical trials focus on chemotherapeutic treatment and allogeneic hematopoietic stem cell transplantation. Therefore, it is important to find appropriate therapeutic methods to prevent the development of ATL or to prolong survival after its occurrence.

In our previous report, we screened 52 agricultural plant samples for their ability to inhibit proliferation in seven kinds of ATL related cell lines to start structure of a study for finding potential drug candidates with the prevention of ATL. We found that an 80% ethanol bitter gourd (*Momordica charantia* L.) seed extract (BGSE) showed an inhibitory effect on the proliferation of ATL-related human leukemia cells [3].

Bitter gourd belongs to the Cucurbitaceae family and is cultivated worldwide as a vegetable crop. The fruit is not only used as a food, but also for its medicinal properties, such as anti-microbial, anti-diabetic, anti-HIV and anti-tumor activities, which were described in a recent review [4]. BGSE has also been reported to have anti-leukemic potential on human acute myelogenous leukemia cells (HL-60) [5]. However, HL-60 and ATL cell lines comprise different types of leukemia cells, and as there have been no reports about bioactive compounds from BGSE active against ATL, the effect of

BGSE on ATL cell proliferation requires elucidation. The aim of the present study was to identify bioactive compounds in BGSE exhibiting activity against ATL.

2. Results and Discussion

2.1. Identification of Active Compounds in BGSE

Figure 1a,b show the HPLC-DAD chromatogram (a) and HPLC-MS-total ion chromatogram (TIC) (b) of BGSE. As shown in Figure 1a, a major peak (peak 1: retention time (RT) = 8.570 min, λ = 270 nm) was detected for BGSE. The TIC showed several peaks (Figure 1b). Figure 1c shows the MS spectrum of the peak with the RT: 8.570 min. in Figure 1b, which shows a deprotonated molecular ion signal at *m/z* 277. This compound was estimated to be $C_{18}H_{29}O_2$ and was proposed to be α-ESA on the basis of reference data [6]. α-ESA was analyzed using the same method. Figure 1d–f shows the HPLC-DAD chromatogram (d), TIC (e) and MS spectrum (f) of α-ESA. α-ESA (50 μg/mL) showed a clear peak at 270 nm (Figure 1d, peak 2) and the same RT of peak 1 in Figure 1a (BGSE). The RT (8.570 min.) of the highest peak in TIC for α-ESA (Figure 1e, peak 2) was the same as peak 1 in Figure 1b. The MS spectrum of α-ESA (Figure 1f) showed the same result as the analysis of BGSE (Figure 1c). From these results, peak 1 in Figure 1 was determined to be α-ESA, and α-ESA is the main compound in BGSE.

We also attempted to quantitatively determine α-ESA in BGSE. Figure 2 shows the α-ESA calibration curve. The calibration curve showed good linearity (R^2 = 0.9977). α-ESA was contained in the concentration range of 34.0 and 34.2 μg/5 mg of freeze-dried bitter gourd seeds (0.68% ± 0.0022%, ±SD, *n* = 5). In our previous report, BGSE suppressed the proliferation of ATL cell lines [3]. The current study shows the presence of α-ESA in BGSE. Tsuzuki *et al.* reported that α-ESA accounted for about 60% of the total fatty acid composition of bitter gourd seed oil [7]. Therefore, α-ESA might have greatly contributed to suppressing the proliferation of ATL cell lines.

2.2. The Inhibitory Effects of Octadecanoic Acid Analogs on ATL Cell Lines

Of the octadecanoic acids, conjugated linoleic acids (CLA), which includes α-ESA, has been acknowledged to have numerous biological activities, such as anti-obesity, anti-diabetes, anti-cancer, anti-arthritis, anti-asthma, and anti-cardiovascular disease effects [8]. However, no study has yet reported the anti-leukemic effects of α-ESA on ATL cell lines. We examined the inhibitory effects of related compounds, seven octadecanoic acid groups, on the proliferation of two types of ATL cell lines (ED and Su9T01). As shown in Table 1, α-ESA (C18:3, n-5), γ-linolenic acid (C18:3, n-6), and α-linolenic acid (C18:3, n-3), which belong to the triunsaturated fatty acid group, substantially inhibited ED cell growth (IC_{50} values of 8.9, 61.3 and 129.9 μM, respectively) and Su9T01 cell growth (IC_{50} values of 29.3, 174.3 and 167.4 μM, respectively). Linoleic acid (C18:2, n-6), which belongs to the diunsaturated fatty acid group, inhibited ED and Su9T01 cell growth (IC_{50} values of 100.7 and 180.1 μM, respectively). In ED cells, these compounds exhibited higher activity than EGCG, which was used as the positive control [9] (IC_{50} = 152.7 μM). On the other hand, in Su9T01 cells, only α-ESA exhibited higher inhibitory activity than EGCG (IC_{50} = 166.0 μM). We cannot calculate the exactly IC_{50} values of monounsaturated fatty acid group (oleic acid (C18:1, n-9) and elaidic acid

(C18:1, n-9)) and saturated fatty acid (stearic acid (C18:0)) in both cell lines. This assay was employed to compare the effects of octadecanoic acids on ED and Su9T01 cell growth; α-ESA showed the highest inhibitory activity. IC_{50} values decreased roughly in proportion to the number of double bonds; therefore, the number of double bonds was an important determinant of anti-proliferation activity. Indeed, the triunsaturated fatty acid group was more potent than EGCG.

Figure 1. HPLC-DAD and MS chromatograms of BGSE and α-ESA. (**a**,**d**) HPLC-DAD chromatograms (270 nm) of BGSE (**a**) and α-ESA (**d**). (**b**,**e**) Total ion chromatograms (TIC) of BGSE (**b**) and α-ESA (**e**). (**c**) MS spectrum (negative-ion spectra) of peak 1 of BGSE in Figure 1b. (**f**) MS spectrum (negative-ion spectra) of peak 2 of α-ESA in Figure 1e.

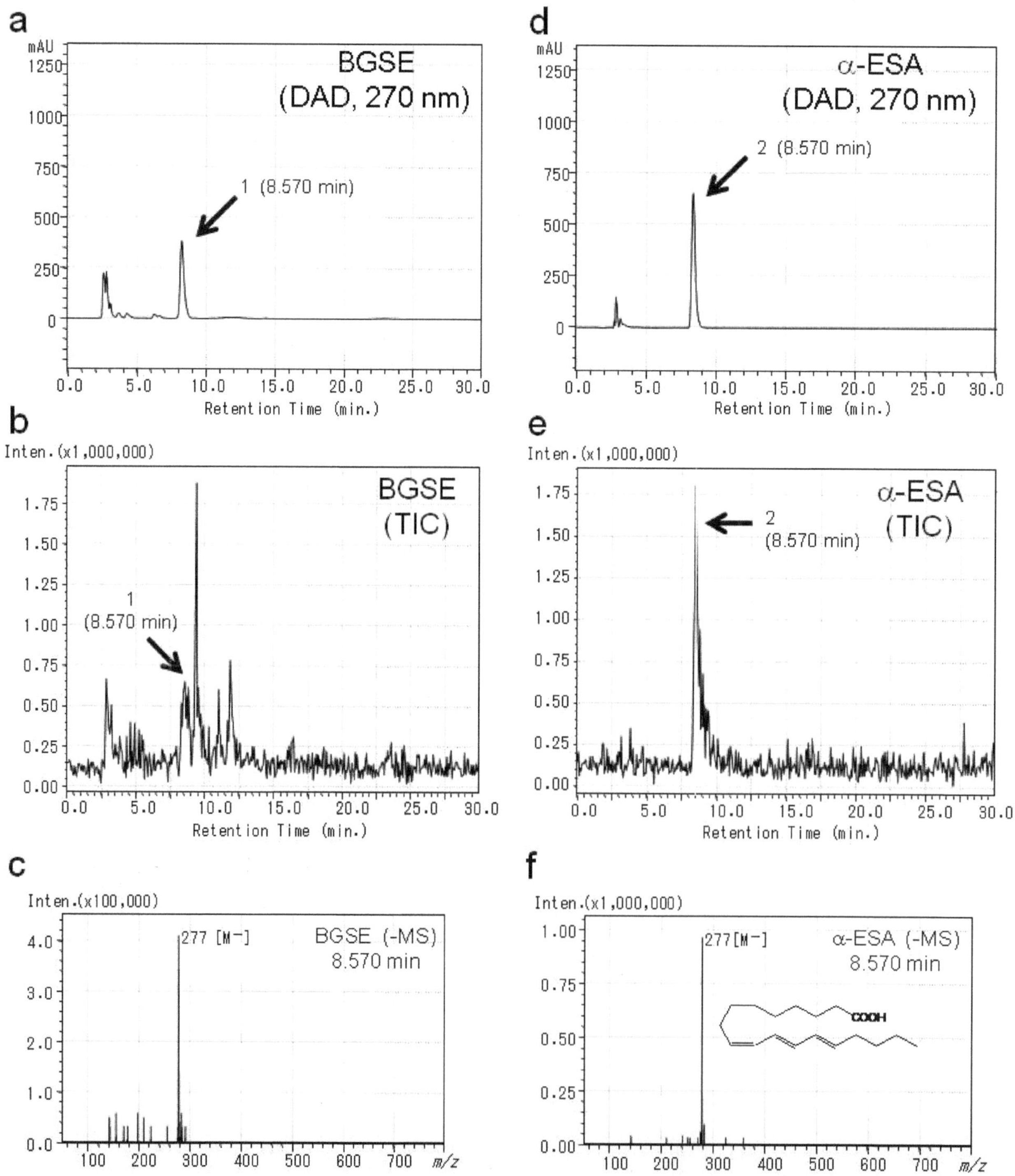

Figure 2. Calibration curve of α-ESA.

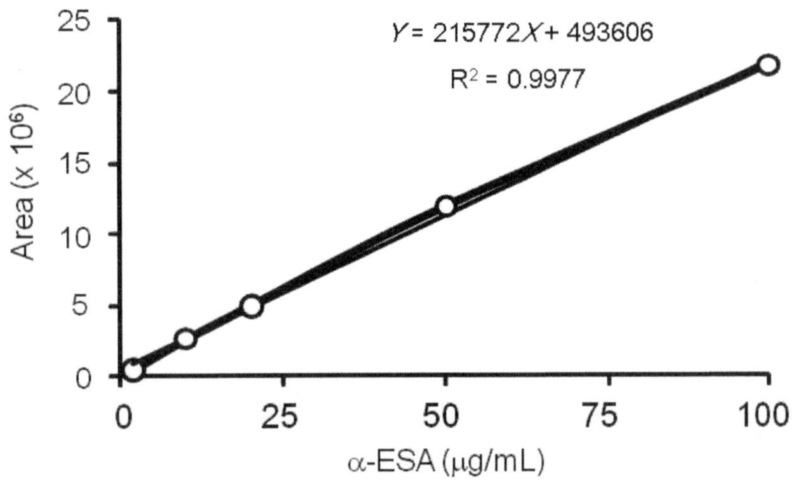

Table 1. Relationship between octadecanoic acid structure and inhibition of adult T-cell leukemia (ATL) cell line proliferation.

Compounds		IC_{50} (μM)	
		ED	**Su9T01**
α-ESA	(C18:3, n-5)	8.9	29.3
γ-linolenic acid	(C18:3, n-6)	61.3	174.3
α-linolenic acid	(C18:3, n-3)	129.9	167.4
linoleic acid	(C18:2, n-6)	100.7	180.1
oleic acid	(C18:1, n-9)	500.0–166.7	500.0–166.7
elaidic acid	(C18:1, n-9)	>500.0	>500.0
stearic acid	(C18:0)	500.0–166.7	>500.0
EGCG	(Positive Control)	152.7	166.0

ATL cells (ED and Su9T01) were incubated for 72 h in RPMI-1640 medium containing each compound. Viable cells were detected using a WST-8 assay kit. The concentration at which cell proliferation is inhibited by 50% compared to untreated control is expressed as IC_{50}.

2.3. The Effect of α-ESA on ATL Cell Line and Phytohemagglutinin-Activated Human Peripheral Blood Mononuclear Cell (PBMC) Proliferation

As shown in Figure 3, we compared the suppressive effect of α-ESA on ED, Su9T01 cells and PBMCs. PBMCs are commonly used as the healthy/normal cell model in comparison to cancer cell lines. Significant differences were observed between ED cells and PBMCs treated at 6, 19 and 56 μM α-ESA ($p < 0.05$). Su9T01 cells and PBMCs treated with between 1 and 500 μM α-ESA were not significantly different. These results confirmed that the α-ESA strongly and selectively inhibited ED cells and moderately inhibited PBMCs, which are healthy normal T cells. α-ESA at 166 and 500 μM significantly decreased proliferation of all cell types. α-ESA showed inhibitory effects on ED, Su9T01 cells and PBMCs (IC_{50} of 8.9, 29.3 and 31.0 μM, respectively).

Figure 3. Effect of α-ESA on ATL cell and PBMC proliferation.

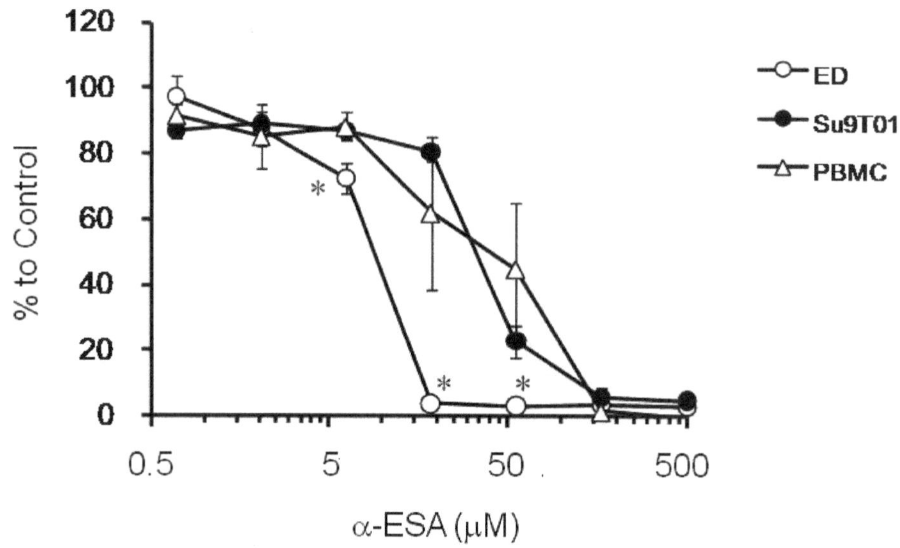

A comparison of the dose-dependent effects of α-ESA on ED, Su9T01 cells and PBMCs. Each mark represents the mean ± SD of three independent tests (Student's *t*-test). Significantly different between ED cells and PBMCs: $p < 0.05$ (*).

Tsuzuki *et al.* and Kobori *et al.* reported that bitter gourd seed oil and its constituent α-ESA had anti-leukemic potential in HL-60 cells [5,10]. While HL-60 and ATL (ED and Su9T01) are leukemia cell lines, they differ in origin and whether they are a result of viral infection. The antiproliferative properties of unsaturated fatty acids are well known. For example, Wendel *et al.* reported that the unsaturated fatty acids (mainly omega-3 fatty acids and derivatives) like conjugated eicosapentaenoic acid as important nutritional adjuvant therapeutics in the management of various human cancer diseases and the impact of nutritional omega-3 fatty acids on cancer prevention [11]. α-ESA also may have similar potential with nutritional function for cancer prevention. The present study is the first to show the inhibitory effects of α-ESA on ATL cells *in vitro*. Specifically, α-ESA showed an inhibitory effect in the rank order: ED cells > Su9T01 cells ≥ PBMCs. Sasaki *et al.* reported that tumor suppressor in lung cancer 1 (*TSLC1*) gene expression was different between ED and Su9T01 cells [12,13]. Therefore, using gene expression analysis, future studies should investigate the regulatory mechanism of *TSLC1* and its DNA methylation, as well as the possible role of α-ESA in the inhibition of ED and Su9T01 proliferation and *TSLC1* expression.

3. Experimental

3.1. Chemicals

α-Eleostearic acid (α-ESA) was obtained from Larodan Fine Chemicals AB, Malmö, Sweden and Cayman Chemical, Ann Arbor, MI, USA. γ-Linolenic acid, linoleic acid, α-linolenic acid, and elaidic acid were purchased from Cayman Chemical. Stearic acid and oleic acid were purchased from Wako, Osaka, Japan. Epigallocatechin-3-gallate (EGCG) was purchased from Nagara Science Co., Gifu, Japan. Ficoll was purchased from GE Healthcare, Uppsala, Sweden. Phytohemagglutinin (M Form) was purchased from Invitrogen, Carlsbad, CA, USA. IL-2 was purchased from R&D Systems,

Minneapolis, MN, USA. A 2-(2-methoxy-4-nitrophenyl)-3-(4-nitrophenyl)-5-(2,4-disulfophenyl)-2*H*-tetrazolium monosodium salt (WST-8) assay kit was purchased from Dojindo, Kumamoto, Japan.

3.2. Identification of Compounds in BGSE

The freeze-dried powder of bitter gourd seeds (5 mg) was extracted with 80% EtOH (0.5 mL) by vortexing for 30 s, followed by centrifugation at 1,500 rpm for 3 min. The supernatant was used for high performance liquid chromatography-diode array detector (HPLC-DAD) and mass spectrometry (MS) analysis. The α-ESA HPLC analysis method was a modification of the methods of Amakura *et al.* and Řezanka *et al.* [6,14]. The HPLC-DAD and MS analysis consisted of a Shimadzu HPLC System (LC-20A Prominence, Shimadzu, Kyoto, Japan) coupled to a SPD-20A (DAD; Shimadzu, Kyoto, Japan) and an LC/MS-ion trap-time of flight (LC/MS-IT-TOF, Shimadzu, Kyoto, Japan) fitted with an atmospheric pressure chemical ionization (APCI) source. HPLC separation was performed on a reverse-phase column (Atlantis T3, 2.1 mm I.D. ϕ100 mm, 3 μm; Waters, Milford, MA, USA). The column was maintained at 40 °C. The mobile phase consisted of eluent A (0.1% acetic acid and MeOH)/eluent B (0.1% acetic acid and 10% MeOH aq.) = 90:10 at a flow rate of 0.10 mL/min. The injection volume was 10 μL. APCI conditions were recorded from m/z = 50 to 400 in negative ion mode. The other MS conditions were as follows: nebulizer N_2 gas, 2.5 L/min; APCI interface temperature, 400.0 °C; curved desolvation line (CDL) temperature, 250.0 °C; heat block temperature, 200.0 °C; detector voltage, 1.80 kV.

3.3. α-ESA Calibration Curve

The α-ESA standard was dissolved in 80% ethanol and serial dilutions were analyzed by HPLC-DAD. α-ESA content was calculated using the following linear equation based on the calibration curve: $Y = 215772X + 493606$, $R^2 = 0.9977$. Y is the area detected by DAD (270 nm), and X is the α-ESA content in μg/mL.

3.4. ATL Cell Proliferation Assay

We used two ATL cell lines (ED and Su9T01) that are highly sensitive to inhibition of cell proliferation, as determined in our previous study [3]. ED cells were kindly provided by Dr. M. Maeda (Kyoto University, Kyoto, Japan) and Su9T01 cells were kindly provided by Dr. N. Arima (Kagoshima University, Kagoshima, Japan). The test compounds were dissolved in dimethyl sulfoxide and subjected to assay screening. The method of ATL assay is described in a previous report [3]. IC_{50} calculation was some curve fitted onto the determined proliferation inhibition points.

3.5. Isolation and Culture of PBMCs

The method of isolation and culture of PBMCs is as follows. Heparinised blood (5 mL) was diluted by adding 5 mL of PBS. The diluted blood samples were divided into four equal parts, loaded on 4 mL of Ficoll and centrifuged at 400 × *g* for 30 min. The PBMC layer was located within the interphase between the Ficoll and plasma. The Ficoll contained the erythrocytes and most of the granulocytes. The plasma was removed using a pipette until ~5 mL above the PBMC interphase. The cells were washed three times with PBS (centrifuged at 200 × *g* for 15 min) and resuspended in RPMI 1640 medium supplemented with 10% foetal bovine serum containing 100 U/mL penicillin G, 100 μg/mL streptomycin, 2 ng/mL IL-2 and 128-fold dilution of phytohemagglutinin to a final cell density of 1×10^6 cells/mL. The PBMC proliferation assay was conducted using the same method as for the ATL proliferation assay [3].

3.6. Statistics

Each experiment was conducted at least three times. All data are expressed as the mean ± standard deviation (SD) of three independent experiments. Statistically significant differences were calculated by Student's *t*-test.

4. Conclusions

α-ESA was shown to be the main bioactive compound in BGSE, and contributes to the inhibition of ED cell differentiation and proliferation without damaging normal cells, leading to the disruption of ATL pathogenesis.

Acknowledgments

We thank Michiyuki Maeda (Kyoto University) and Naomichi Arima (Kagoshima University) for supplying the cell lines. We also thank Yuuki Maeda and Naomi Makisumi for their excellent technical assistance. This work was supported by a Grant-in-Aid from the Collaboration of Regional Entities for the Advancement of Technological Excellence (CREATE) from the Japanese Science and Technology Agency.

Conflicts of Interest

The authors declare no conflict of interest.

References

1. Arisawa, K.; Soda, M.; Endo, S.; Kurokawa, K.; Katamine, S.; Shimokawa, I.; Koba, T.; Takahashi, T.; Saito, H.; Doi, H.; *et al.* Evaluation of adult T-cell leukemia/lymphoma incidence and its impact on non-Hodgkin lymphoma incidence in southwestern Japan. *Int. J. Cancer* **2000**, *85*, 319–324.
2. Proietti, F.A.; Carneiro-Proietti, A.B.; Catalan-Soares, B.C.; Murphy, E.L. Global epidemiology of HTLV-I infection and associated diseases. *Oncogene* **2005**, *24*, 6058–6068.
3. Kai, H.; Akamatsu, E.; Torii, E.; Kodama, H.; Yukizaki, C.; Sakakibara, Y.; Suiko, M.; Morishita, K.; Kataoka, H.; Matsuno, K. Inhibition of proliferation by agricultural plant extracts in seven human adult T-cell leukaemia (ATL)-related cell lines. *J. Nat. Med.* **2011**, *65*, 651–655.
4. Fang, E.F.; Ng, T.B. Bitter gourd (*Momordica charantia*) is a cornucopia of health: A review of its credited antidiabetic, anti-HIV, and antitumor properties. *Curr. Mol. Med.* **2011**, *11*, 417–436.

5. Soundararajan, R.; Prabha, P.; Rai, U.; Dixit, A. Antileukemic Potential of *Momordica charantia* Seed Extracts on Human Myeloid Leukemic HL60 Cells. *J. Evid.-Based Complement. Alternat. Med.* **2012**, doi:10.1155/2012/732404.
6. Amakura, Y.; Kondo, K.; Akiyama, H.; Ito, H.; Hatano, T.; Yoshida, T.; Maitani, T. Characteristic long-chain fatty acid of *Pleurocybella porrigens*. *Shokuhin Eiseigaku Zasshi* **2006**, *47*, 178–181.
7. Tsuzuki, T.; Tokuyama, Y.; Igarashi, M.; Miyazawa, T. Tumor growth suppression by alpha-eleostearic acid, a linolenic acid isomer with a conjugated triene system, via lipid peroxidation. *Carcinogenesis* **2004**, *25*, 1417–1425.
8. Ruiz, R.A.; Reglero, G.; Ibañez, E. Recent trends in the advanced analysis of bioactive fatty acids. *J. Pharm. Biomed. Anal.* **2012**, *51*, 305–326.
9. Li, H.C.; Yashiki, S.; Sonoda, J.; Lou, H.; Ghosh, S.K.; Byrnes, J.J.; Lema, C.; Fujiyoshi, T.; Karasuyama, M.; Sonoda, S. Green tea polyphenols induce apotosis *in vitro* in peripheral blood T lymphocytes of adult T-cell leukemia patients. *Jpn. J. Cancer Res.* **2000**, *91*, 34–40.
10. Kobori, M.; Ohnishi-Kameyama, M.; Akimoto, Y.; Yukizaki, C.; Yoshida, M. α-eleostearic acid and its dihydroxy derivative are major apoptosis-inducing components of bitter gourd. *J. Agric. Food Chem.* **2008**, *56*, 10515–10520.
11. Wendel, M.; Heller, A.R. Anticancer actions of omega-3 fatty acids—Current state and future perspectives. *Anticancer Agents Med. Chem.* **2009**, *9*, 457–470.
12. Sasaki, H.; Nishikata, I.; Shiraga, T.; Akamatsu, E.; Fukami, T.; Hidaka, T.; Kubuki, Y.; Okayama, A.; Hamada, K.; Okabe, H.; *et al.* Overexpression of a cell adhesion molecule, TSLC1, as a possible molecular marker for acute-type adult T-cell leukemia. *Blood* **2005**, *105*, 1204–1213.
13. Watanabe, M.; Nakahata, S.; Hamasaki, M.; Saito, Y.; Kawano, Y.; Hidaka, T.; Yamashita, K.; Umeki, K.; Taki, T.; Taniwaki, M.; *et al.* Downregulation of CDKN1A in adult T-cell leukemia/lymphoma despite overexpression of CDKN1A in human T-lymphotropic virus 1-infected cell lines. *J. Virol.* **2010**, *84*, 6966–6977.
14. Řezanka, T.; Votruba, J. Analysis of Fatty Acids by APCI-MS. In *Modern Methods for Lipid Analysis by Liquid Chromatography: Mass Spectrometry and Related Techniques*; William, C.B., Ed.; AOCS Press: Urbana, IL, USA, 2005; pp. 242–275.

2

Annexin-Mediated Calcium Signalling in Plants

Julia M. Davies

Department of Plant Sciences, University of Cambridge, Cambridge CB2 3EA, UK;
E-Mail: jmd32@cam.ac.uk

Abstract: Calcium-permeable channels underpin elevations of free calcium that encode specific signals in stress adaptation, development and immunity. Identifying the genes encoding these channels remains a central goal of plant signalling research. Evidence now suggests that members of the plant annexin family function as unconventional calcium-permeable channels, with roles in development and stress signalling. Arabidopsis annexin 1 mediates a plasma membrane calcium-permeable conductance in roots that is activated by reactive oxygen species. Recombinant annexin 1 forms a very similar conductance in planar lipid bilayers, indicating that this protein could facilitate the *in vivo* conductance directly. The annexin 1 mutant is impaired in salinity-induced calcium signalling. Protein–protein interactions, post-translational modification and dynamic association with membranes could all influence annexin-mediated calcium signalling and are reviewed here. The prospect of annexins playing roles in calcium signalling events in symbiosis and immunity are considered.

Keywords: annexin; *Arabidopsis*; calcium; channel; *Medicago*; signal; stress

---

1. Introduction

Calcium influx to the cytosol (from the apoplast or from organellar stores) is central to elevation of cytosolic free Ca^{2+} ($[Ca^{2+}]_{cyt}$) as a second messenger in developmental, stress and immune signalling [1]. Elevation of free Ca^{2+} within some organelles could also have a signalling role. Gradually, the Ca^{2+}-permeable channels involved in the stimulus-specific, transient free Ca^{2+} elevations or oscillations are being identified (reviewed by [2,3]). At the vacuole of *Arabidopsis thaliana* and rice, TPC1 (Two Pore Channel 1) would be capable of releasing Ca^{2+} to the cytosol [4], although modelling studies suggest that it would not be a component of guard cell Ca^{2+} signalling for aperture control [5]. At the plasma membrane, members of the Glutamate Receptor-Like (GLR) family of ion channel sub-units mediate Ca^{2+} influx into pollen tubes and root cells [6,7]. In pollen tubes, they co-reside with members of the Cyclic Nucleotide-Gated Channel (CNGC) family [8]. AtCNGC2 is the best studied of the family and lies downstream of specific receptors for defence responses [9,10]. All of these channels are "conventional" in that each gene encodes an integral, trans-membrane spanning subunit with a pore-forming loop that is targeted to a specific membrane and is most likely united with other subunits to form a functional channel.

Recent studies have shown that some Ca^{2+} influx pathways may not be formed by conventional channels. There is scope for passive Ca^{2+} transport mediated by annexins [11,12]. These small amphipathic proteins are distributed throughout cells (reviewed by [13]) and can be transported within the plant via the phloem [14]. Expression can be regulated by Ca^{2+} [15]. There is now strong evidence for plant annexins' forming Ca^{2+}-permeable transport pathways across bilayers *in vitro*. Results from an annexin loss of function mutant are consistent with an *in vivo* Ca^{2+} transport function and further studies are now needed to establish how annexins could directly mediate Ca^{2+} transport in native membranes. It is possible that annexins could be recruited directly to membranes, independently of vesicle delivery, to operate in stimulus-specific signalling. This short review will introduce this family of Ca^{2+}-binding proteins and address what is known about their role in Ca^{2+} signalling in plants.

2. Ubiquitous Annexins

Genome studies have revealed that higher plants harbour multi-gene annexin families; eight in *Arabidopsis*, ten in rice, and twenty-three in soybean [16,17]. Excellent reviews by Clark *et al.* [16] and Jami *et al.* [18] address their phylogeny and evolution. These small (32 to 42 kDa) proteins are expressed throughout the higher plant body, with expression varying with development and environmental conditions, including light, water availability, temperature, salinity, acid rain, gravity, metal stress, mechanical stress, presence of microbes and nutrient deprivation [13,19–26]. Transcriptional regulators are now being identified, such as MYB98 and UPB1 in *Arabidopsis thaliana* [27,28]. Analyses of expression and protein abundance have revealed widespread distribution of annexins through the plant, with greater abundance at growth points such as root hair apices (reviewed by [13,16]). Distribution through the plant via the phloem also appears likely [14]. Unlike conventional transport proteins, an annexin can ostensibly exist in the cytosol or extracellular matrix, in addition to being membrane associated or inserted. The clearest example is *Arabidopsis* annexin 1 (AtANN1; At1g35720). In addition to a predominant presence in the cytosol, proteomic, immunolocalisation, radiolabelling and GFP studies have identified AtANN1 at the plasma membrane (as an integral protein), ER, vacuole, mitochondria, chloroplast, in phloem exudate and cell wall [14,29,30]. How an annexin becomes extracellular is unknown but the AtANN1 sequence is consistent with its being a non-classical secreted protein [31] and it also harbours a diacidic motif that should target it to the plasma membrane [13]. By analogy with animal annexins, export could also be via exocytosis or ABC transporters (reviewed by [13]). Extracellular animal annexin function can be through receptor binding [32] but this remains to be explored for plant annexins.

The mechanistic basis of membrane association is far better understood but still lags behind that of animal annexins. Plant annexins contain up to four "annexin repeats" that would facilitate reversible

Ca^{2+}-dependent binding to negatively charged phospholipid head groups [33,34]. Half maximal binding requires nanomolar to millimolar Ca^{2+} (reviewed by [13]). Membrane binding may also involve the *N*-terminus. Ca^{2+}-independent binding to lipids is possible at neutral and acidic pH [31,35,36]. In common with animal annexins, plant annexins can fully or partially insert into membranes, with clear examples coming from wheat and *Arabidopsis* [35,37]. Some animal annexins support Ca^{2+} channel-like behaviour *in vitro* by inserting into or associating with the bilayer, with transport activity regulated by ATP, GTP, peroxide, pH and voltage (reviewed by [13,38]. Given the structural similarity of plant annexins to their animal counterparts, including the conservation of salt bridges implicated in channel selectivity and regulation [13,31,38], it is reasonable to hypothesise that plant annexins capable of membrane association or insertion could act as Ca^{2+}-permeable channels *in vivo* and so have a role in Ca^{2+} signalling.

3. Ca^{2+} Transport by Plant Annexins

The first indication that plant annexins could form Ca^{2+}-permeable transport pathways came from the incorporation of (recombinant) *Capsicum annuum* CaANN24 into vesicles containing a Ca^{2+} indicator dye [33]. Since then, purified native *Zea mays* annexins ZmANN33/35 were found to increase $[Ca^{2+}]_{cyt}$ when added to the extracellular membrane face of *Arabidopsis* root protoplasts as a bioassay [31]. This indicated that extracellular annexins could somehow modulate $[Ca^{2+}]_{cyt}$ but whether this was by directly forming a Ca^{2+} influx pathway or through activation of other channels was not determined [31]. More tellingly, ZmANN33/35 formed a Ca^{2+}- and K^+-permeable conductance when added to the equivalent of the cytosolic face of a planar lipid bilayer, designed to act as a plasma membrane mimetic [31]. This conductance was blocked by the cation channel blocker Gd^{3+} present at the equivalent extracellular face of the bilayer, indicating that a trans-bilayer conductance had been formed by the annexins [31]. Incubating the annexins with their cognate antibody prevented the formation of the conductance. The incorporation of malondialdehyde (MDA) into the planar lipid bilayer to mimic lipid peroxidation caused a profound change in the way the ZmANN33/35 conductance was regulated by voltage. MDA forms in membranes during stress responses known to involve reactive oxygen species (ROS) and $[Ca^{2+}]_{cyt}$ elevation [39]. In control conditions the annexin-mediated conductance increased in a linear fashion as voltage became more negative (hyperpolarised). In contrast, MDA restricted annexin Ca^{2+} transport activity to more hyperpolarised voltages and this could in turn have implications for a resultant $[Ca^{2+}]_{cyt}$ signal *in vivo* [31,39]. The mechanism for this change in voltage sensitivity is also unknown but the channel-forming animal annexin A5 can bind to MDA and is implicated in Ca^{2+} influx across the plasma membrane in response to hydrogen peroxide [40,41]. Perhaps binding of one or both of the *Zea* annexins to MDA effected voltage regulation.

Work on ZmANN33/35 in bilayers established that the conductance was selective for K^+ over Ca^{2+} [31]. Single channel behaviour (discrete, step changes in current translocated across the bilayer) has rarely been observed but a single channel conductance of 17 pS was recorded in MDA-containing bilayers [38]. Recombinant *Medicago truncatula* Annexin 1 (MtANN1) has recently been shown to support single channel activity in planar lipid bilayers [42]. When transporting K^+, three different single channel conductances were observed; 16pS, 135 pS and 329 pS depending on the amount of annexin present. This is similar to channel behaviour shown by animal annexins in bilayers, where amount of protein is linked to level of oligomerization, association with the bilayer and channel characteristics such as conductance and voltage dependence [13,29,38,42,43]. Testing for Ca^{2+} transport by MtANN1 is now feasible.

Further progress has been made on the transport activity and function of the predominantly abundant annexin of Arabidopsis, AtANN1. Recombinant AtANN1 was first reported to form a K^+ conductance in planar lipid bilayers by Gorecka *et al.* [36], with activity promoted by acidic pH. Exposure to Ca^{2+} was found to prevent transport activity unless the planar lipid bilayer was itself exposed to copper and ascorbate to generate hydroxyl radicals [11]. In these experiments, AtANN1 was present at the cytosolic face of a plasma membrane mimetic bilayer and hydroxyl radicals ($OH^\bullet$) were generated at the extracellular face. $OH^\bullet$ are the most potent and short-lived of the ROS. Why Ca^{2+} was inhibitory to AtANN1 transport function and how $OH^\bullet$ overcame this now need to be elucidated. In common with ZmANN33/35, Gd^{3+} at the "extracellular" bilayer face blocked the AtANN1-mediated conductance thus indicating that a trans-bilayer transport pathway had been formed and conductance formation was prevented by incubation with anti-AtANN1 antibody [11]. Analysis of ionic selectivity has revealed that although only modestly permeable to Ca^{2+}, the $OH^\bullet$-activated AtANN1 conductance discriminates between K^+ and Na^+ very well; the Ca^{2+}:K^+ selectivity ratio is 0.64, Ca^{2+}:Na^+ is 11 and the K^+:Na^+ is 18 [11,12].

Plant annexins have been found to have *in vitro* ATPase and GTPase activity (reviewed by [13,44]. AtANN1 has been identified *in vitro* as an ATP-binding protein [45]. This has led to the proposal that it may be involved in $[Ca^{2+}]_{cyt}$ elevations caused by extracellular ATP [13,46]. Both extracellular ATP and ADP cause transient elevation of $[Ca^{2+}]_{cyt}$ in plant cells and help regulate growth, stress and immune responses [47–49]. However, plant genomes do not contain the equivalent genes encoding animal ATP/ADP receptors [49]. The capacity of AtANN1 to be extracellular, bind ATP and form a Ca^{2+} transport route makes it a candidate for the plant's functional equivalent of those receptors [13,46]. Although there are still no reports on AtANN1, recently the Na^+ transport activity of recombinant MtANN1 in planar lipid bilayers was reported to be promoted by ATP (data not shown in [42]).

While bilayer studies have clearly shown the capacity of annexins to translocate Ca^{2+} and K^+, few studies have addressed *in vivo* transport function. As, in common with animal annexins, plant annexins are firmly implicated in exocytosis [50], analysis of loss of function mutants may not yield clear-cut results. The absence of a conductance could be due to a failure in annexin-mediated exocytotic delivery of a channel subunit to a membrane rather than a failure in annexin-mediated ion transport itself. In fairness, there are caveats also to the interpretation of conventional channel mutants; for example, loss of transport function could be due to a pleiotropic effect of the mutation. The safeguard is to examine the protein's transport activity *in vitro*, without the confounding effects of a cellular expression system, whether native or heterologous. Does the *in vitro* transport match that of the membrane?

4. *In Vivo* Activity and Functions of AtANN1

The root epidermal and root hair apical plasma membrane of *Arabidopsis* contain a hyperpolarisation-activated Ca^{2+}-permeable channel conductance that is activated by extracellular hydroxyl radicals ($OH^•$) and is involved in growth [11,51,52]. Using patch clamp electrophysiology, an *Atann1* knockout mutant was found to lack this Ca^{2+} conductance in both epidermal and root hair apical plasma membrane, with activity restored by complementation. The transport characteristics of the $OH^•$-activated Ca^{2+} conductance generated by recombinant AtANN1 in planar lipid bilayers agree well with that of the native membrane, strongly supporting AtANN1's direct formation of this transport pathway. Root epidermal protoplasts from the mutant were significantly impaired in their ability to elevate $[Ca^{2+}]_{cyt}$ in response to extracellular $OH^•$, consistent with a Ca^{2+} transport function for AtANN1 [11]. Mutant roots and root hairs were significantly shorter than wild type, consistent with impaired Ca^{2+} uptake [11].

Recent work has shown that AtANN1 is involved in root $[Ca^{2+}]_{cyt}$ elevation in response to hydrogen peroxide, using aequorin as a $[Ca^{2+}]_{cyt}$ reporter [53]. H_2O_2 evokes a markedly different $[Ca^{2+}]_{cyt}$ response to extracellular $OH^•$ both in whole roots and root epidermal protoplasts, indicating that different ROS can generate specific $[Ca^{2+}]_{cyt}$ signals. In addition to having a role in H_2O_2-induced Ca^{2+} influx, AtANN1 was implicated in mediating Ca^{2+} release from intracellular stores when plasma membrane influx was blocked by Gd^{3+}. Although this may not be physiologically relevant, it helps make some sense of AtANN1's reported associations with endomembranes and shows that it can respond to perturbation of $[Ca^{2+}]_{cyt}$ homeostasis. With no block of influx by Gd^{3+}, loss of AtANN1 function perturbed and diminished the H_2O_2-induced $[Ca^{2+}]_{cyt}$ signal leading to impaired transcription of *Glutathione-S-Transferase1 Tau 1* (*GST1*; [53]). This upregulation of *AtGST1* by H_2O_2 was shown previously to be dependent on Ca^{2+} influx [54]. The mode of AtANN1's action in this system is unknown but H_2O_2 causes AtANN1 to dimerise *in vitro* [55].

As an $OH^•$-activated plasma membrane Ca^{2+} conductance, AtANN1 is expected to operate downstream of plasma membrane NADPH oxidases, the activity of which can ultimately source extracellular $OH^•$ [52,56]. Specifically, AtANN1 is likely to operate with the NADPH oxidase encoded by AtRBOHC (Respiratory Burst Oxidase Homologue C), which operates in root growth and salinity stress signalling [52,57]. The ROS sourced by AtRBOHC stabilise transcript for the plasma membrane Na^+/H^+ antiporter, SOS1 that is fundamental to resisting salinity stress and this stabilisation also requires Ca^{2+} influx [57]. The *Atann1* loss of function mutant fails to activate the root epidermal plasma membrane Ca^{2+} influx conductance in response to salinity stress (Figure 1) [12].

Moreover, AtANN1 underpins the salinity induced $[Ca^{2+}]_{cyt}$ elevation in root epidermal protoplasts that requires oxidation, consistent with its acting as an ROS-activated Ca^{2+} influx conductance [12]. In this respect, the low Na^+ permeability of recombinant AtANN1 in planar lipid bilayers makes biological sense because it could act downstream of AtRBOHC to amplify the $[Ca^{2+}]_{cyt}$ signal without exposing the root to further Na^+ influx. There is a profound NaCl-induced recruitment of AtANN1 to membranes [57] and while some of this may be for Ca^{2+} signalling, it may be an attempt by the plant to inhibit Na^+ ingress and K^+ loss. Loss of AtANN1 function results in significantly increased Na^+ influx and K^+ efflux from roots [12]. How AtANN1 functions as a negative regulator of these transport reactions and the identities of these important transport proteins now need to be determined. Downstream of the impaired $[Ca^{2+}]_{cyt}$ signal in the *Atann1* mutant was a failure to increase *AtSOS1* transcription and a significant reduction in the production of secondary roots in response to salinity stress [12].

Figure 1. Functional NADPH oxidase/annexin unit in calcium signalling. Studies with *Arabidopsis* suggest that annexin 1 (AtANN1) can function downstream of a plasma membrane NADPH oxidase in salinity stress. In roots, Na^+ entry across the plasma membrane (PM) causes elevation of cytosolic free Ca^{2+} which could activate NADPH oxidase through the latter's EF hands (helix-loop-helix structural domains usually involved in Ca^{2+} binding). This would result in extracellular ROS production that would activate AtANN1-mediated Ca^{2+} influx to promote SOS1 transcription and secondary root formation [11,12,58].

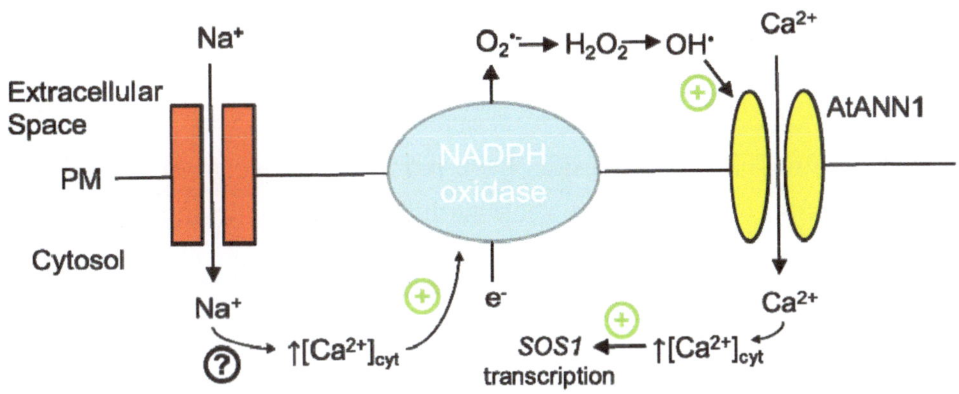

5. Regulation of Annexin Positioning

Positioning within the cell or extracellularly will most certainly prove to be of fundamental importance to annexin function in Ca^{2+} signalling, whether acting as a transport pathway or not. For animal annexins, contact with membranes is regulated by various factors including pH, voltage, $[Ca^{2+}]_{cyt}$, membrane curvature and specificity of binding to phospholipid headgroups (reviewed by [13]). Cold, salinity and mechanical stress cause repositioning of plant annexins, often to membranes [35,57,59,60] while infection by *Phytophthora* causes secretion of a potato annexin [21].

Recently, AtANN1 has been detected in a detergent resistant plasma membrane "nanodomain" involved in mesophyll ABA signalling [61]. Its role there is unknown and while it could directly contribute to ABA-induced $[Ca^{2+}]_{cyt}$ signalling as a transporter, work on animal annexins has shown involvement in formation of such domains [62]. MtANN2 has also been recovered from a detergent resistant plasma membrane domain, where it resided with an NADPH oxidase [63]. Detergent-resistant domains hold NADPH oxidases at the apex of pollen tubes [64] and this may also be the case for root hairs. The co-localisation of NADPH oxidases with annexins as ROS-activated Ca^{2+} transporters would afford tight spatial specificity of $[Ca^{2+}]_{cyt}$ elevation in polar growth of pollen tubes and root hairs. Lipid composition of the extracellular face of the plasma membrane could recruit extracellular annexins to act in $[Ca^{2+}]_{cyt}$ signalling. During plant cell death, phosphatidylserine everts to the extracellular leaflet of the plasma membrane [65,66]. The ability of extracellular ZmANN33/35 to elevate $[Ca^{2+}]_{cyt}$ has been hypothesised to be involved in cell death, which would be consistent with the ability of annexins to bind phosphatidylserine [31].

6. Interacting Proteins

Interaction with other proteins has been reviewed by [13] but the bearing of such interactions on Ca^{2+}-mediated signalling remains to be determined. Annexin-annexin interaction could influence a signalling function. Peroxide as a key component of ROS signalling, causes *in vitro* oligomerisation of ZmANN33/35 [44] and dimerisation of AtANN1 [55]. AtANN1 can interact with AtANN4, possibly to function in drought stress, but the impact of this on AtANN1's role in $[Ca^{2+}]_{cyt}$ signalling remains to be tested [67]. Recently, sea cotton (*Gossypium barbadense*) annexins 5 and 6 have been shown to form homodimers and heterodimers [68]. GbANN6 was found at the plasma membrane and nucleolus when heterologously expressed and was also found to bind F-actin. Tomato and mimosa annexins also bind actin while annexin expression is now clearly involved in cotton fibre elongation [68,69]. Expression of *GbANN6* in *Arabidopsis* increased root cell length, which was positively correlated with amount and position of F-actin [68]. These results have led to the proposal of GbANN6's acting as a scaffolding protein [68]. It will be interesting to see whether annexin-actin interaction has any bearing on actin regulation of the plasma membrane hyperpolarisation-activated Ca^{2+} channels involved in pollen viability and growth [69,70].

Interaction with C2 domain-containing proteins via an K/H/RGD motif has been proposed, which would implicate annexins in regulation of such signalling components as phospholipases [13,16,18,71]. *In vitro*, *Capsicum* annexin p35 inhibited porcine pancreatic phospholipase (PLA) A_2 [72]. A tomato annexin was recently found to interact with a Universal Stress Protein in drought responses [73]. Finally, data from Huang *et al.* [68] eliminated several CDPKs, CIPKs and a wall-associated kinase as interacting partners for GbANN6. Such studies will be of significant value in elucidating the cellular functions of annexins.

7. Post-Translational Modifications

Function in signalling may also be regulated by post-translational modification. Conservation of two cysteine residues affords both S-nitrosylation and S-glutathionylation of AtANN1, with S-glutathionylation triggered by ABA [74,75]. This modification would impede Ca^{2+}-mediated membrane association [75] and perhaps help terminate a Ca^{2+} signal. Participation of annexins in NO-regulated Ca^{2+} signalling now needs to be tested for.

Phosphorylation of plant annexins has been reported, with that of a *Brassica napus* annexin evident at the N terminal of the second annexin repeat [76]. Rice annexins interact with a MAPKK and Ste20-related protein kinase [77] while *Gossypium hirsutum* GhANN1 is phosphorylated by a plasma membrane-associated kinase [78]. AtANN1 can undergo phosphorylation by SnRK2s in ABA signalling [79] and, in common with annexins of other species, its transcript is upregulated by ABA [13,17,57,75]. Exactly how AtANN1's phosphorylation fits into drought or salinity-induced signalling is unknown but it would be useful to re-examine the transport characteristics of recombinant AtANN1 with phosphorylation as a variable. Phosphorylation enhances AtANN1's very weak *in vitro* peroxidase activity [55] but this activity could be due to contaminating proteins [16]. Exogenous H_2O_2 strongly suppresses *AtANN1* transcription in roots, most probably through the UPBEAT1 (AtUPB1) transcription factor [28,53]. However, H_2O_2 upregulates peroxidase activity [80] and so it appears that

AtANN1's main function in response to this ROS is as a component of Ca^{2+} transport rather than to de-toxify ROS [53].

8. Conclusions

With the capacity of annexins to modulate Ca^{2+} transport and cytosolic Ca^{2+} tested, the range of annexin involvement in calcium signalling now needs to be explored. The association of annexins with intracellular membranes means that research should not be limited just to cytosolic calcium, but needs to be extended to mitochondria, chloroplasts and the nucleus. There is a clear need to answer the longstanding question of whether annexins such as MtANN1 function in symbiotic nuclear calcium signalling [81–84]. Similarly, the role of annexins in immunity requires further attention. More tools are now available with which to test for annexin function in Ca^{2+} signalling and it is hoped that more plant researchers will take up this challenge.

Acknowledgments

This work was funded by the University of Cambridge. I thank past and present members of the lab and our collaborators for their work on annexins.

Conflicts of Interest

The author declares no conflict of interest.

References

1. McAinsh, M.R.; Pitman, J.K. Shaping the calcium signature. *New Phytol.* **2009**, *181*, 275–294.
2. Jammes, F.; Hu, H.-C.; Villiers, F.; Boueten, R.; Kwak, J.M. Calcium-permeable channels in plants. *FEBS J.* **2011**, *278*, 4262–4276.
3. Swarbreck, S.M.; Colaço, R.; Davies, J.M. Update on plant calcium-permeable channels. *Plant Physiol.* **2013**, *163*, 1514–1522.
4. Dadacz-Narloch, B.; Kimura, S.; Kurusu, T.; Farmer, E.E.; Becker, D.; Kuchitsu, K.; Hedrich, R. On the cellular site of two-pore channel TPC1 action in the Poaceae. *New Phytol.* **2013**, *200*, 663–674.
5. Chen, Z.-H.; Hills, A.; Bätz, U.; Amtmann, A.; Lew, V.L.; Blatt, M.R. Systems dynamic modelling of the stomatal guard cell predicts emergent behaviors in transport, signalling, and volume control. *Plant Physiol.* **2012**, *159*, 1235–1251.
6. Michard, E.; Lima, P.T.; Borges, F.; Silva, A.C.; Portes, M.T.; Carvalho, J.E.; Gilliham, M.; Liu, L.H.; Obermeyer, G.; Feijó, J. Glutamate receptor-like genes form Ca^{2+} channels in pollen tubes and are regulated by pistil D-serine. *Science* **2011**, *332*, 434–437.
7. Vincill, E.D.; Bieck, A.M.; Spalding, E.P. Ca^{2+} conduction by an amino acid-gated ion channel related to glutamate receptors. *Plant Physiol.* **2012**, *159*, 40–46.
8. Tunc-Ozmedir, M.; Rato, C.; Brown, E.; Rogers, S.; Mooneyham, A.; Frietsch, S.; Myers, C.T.; Poulsen, L.R.; Malhó, R.; Harper, J.F. Cyclic nucleotide gated channels 7 and 8 are essential for male reproductive fertility. *PLoS One* **2013**, *8*, e55277.

9. Ma, Y.; Zhao, Y.; Walker, R.K.; Berkowitz, G.A. Molecular steps in the immune signaling pathway evoked by plant elicitor peptides: Ca^{2+}-dependent protein kinases, nitric oxide, and reactive oxygen species are downstream from the early Ca^{2+} signal. *Plant Physiol.* **2012**, *163*, 1459–1471.
10. Ma, Y.; Walker, R.K.; Zhao, Y.; Berkowitz, G.A. Linking ligand perception by PEPR pattern recognition receptors to cytosolic Ca^{2+} elevation and downstream immune signaling in plants. *Proc. Natl. Acad. Sci. USA* **2013**, *109*, 19852–19857.
11. Laohavisit, A.; Shang, Z.; Rubio, L.; Cuin, T.A.; Very, A.-A.; Wang, A.H.; Mortimer, J.C.; Macpherson, N.; Coxon, K.M.; Battey, N.H.; *et al. Arabidopsis* annexin1 mediates the radical-activated plasma membrane Ca^{2+}- and K^{+}-permeable conductance in root cells. *Plant Cell* **2012**, *24*, 1522–1533.
12. Laohavisit, A.; Richards, S.L.; Shabala, L.; Chen, C.; Colaço, R.; Swarbreck, S.M.; Shaw, E.; Dark, A.; Shabala, S.; Shang, Z.-L.; *et al.* Salinity-induced calcium signaling and root adaptation in *Arabidopsis* require the regulatory protein annexin1. *Plant Physiol.* **2013**, *163*, 253–262.
13. Laohavisit, A.; Davies, J.M. Annexins. *New Phytol.* **2011**, *189*, 40–53.
14. Guelette, B.S.; Benning, U.F.; Hoffmann-Benning, S. Identification of lipids and lipid-binding proteins in phloem exudates from *Arabidopsis thaliana*. *J. Exp. Bot.* **2012**, *63*, 3603–3616.
15. Li, B.; Li, D.-D.; Zhang, J.; Xia, H.; Wang, X.-L.; Li, Y.; Li, X.-B. Cotton AnnGh3 encoding an annexin protein is preferentially expressed in fibers and promotes initiation and elongation of leaf trichomes in transgenic *Arabidopsis*. *J. Integr. Plant Biol.* **2013**, *55*, 902–916.
16. Clark, G.B.; Morgan, R.O.; Fernandez, M.-P.; Roux, S.J. Evolutionary adaptation of plant annexins has diversified their molecular structures, interactions and functional roles. *New Phytol.* **2012**, *196*, 695–712.
17. Feng, Y.M.; Wei, X.K.; Liao, W.X.; Huang, L.H.; Zhang, H.; Liang, S.C.; Peng, H. Molecular analysis of the annexin gene family in soybean. *Biol. Plant.* **2013**, *57*, 655–662.
18. Jami, S.K.; Clark, G.B.; Ayele, B.T.; Ashe, P.; Kirti, P.B. Genome-wide comparative analysis of annexin superfamily in plants. *PLoS One* **2012**, *7*, e47801.
19. Liu, T.-W.; Fu, B.; Niu, L.; Chen, J.; Wang, W.-H.; He, J.-X.; Pei, Z.-M.; Zheng, H.-L. Comparative proteomic analysis of proteins in response to simulated acid rain in *Arabidopsis*. *J. Proteome Res.* **2011**, *10*, 2579–2589.
20. Chu, P.; Chen, H.; Zhou, Y.L.; Li, Y.; Ding, Y.; Jiang, L.; Tsang, E.W.T.; Wu, K.; Huang, S. Proteomic and functional analyses of *Nelumbo nucifera* annexins involved in seed thermotolerance and germination vigor. *Planta* **2012**, *235*, 1271–1288.
21. Fernandez, M.B.; Pagano, M.R.; Daleo, G.R.; Guevara, M.G. Hydrophobic proteins secreted into the apoplast may contribute to resistance against *Phytophthora infestans* in potato. *Plant Physiol. Biochem.* **2012**, *60*, 59–66.
22. Urbany, C.; Colby, T.; Stich, B.; Schmidt, L.; Schimdt, J.; Gebhardt, C. Analysis of natural variation of the potato tuber proteome reveals novel candidate genes for tuber bruising. *J. Proteome Res.* **2012**, *11*, 703–716.
23. Limpens, E.; Moling, S.; Hooiveld, G.; Pereira P.A.; Bisseling, T.; Becker, J.D.; Küster, H. Cell-and tissue-specific transcriptome analyses of *Medicago truncatula* root nodules. *PLoS One* **2013**, *8*, e64377.

24. Yacoubi, R.; Job, C.; Belghazi, M.; Chaibi, W.; Job, D. Proteomic analysis of the enhancement of seed vigour in osmoprimed alfalfa seeds germinated under salinity stress. *Seed Sci. Res.* **2013**, *23*, 99–110.
25. Zhang, Y.; Xu, L.; Zhu, X.; Gong, Y.; Xiang, F.; Sun, X.; Liu, L. Proteomic analysis of heat stress response in leaves of radish (*Raphanus sativus* L.). *Plant Mol. Biol. Report.* **2013**, *31*, 195–203.
26. Zhou, M.-L.; Yang, X.-B.; Zhang, Q.; Zhou, M.; Zhao, E.-Z.; Tang, Y.-X.; Zhu, X.-M.; Shao, J-R.; Wu, Y.-M. Induction of annexin by heavy metals and jasmonic acid in *Zea mays*. *Funct. Integr. Genomics* **2013**, *13*, 241–251.
27. Punwani, J.A.; Rabiger, D.S.; Drews, G.N. MYB98 positively regulates a battery of synergid-expressed genes encoding filiform apparatus-localized proteins. *Plant Cell* **2007**, *19*, 2557–2568.
28. Tsukagoshi, H.; Busch, W.; Benfey, P.N. Transcriptional regulation of ROS controls transition from proliferation to differentiation in the root. *Cell* **2010**, *143*, 606–616.
29. Laohavisit, A.; Davies, J.M. Multifunctional annexins. *Plant Sci.* **2009**, *177*, 532–539.
30. Obata, T.; Matthes, A.; Koszior, S.; Lehmann, M.; Araujo, W.L.; Bock, R.; Sweetlove, L.J.; Fernie, A.R. Alteration of mitochondrial protein complexes in relation to metabolic regulation under short-term oxidative stress in *Arabidopsis* seedlings. *Phytochemistry* **2011**, *72*, 1081–1091.
31. Laohavisit, A.; Mortimer, J.C.; Demidchik, V.; Coxon, K.M.; Stancombe, M.A.; Macpherson, N.; Brownlee, C.; Hofmann, A.; Webb, A.A.R.; Miedema, H.; *et al. Zea mays* annexins modulate cytosolic free Ca^{2+} and generate a Ca^{2+}-permeable conductance. *Plant Cell* **2009**, *21*, 479–493.
32. Swisher, J.F.; Burton, N.; Bacot, S.M.; Vogel, S.N.; Feldman, G.M. Annexin A2 tetramer activates human and murine macrophages through TLR4. *Blood* **2010**, *115*, 549–558.
33. Hofmann, A.; Proust, J.; Dorowski, A.; Schantz, R.; Huber, R. Annexin 24 from *Capsicum annuum*. X-ray structure and biochemical characterization. *J. Biol. Chem.* **2000**, *275*, 8072–8082.
34. Hu, N.-J.; Yusof, A.M.; Winter, A.; Osman, A.; Reeve, A.K.; Hofmann, A. The crystal structure of calcium-bound annexin Gh1 from *Gossypium hirsutum* and its implications for membrane binding mechanisms of plant annexins. *J. Biol. Chem.* **2008**, *283*, 18314–18322.
35. Breton, G.; Vazquez-Tello, A.; Danyluk, J.; Sarhan, F. Two novel intrinsic annexins accumulate in wheat membranes in response to low temperature. *Plant Cell Physiol.* **2000**, *41*, 177–184.
36. Gorecka, K.M.; Thouverey, C.; Buchet, R.; Pikula, S. Potential role of annexin AtANN1 from *Arabidopsis thaliana* in pH-mediated cellular response to environment stimuli. *Plant Cell Physiol.* **2007**, *48*, 792–803.
37. Santoni, V.; Rouquie, D.; Doumas, P.; Mansion, M.; Boutry, M.; Degand, H.; Dupree, P.; Packman, L.; Sherrier, J.; Prime, T.; *et al.* Use of a proteome strategy for tagging proteins present at the plasma membrane. *Plant J.* **1998**, *16*, 633–641.
38. Laohavisit, A.; Davies, J.M. Annexins. In *Coding and Decoding of Calcium Signals in Plants*; Springer-Verlag: Heidelberg, Germany, 2011; pp. 111–128.
39. Laohavisit, A.; Brown, A.T.; Cicuta, P.; Davies, J.M. Annexins: Components of the calcium and reactive oxygen signaling network. *Plant Physiol.* **2010**, *152*, 1824–1829.
40. Kubista, H.; Hawkins, T.E.; Moss, S.E. Annexin V mediates a peroxide-induced Ca^{2+}-influx in B-cells. *Curr. Biol.* **1999**, *9*, 1403–1406.

41. Balasubramanian, K.; Bevers, E.M.; Willems, G.M.; Schroit, A.J. Binding of annexin V to membrane products of lipid peroxidation. *Biochemistry* **2001**, *40*, 8672–8676.
42. Kodavali, P.K.; Skowronek, K.; Koszela-Piotrowska, I.; Strzelecka-Kiliszek, A.; Pawlowski, K.; Pikula, S. Structural and functional characterization of annexin 1 from *Medicago truncatula*. *Plant Physiol. Biochem.* **2013**, *73*, 56–62.
43. Neumann, E.; Siemens, P.M.; Toensing, K. Electroporative fast pore-flickering of the annexin V-lipid surface complex, a novel gating concept for ion transport. *Biophys. Chem.* **2000**, *86*, 203–220.
44. Mortimer, J.C.; Coxon, K.M.; Laohavisit, A.; Davies, J.M. Heme-independent soluble and membrane-associated peroxidase activity of a *Zea mays* annexin preparation. *Plant Signal. Behav.* **2009**, *4*, 428–430.
45. Ito, J.; Heazlewood, J.L.; Millar, A.H. Analysis of the soluble ATP-binding proteome of plant mitochondria identifies new proteins and nucleotide triphosphate interactions within the matrix. *J. Proteome Res.* **2006**, *5*, 3459–3469.
46. Shang, Z.; Laohavisit, A.; Davies, J.M. Extracellular ATP activates an *Arabidopsis* plasma membrane Ca^{2+}-permeable conductance. *Plant Signal. Behav.* **2009**, *4*, 985–991.
47. Demidchik, V.; Nichols, C.; Dark, A.; Oliynyk, M.; Glover, B.J.; Davies, J.M. Is ATP a signaling agent in plants? *Plant Physiol.* **2003**, *133*, 456–461.
48. Demidchik, V.; Shang, L.; Shin, R.; Thompson, E.; Rubio, L.; Laohavisit, A.; Mortimer, J.C.; Chivasa, S.; Slabas, A.R.; Glover, B.J.; *et al.* Plant extracellular ATP signalling by plasma membrane NADPH oxidase and Ca^{2+} channels. *Plant J.* **2009**, *58*, 903–913.
49. Clark, G.; Roux, S.J. Apyrases, extracellular ATP and the regulation of growth. *Curr. Opin. Plant Biol.* **2011**, *14*, 700–706.
50. Carroll, A.D.; Moyen, C.; van Kesteren, P.; Tooke, F.; Battey, N.H.; Brownlee, C. Ca^{2+}, annexins, and GTP modulate exocytosis from maize root cap protoplasts. *Plant Cell* **1998**, *10*, 1267–1276.
51. Demidchik, V.; Shabala, S.N.; Coutts, K.B.; Tester, M.A.; Davies, J.M. Free oxygen radicals regulate plasma membrane Ca^{2+} & K^{+}-permeable channels in plant root cells. *J. Cell Sci.* **2003**, *116*, 81–88.
52. Foreman, J.; Demidchik, V.; Bothwell, J.H.F.; Mylona, P.; Miedema, H.; Torres, M.A.; Linstead, P.; Costa, S.; Brownlee, C.; Jones, J.D.G.; *et al.* Reactive oxygen species produced by NADPH oxidase regulate plant cell growth. *Nature* **2003**, *422*, 442–446.
53. Richards, S.L.; Laohavisit, A.; Mortimer, J.C.; Shabala, L.; Swarbreck, S.M.; Shabala, S.; Davies, J.M. Annexin 1 regulates the H_2O_2-induced calcium signature in *Arabidopsis thaliana* roots. *Plant J.* **2013**, *77*, 136–145.
54. Rentel, M.C.; Knight, M.R. Oxidative stress-induced calcium signaling in *Arabidopsis*. *Plant Physiol.* **2004**, *135*, 1471–1479.
55. Gorecka, K.M.; Konopka-Postupolska, D.; Hennig, J.; Buchet, R.; Pikula, S. Peroxidase activity of annexin 1 from *Arabidopsis thaliana*. *Biochem. Biophys. Res. Commun.* **2005**, *336*, 868–875.
56. Renew, S.; Heyno, E.; Schopfer, P.; Liszkay, A. Sensitive detection and localization of hydroxyl radical production in cucumber roots and Arabidopsis seedlings by spin trapping electron paramagnetic resonance spectroscopy. *Plant J.* **2005**, *44*, 342–347.

57. Lee, S.; Lee, E.J.; Yang, E.J.; Lee, J.E.; Park, A.R.; Song, W.H.; Park, O.K. Proteomic identification of annexins, calcium-dependent membrane binding protein that mediate osmotic stress and abscisic acid signal transduction in *Arabidopsis*. *Plant Cell* **2004**, *16*, 1378–1391.
58. Chung, J.S.; Zhu, J.K.; Bressan, R.A.; Hasegawa, P.M.; Shi, H.Z. Reactive oxygen species mediate Na^{+}-induced SOS1 mRNA stability in *Arabidopsis*. *Plant J.* **2008**, *53*, 554–565.
59. Thonat, C.; Mathieu, C.; Crevecoeur, M.; Penel, C.; Gaspar, T.; Boyer, N. Effects of a mechanical stimulation of localization of annexin-like proteins in *Bryonia dioica* internodes. *Plant Physiol.* **1997**, *114*, 981–988.
60. Clark, G.B.; Rafati, D.S.; Bolton, R.J.; Dauwalder, M.; Roux, S.J. Redistribution of annexin in gravistimulated pea plumules. *Plant Physiol. Biochem.* **2000**, *38*, 937–947.
61. Demir, F.; Horntrich, C.; Blachutzik, J.O.; Scherzer, S.; Reinders, Y.; Kierszniowska, S.; Schulze, W.X.; Harms, G.S.; Hedrich, R.; Geiger, D.; *et al. Arabidopsis* nano-domain-delimited ABA signalling pathway regulates the anion channel SLAH3. *Proc. Natl. Acad. Sci. USA* **2013**, *110*, 8296–8301.
62. Chasserot-Golaz, S.; Vitale, N.; Umbrecht-Jenck, E.; Knight, D.; Gerke, V.; Bader, M.F. Annexin 2 promotes the formation of lipid microdomains required for calcium-regulated exocytosis of dense-core vesicles. *Mol. Biol. Cell* **2005**, *16*, 1108–1119.
63. Lefebvre, B.; Furt, F.; Hartmann, M.-A.; Michaelson, L.V.; Carde, J.-P.; Sargueil-Boiron, F.; Rossignol, M.; Napier, J.A.; Cullimore, J.; Bessoule, J.-J.; *et al.* Characterization of lipid rafts from *Medicago truncatula* root plasma membranes: A proteomic study reveals the presence of a raft-associated redox system. *Plant Physiol.* **2007**, *144*, 402–418.
64. Liu, P.; Li, R.-L.; Zhang, L.; Wang, Q.-L.; Niehaus, K.; Baluska, F.; Samaj, J.; Lin, J.-X. Lipid microdomain polarization is required for NADPH oxidase-dependent ROS signaling in *Picea meyeri* pollen tube tip growth. *Plant J.* **2009**, *60*, 303–313.
65. Ning, S.-B.; Song, Y.-C.; van Damme, P. Characterization of the early stages of programmed cell death in maize root cells by using comet assay and the combination of cell electrophoresis with annexin binding. *Electrophoresis* **2002**, *23*, 2096–2102.
66. Reina-Pinto, J.J.; Voisin, D.; Kurdyukov, S.; Faust, A.; Haslam, R.P.; Michaelson, L.V.; Efremova, N.; Franke, B.; Schreiber, L.; Napier, J.A.; *et al.* Misexpression of FATTY ACID ELONGATION1 in the *Arabidopsis* epidermis induces cell death and suggests a critical role for phospholipase A2 in this process. *Plant Cell* **2009**, *21*, 1252–1272.
67. Huh, S.M.; Noh, E.Y.; Kim, H.G.; Jeon, B.W.; Bae, K.; Hu, H.C.; Kwak, J.M.; Park, O.K. *Arabidopsis* annexins AnnAt1 and AnnAt4 interact with each other and regulate drought and salt stress responses. *Plant Cell Physiol.* **2010**, *51*, 1499–1514.
68. Huang, Y.; Wang, J.; Zhang, L.; Zuo, K. A cotton annexin protein AnxGb6 regulates fiber elongation through its interaction with actin 1. *PLoS One* **2013**, *8*, e66160.
69. Wang, Y.F.; Fan, L.M.; Zhang, W.Z.; Zhang, W.; Wu, W.H. Ca^{2+}-permeable channels in the plasma membrane of *Arabidopsis* pollen are regulated by actin microfilaments. *Plant Physiol.* **2004**, *136*, 3892–3904.
70. Wu, J.-Y.; Jin, C.; Qu, H.-Y.; Tao, S.-T.; Xu, G.-H.; Wu, J.; Wu, H.-Q.; Zhang, S.-L. Low temperature inhibits pollen viability by alteration of actin cytoskeleton and regulation of pollen plasma membrane ion channels in *Pyrus pyrifolia*. *Environ. Exp. Bot.* **2012**, *78*, 70–75.

71. Morgan, R.O.; Martin-Almedina, S.; Garcia, M.; Jhoncon-Kooyip, J.; Fernandez, M.P. Deciphering function and mechanism of calcium-binding proteins from their evolutionary imprints. *Biochim. Biophys. Acta* **2006**, *1763*, 1238–1249.
72. Hoshino, T.; Mizutani, A.; Chida, M.; Hidaka, H.; Mizutani, J. Plant annexins form homodimer during Ca^{2+}-dependent liposome aggregation. *Biochem. Mol. Biol. Int.* **1995**, *35*, 749–755.
73. Loukehaich, R.; Wang, T.; Ouyang, B.; Ziaf, K.; Li, H.; Zhang, J.; Lu, Y.; Ye, Z. SpUSP, an annexin-interacting universal stress protein, enhances drought tolerance in tomato. *J. Exp. Bot.* **2012**, *63*, 5593–5606.
74. Lindermayr, C.; Saalbach, G.; Durner, J. Proteomic identification of S-nitrosylated proteins in *Arabidopsis*. *Plant Physiol.* **2005**, *137*, 921–930.
75. Konopka-Postupolska, D.; Clark, G.; Goch, G.; Debski, J.; Floras, K.; Cantero, A.; Fiolek, B.; Roux, S.; Hennig, J. The role of annexin 1 in drought stress in *Arabidopsis*. *Plant Physiol.* **2009**, *150*, 1394–1410.
76. Agrawa, G.K.; Thelen, J.J. Large scale identification and quantitative profiling of phosphoproteins expressed during seed filling in oilseed rape. *Mol. Cell. Proteomics* **2006**, *5*, 2044–2059.
77. Rohila, J.S.; Chen, M.; Chen, S.; Chen, J.; Cerny, R.; Dardick, C.; Canlas, P.; Xu, X.; Gribskov, M.; Kanrar, S.; *et al.* Protein-protein interactions of tandem affinity purification-tagged protein kinases in rice. *Plant J.* **2006**, *46*, 1–13.
78. Andrawis, A.; Solomon, M.; Delmer, D.P. Cotton fibre annexins: A potential role in the regulation of callose synthase. *Plant J.* **1993**, *3*, 763–772.
79. Wang, P.; Xue, L.; Batelli, G.; Lee, S.; Hou, Y.-J.; van Oosten, M.J.; Zhang, H.; Tao, W.A.; Zhu. J.-K. Quantitative phosphoproteomics identifies SnRK2 protein kinase substrates and reveals the effectors of abscisic acid action. *Proc. Natl. Acad. Sci. USA* **2013**, *110*, 11205–11210.
80. Yang, Y.L.; Zhang, Y.Y.; Lu, J.; Zhang, H.; Liu, Y.; Shi, R.X. Exogenous H_2O_2 increased catalase and peroxidase activities and proline content in *Nitraria tangutorum* callus. *Biol. Plant.* **2012**, *56*, 330–336.
81. Kovács, I.; Ayaydin, F.; Oberschall, A.; Ipacs, I.; Bottka, S.; Pongor, S.; Dudits, D.; Toth, E.C. Immunolocalization of a novel annexin-like protein encoded by a stress and abscisic acid responsive gene in alfalfa. *Plant J.* **1998**, *15*, 185–197.
82. De Carvalho-Niebel, F.; Lescure, N.; Cullimore, J.V.; Gamas, P. The *Medicago truncatula MtAnn1* gene encoding an annexin is induced by nod factors and during the symbiotic interaction with *Rhizobium meliloti*. *Mol. Plant Microbe Interact.* **1998**, *11*, 504–513.
83. De Carvalho-Niebel, F.; Timmers, A.C.J.; Chabaud, M.; Defaux-Petras, A.; Barker, D.G. The Nod factor-elicited annexin *MtAnn1* is preferentially localised at the nuclear periphery in symbiotically activated root tissues of *Medicago truncatula*. *Plant J.* **2002**, *32*, 343–352.
84. Talukdar, T.; Gorecka, K.M.; de Carvalho-Niebel, F.; Downie, J.A.; Cullimore, J.; Pikula, S. Annexins—Calcium- and membrane-binding proteins in the plant kingdom. Potential role in nodulation and mycorrhization in *Medicago truncatula*. *Acta Biochim. Pol.* **2009**, *56*, 199–210.

3

Accumulation of TIP2;2 Aquaporin during Dark Adaptation Is Partially PhyA Dependent in Roots of *Arabidopsis* Seedlings

Yumi Uenishi [1,†], Yukari Nakabayashi [2,‡], Ayako Tsuchihira [3], Mari Takusagawa [1,§], Kayo Hashimoto [1], Masayoshi Maeshima [3] and Kumi Sato-Nara [2,*]

[1] Graduate School of Humanities and Sciences, Nara Women's University, Nara 630-8506, Japan; E-Mail: mak_hashimoto@cc.nara-wu.ac.jp

[2] Faculty of Science, Nara Women's University, Nara 630-8506, Japan

[3] Graduate School of Bioagricultural Sciences, Nagoya University, Nagoya 464-8601, Japan; E-Mails: tsuchihira.ayako@f.mbox.nagoya-u.ac.jp (A.T.); maeshima@agr.nagoya-u.ac.jp (M.M.)

[†] Present address: NIPPON Funmatsu Yakuhin CO., Ltd, Osaka 541-0045, Japan; E-Mail: uenishi@nfy.co.jp.

[‡] Present address: Graduate School of Information Science, Nagoya University, Nagoya 464-8601, Japan; E-Mail: nakabayashi.yukari@a.mbox.nagoya-u.ac.jp.

[§] Present address: Graduate School of Science, Kyoto University, Kyoto 606-8502, Japan; E-Mail: mtakusagawa@pmg.bot.kyoto-u.ac.jp.

* Author to whom correspondence should be addressed; E-Mail: kumisn@cc.nara-wu.ac.jp

---

Abstract: Light regulates the expression and function of aquaporins, which are involved in water and solute transport. In *Arabidopsis thaliana*, mRNA levels of one of the aquaporin genes, *TIP2;2*, increase during dark adaptation and decrease under far-red light illumination, but the effects of light at the protein level and on the mechanism of light regulation remain unknown. Numerous studies have described the light regulation of aquaporin genes, but none have identified the regulatory mechanisms behind this regulation via specific photoreceptor signaling. In this paper, we focus on the role of phytochrome A (phyA) signaling in the regulation of the TIP2;2 protein. We generated *Arabidopsis* transgenic plants expressing a TIP2;2-GFP fusion protein driven by its own promoter, and showed several differences in TIP2;2 behavior between wild type and the *phyA* mutant. Fluorescence of TIP2;2-GFP protein in the endodermis of roots in the wild-type seedlings increased during dark adaptation, but not in the *phyA* mutant. The amount of the TIP2;2-GFP protein in wild-type seedlings decreased rapidly under far-red light illumination, and a delay in reduction of TIP2;2-GFP was observed in the *phyA* mutant. Our results imply that phyA, cooperating with other photoreceptors, modulates the level of TIP2;2 in *Arabidopsis* roots.

Keywords: *Arabidopsis thaliana*; aquaporin; dark adaptation; endodermis; light regulation; phytochrome; root; water transport

---

1. Introduction

As a consequence of water absorption during development, vacuoles occupy more than 90% of the total volume of plant cells [1]. Vacuoles play multifaceted roles, including accumulation of useful substances, recycling of cell components, regulation of turgor pressure, sequestration of toxic ions, and detoxification of xenobiotics [2,3]. Furthermore, they have a space-filling function. Vacuolar transporters play essential roles in these functions [4]. One group of aquaporins, tonoplast intrinsic proteins (TIPs), serves as channels to pass water and small neutral molecules through the tonoplast [5–7]. TIP (TP25 in *Phaseolus vulgaris*) was the first aquaporin reported in plants [8]. In *Arabidopsis thaliana*, ten TIP members are classified into five subgroups, TIP1 to TIP5: these include three TIP1s, three TIP2s, two TIP3s, one TIP4;1 and one TIP5;1 [9]. In addition to water [10–15], TIPs are permeable to ammonia [16,17], urea [18], H_2O_2 [19] and glycerol [20]. For example, although both TIP1;1 and TIP2;1 are permeable to water [15] and urea [18], TIP1;1 is also permeable to H_2O_2 [19] and TIP2;1 is permeable to ammonia [16]. Therefore, each TIP member appears to allow permeation of specific substrate molecules and to play a specific physiological role. The diversity of physiological roles of TIPs is supported by earlier observations that different types of vacuoles contain distinct members of individual TIPs [21,22]. The tonoplast of storage vacuoles contains δ-TIP (presently TIP2), while the lytic vacuoles mainly contain γ-TIP (TIP1) [21]. Three types of TIPs, α-TIP (TIP3), TIP2, and sometimes TIP1, coexist in protein storage vacuoles of seeds [21]. The coexistence and distinct patterns of tissue- and organ-specific expression of TIP members have also been observed in roots and leaves of *Arabidopsis* using fluorescent reporters [23–25]. Thus, the expression of TIP isoforms is controlled by developmental cues associated with a specific vacuolar differentiation pattern [26].

External cues such as light and environmental stresses also regulate the expression of *TIPs* [14,27–31]. The expression of several *TIP2* genes is affected by light: *SunTIP7 and SunTIP20* in *Helianthus* leaves are up-regulated by light [28], while expression of others such as *DIP* in *Antirrhinum* cotyledons [32] and *TIP2;2* in *Arabidopsis* roots [31] is increased during the dark period. The modulation of *TIP2s*' expression by light is probably involved in light regulation of vacuolar functions. However, the mechanism of light regulation of TIP2 and its relationship with vacuolar functions have not been studied with respect to aspects of developmental regulation and stress responses.

The modification of vacuolar functions by increased levels of specific TIPs promotes water permeability in cells and growth of cells and organs. When tomato *SlTIP2;2* is transiently expressed in mesophyll cells in *Arabidopsis*, water permeability of the protoplasts increases [33]. Over-expression of *Panax ginseng PgTIP1* in *Arabidopsis* plants promotes the growth of leaves, roots and seeds [34]. Overexpression of *NtTIP1;1* promotes regeneration, enlargement, and division of tobacco BY-2 cells [35]. Therefore, modulation of *TIP* expression during adaptation to light conditions is probably connected with regulation of water transport and growth in plants through the regulation of vacuolar functions.

Plants monitor light in the environment via multiple photoreceptors such as phytochromes, cryptochromes and phototropins [36,37]. The phytochrome (phy) family encodes red/far-red (FR) light photoreceptors, and *Arabidopsis* has five members, phyA through phyE [38]. Phytochromes generally mediate reversible red/FR light responses [36,39,40]. Furthermore, phyA mediates the very-low-fluence response to broad-spectrum light as a highly sensitive light "antenna" [40,41] as well as the high-irradiation response to FR light [42]. phyA is involved in the regulation of growth and development of not only aerial parts [39] but also roots, affecting root elongation [43], root hair formation [44], and red light-induced root phototropism [45]. One of our research interests is why and how light regulates the growth and development of roots underground.

Our previous study focusing on phyA regulation of gene expression in roots indicated that transcripts of several aquaporin genes including *TIP1;1*, *TIP1;2* and *TIP2;2* increase during dark adaptation and that phyA is involved in regulation of these genes [31]. One of the *TIP* genes strongly expressed in the roots [46], *TIP2;2*, was greatly upregulated during dark adaptation [31]. Light regulation of *TIP2;2* expression could affect the growth and metabolism of roots as a consequence of modification of vacuolar function. However, the effects of light and phyA signaling on TIP2;2 protein levels have not been studied. In this paper, we focused on the roles of light and phyA signaling on the regulation of TIP2;2 protein. We generated transgenic *Arabidopsis* plants expressing a TIP2;2-GFP fusion protein driven by its own promoter, and compared tissue and cellular localization of TIP2;2-GFP in wild-type and *phyA* mutant cells. We also analyzed the protein levels in response to dark conditions and FR illumination and found several differences in gene expression and protein accumulation of TIP2;2 between wild-type and *phyA* mutant cells.

2. Results and Discussion

2.1. Effects of phyA Mutation on Fluorescent Intensity of TIP2;2-GFP under Light Conditions

In order to examine differences in protein levels of TIP2;2 under light and dark conditions, and between wild-type and *phyA* mutant, we generated transgenic *Arabidopsis* expressing a *TIP2;2-GFP* fusion gene driven by its own promoter (Supplementary Figure S1). RT-PCR and immunoblotting with anti-GFP antibodies confirmed that the TIP2;2-GFP fusion protein is expressed in the transgenic lines (Supplementary Figure S1). Using a confocal laser scanning microscope (LSM), GFP fluorescence was mainly observed on the tonoplast not the plasma membrane, in root cells of all lines (Supplementary Figure S1). Therefore, the exogenous TIP2;2-GFP might be correctly localized in the tonoplast in all transgenic lines. Approximately 1.9 kb of the 5' upstream region, which was used as the *TIP2;2* promoter, contained 521 *cis*-elements based on searching an open database of *cis*-elements, PLACE (plant *cis*-acting regulatory DNA elements) [47]. Of these, 65 *cis*-elements are involved in light regulation (Supplementary Table S1), including the G-box, to which phytochrome interacting factors bind [48], and the conserved I-box sequence found upstream of many light-regulated genes [49]. Other *cis*-elements contained in the sequence were nine sugar-repressive elements [50], eight ABA-responsive elements [51] and 14 auxin-responsive elements [52] (Supplementary Table S1).

2.1.1. Tissue-Specific Localization of TIP2;2-GFP in the Wild Type and *phyA* Mutant

First, to examine differences between two ecotypes, Col and L*er*, and between the wild type (L*er*) and *phyA* mutant, we analyzed the tissue specificity of TIP2;2-GFP expression in roots of light-grown seedlings of Col, L*er* and *phyA* lines. TIP2;2-GFP was detected in the pericycle, endodermis, cortex and epidermis of the mature region of the roots, but not in immature tissues such as the root apical meristem or the developing lateral root primordium (Figure 1).

Figure 1. Tissue-specific expression of TIP2;2-GFP in roots of *Arabidopsis* seedlings. Fluorescence (**a**,**c**,**e**,**g**,**i**) and transmitted light (**b**,**d**,**f**,**h**,**j**) images of roots of transgenic Col plants expressing a TIP2;2-GFP fusion protein driven by its own promoter. Seedlings were grown for 6 days (**a**–**h**) or 11 days (**i**,**j**) in a 16 h light/8 h dark cycle. (**a**–**h**) Mature region (**a**,**b**), border region of the elongation to mature zones (**c**,**d**), root tip (**e**,**f**), and lateral root primordium (**g**,**h**) of the primary root. (**i**,**j**) Growing lateral roots with bright fluorescence. co, cortical cell; en, endodermal cell; ep, epidermal cell; lp, lateral root primordium; lr, lateral root; mr, main root; p, pericycle cell; rm, root apical meristem. Bars, 50 μm.

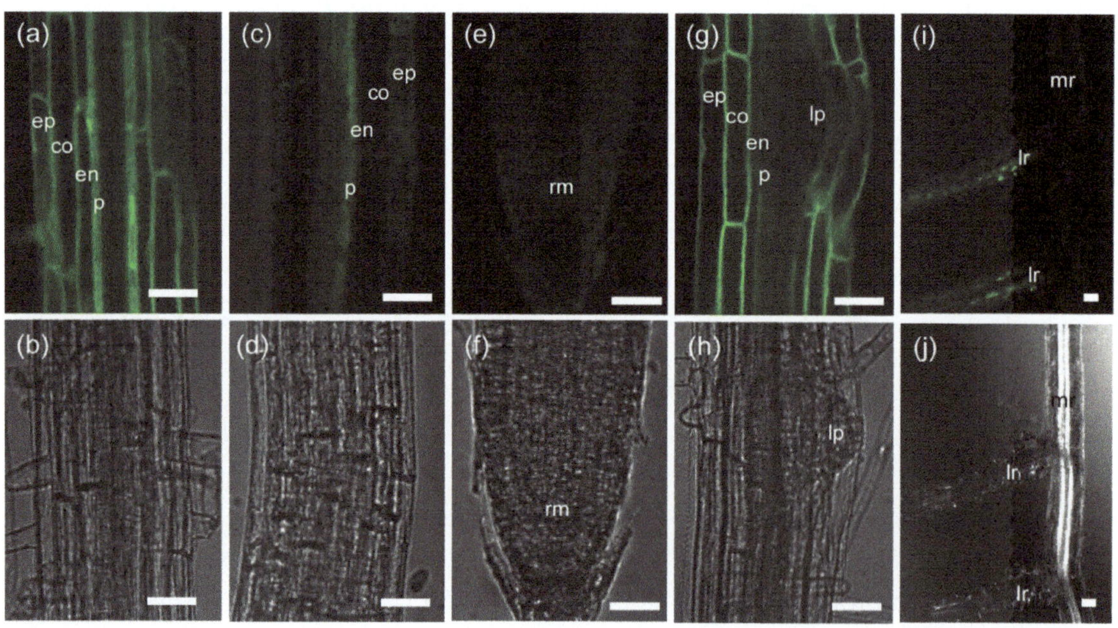

The strongest fluorescence was detected in the mature region near the differentiation zone, and fluorescence weakened at a more basal region near the hypocotyl. Furthermore, a region of the growing lateral roots in the light-grown plants showed strong fluorescence (Figure 1i). Similar patterns were observed in the roots of all transgenic lines (3 Col, 3 L*er*, and 4 *phyA* background-lines) except for slight differences in the timing of when the tissues began expression and in the fluorescent intensity of the inner tissues, as described below. In Col seedlings, TIP2;2-GFP was first expressed in the pericycle of the border region of the elongation to mature zones and expression expanded to the outer region (Figure 1c). However, in the seedlings with L*er* and *phyA* backgrounds, TIP2;2-GFP was first expressed in the epidermis of the elongation zone (data not shown). This expression pattern resembled that in a previous study reporting that TIP2;2-YFP is first expressed in the cortex and epidermis and then expands to the pericycle during development [23]; these observations of *TIP2;2-YFP* transgenic plants in the Col background differ from ours in that TIP2;2-GFP expression in Col roots started in the pericycle and expanded to the epidermis. This difference may be due to the length of the promoter: Gattolin *et al.* [23] used approximately 1.4 kb of the upstream region of the *TIP2;2* gene as its own promoter, and we used about 1.9 kb. Based on a PLACE database search [47], there are about 160 possible *cis*-elements (Supplementary Table S1), including some involved in light and hormone regulation, in the approximately 550 bp of the upstream region that we included and Gattolin *et al.* [23] did not. Furthermore, based on the differing expression patterns of TIP2;2-GFP in the L*er* and Col backgrounds, the upstream region we used may contain some elements differing between ecotypes. However, additional experiments are necessary to clarify the cause of different tissue-specific localization of the fusion protein between ecotypes, or in our TIP2;2-GFP lines and their TIP2;2-YFP lines.

2.1.2. Fluorescence of TIP2;2-GFP in Light-Grown Wild-Type and *phyA* Mutant Plants

Next, we examined the effects of the *phyA* mutation on TIP2;2-GFP expression. In the roots of light-grown seedlings, the fluorescent intensity of three TIP2;2-GFP/*phyA* lines was higher than that of two TIP2;2-GFP/L*er* lines (Figure 2). This difference in fluorescent intensity was consistent with immunoblotting results (Figure 3a) and other experiments of fluorescent intensity (Figure 4).

Figure 2. Relative fluorescent intensity in the roots of transgenic *Arabidopsis* expressing a *TIP2;2-GFP* fusion gene in plants of wild-type (L*er*) and *phyA* mutant background. (**a**) Differing fluorescent intensity of TIP2;2-GFP in the roots of light-grown L*er* and *phyA* mutant plants. Plants were grown on MS-agar medium for seven days in a 16 h light/8 h dark cycle. All images were acquired at approximately 5 mm from the root tip by LSM under the same conditions on the same day. Brightness and contrast of images were not modified. Bars, 50 μm; (**b**) Relative fluorescent intensity in the roots of light-grown L*er* and *phyA* seedlings expressing a *TIP2;2-GFP* fusion gene. Relative fluorescent intensity is shown as the fold difference relative to the mean intensity of the L*er*5-1 line. Results are means ± SD of five independent plants. Asterisks indicate statistical differences when compared with L*er*5-1 using Welch's *t*-test (* $p < 0.05$).

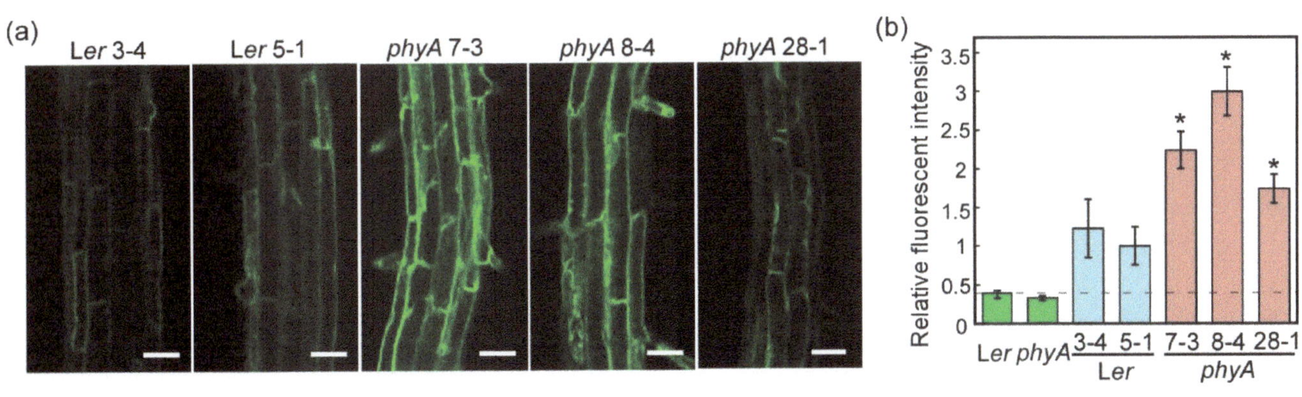

Figure 3. Light-dependent changes in aquaporin abundance in roots of L*er* seedlings and *phyA* mutants during dark adaptation and FR illumination. Aliquots of solubilized crude membrane proteins were immunoblotted with anti-GFP and anti-γ-VM23 (TIP1) antibodies. After detection of TIP2;2-GFP, the same blot was used for TIP1 detection. The background lines were L*er* and *phyA*, and transgenic plants are identified by line number (e.g., L*er*3-4). (**a**) Immunoblots of TIP2;2-GFP in the roots of 7-day light-grown seedlings; (**b**) Diagrams of growth conditions. Seedlings were grown under 12 h irradiation with white light for three days and were then grown continuously in the light (12L12D, L) or dark (D) for another three days. Dark-adapted seedlings were illuminated by FR light for 1, 2 and 6 h; (**c**) Immunoblots of TIP2;2-GFP during dark adaptation and FR illumination. Growth conditions are shown in (**b**).

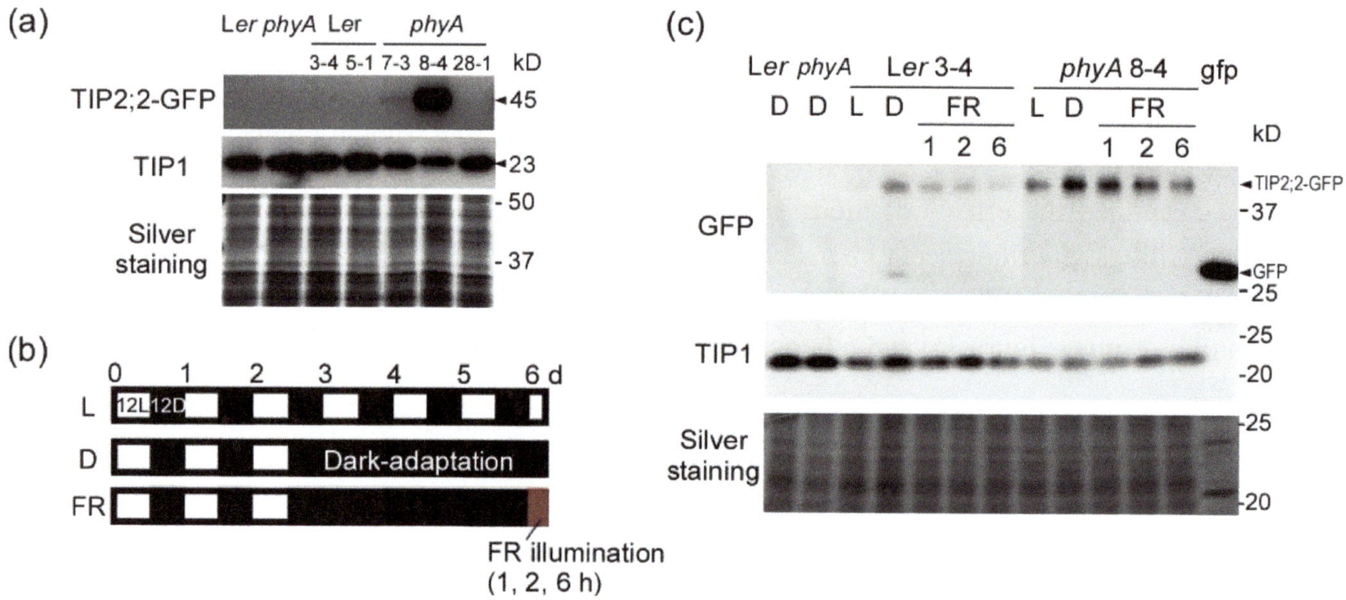

We randomly selected *TIP2;2-GFP* transgenic lines in the L*er* and *phyA* backgrounds, but *phyA* lines showed a tendency to be brighter than L*er* lines. To evaluate positional effects due to the insertion region of the transgene, we crossed the brightest line, TIP2;2-GFP/*phyA*8-4, with wild type (L*er*) and observed the fluorescent intensity of the progeny. The segregation ratio of the F_2 lines was *PHYA*: *phyA* = 3:1 (Table 1, $\chi^2 = 0.00456$) and *phyA tip2;2-gfp*$^-$: other phenotypes = 1:15 ($\chi^2 = 1.54$). The segregation ratio of the F_2 lines obtained by backcrossing TIP2;2-GFP/*phyA*8-4 to *phyA-201* was *phyA tip2;2-gfp*$^-$: *phyA TIP2;2-GFP*$^+$ = 1:3 ($\chi^2 = 1.75$). Therefore, insertion of the *TIP2;2-GFP* transgene in TIP2;2-GFP/*phyA*8-4 can be considered a single locus unlinked to the *phyA* locus. Under stereoscopic observation, all of the plants with intense green fluorescence had the *phyA* phenotype. More than 60% of the F_2 progeny with the wild-type phenotype displayed clear or detectable green fluorescence. We selected homozygote lines of *TIP2;2-GFP* and *PHYA* from the F_3 progeny, and named one of the lines TIP2;2-GFP/L*er*8-4.

Figure 4. Increase in TIP2;2-GFP protein levels during dark adaptation. (**a**) Fluorescent images of TIP2;2-GFP in the roots of light- and dark-grown TIP2;2-GFP/Col46-5 plants. Seedlings grown for seven days in the light were then grown continuously in the light or dark for another four days. Black and white bars, 50 μm; (**b**) Relative fluorescent intensity of TIP2;2-GFP in the main roots and lateral roots of light- and dark-grown TIP2;2-GFP/Col46-5 plants. The main roots and lateral roots of seven plants were scanned and the brightness of the GFP fluorescence was quantified. Average values for each root were plotted. Scanned areas: Root tip, area approximately 0.5 to 1.2 mm from the root tip of the main root; MR, area approximately 5 to 10 mm from the root tip of the main root; LR, area approximately 0.4 to 1 mm from the root tip of the young lateral root (approximately 0.6 to 1.2 mm long). Asterisks indicate statistical differences found using the Mann-Whitney U-test (* $p < 0.05$); (**c**) Comparison of TIP2;2-GFP signals in the roots of light- and dark-grown TIP2;2-GFP transgenic plants with a L*er* or *phyA* background. Seedlings grown for four days in the light were then grown continuously in the light or dark for three days. The central regions, which contain the central cylinder and protoxylem, of main roots were scanned at approximately 5 mm from the root tip and the brightness of the GFP fluorescence was quantified. Results are means ± SD of seven independent plants. Asterisks indicate statistical differences found using Student's *t*-test (* $p < 0.05$).

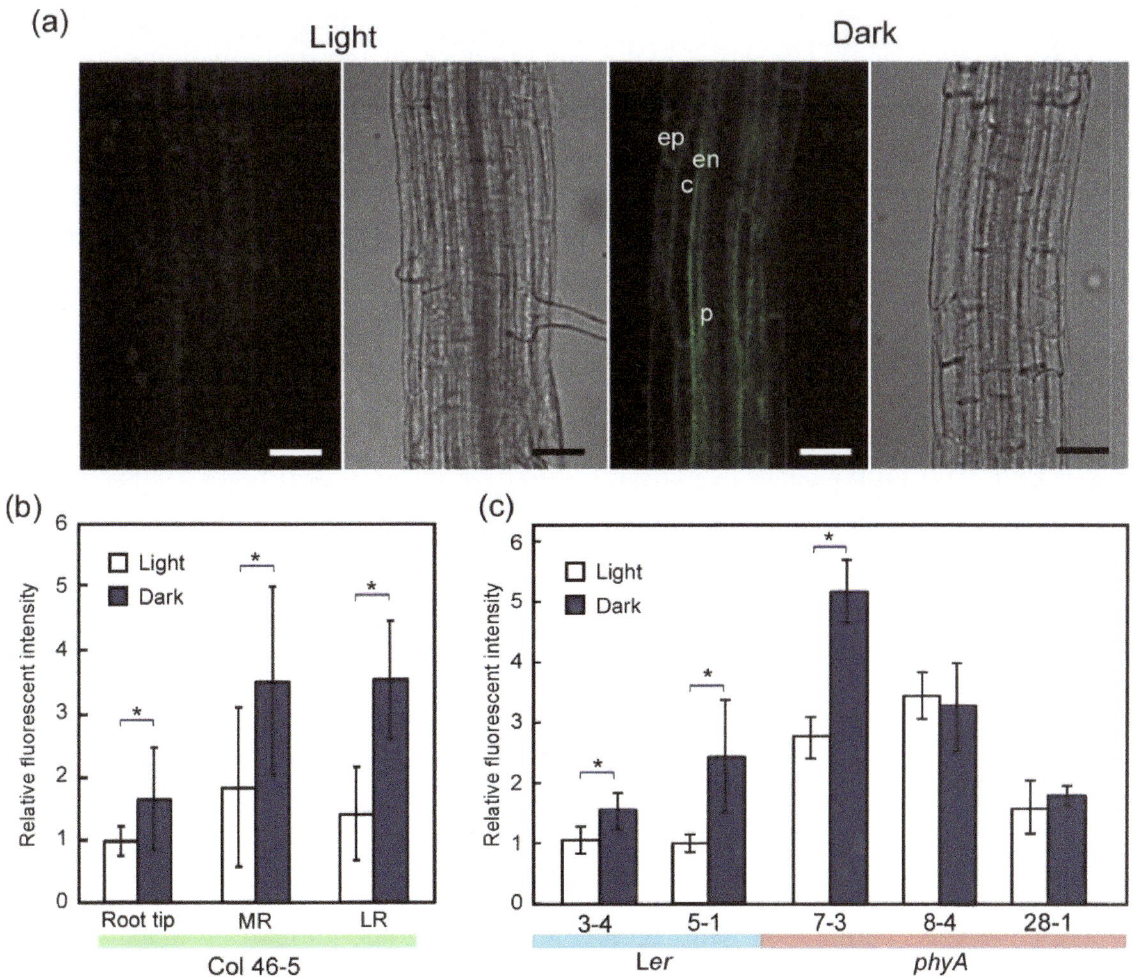

Table 1. Segregation ratios of wild-type and *phyA* mutated F_2 progeny from TIP2;2-GFP/*phyA*8-4 plants crossed with L*er* plants or *phyA-201* plants.

F_2 plant	Number of plants classified with *phyA* mutation and fluorescent intensity *[1] *PHYA*			*phyA*			χ^2 values *[2] Expected frequency
	++	+	−	++	+	−	
TIP2;2-GFP/ *phyA* 8-4 × L*er*	0	37	18	2	14	2	0.00456 [PHYA]: [phyA] = 3:1 1.54 [*PHYA* or *GFP*$^+$]: [*phyA gfp*$^-$] = 15:1
TIP2;2-GFP/ *phyA* 8-4 × *phyA*	0	0	0	2	18	3	1.75 [*GFP*$^+$]:[*gfp*$^-$] = 3:1

*[1] Fluorescent intensities of roots classified into strong (++), detectable (+) and undetectable (−) under stereoscopic observation; *[2] All the calculated values of χ^2 were less than the value for $\chi^2_{0.95}$ = 3.84 with one degree of freedom, suggesting that each expected frequency is acceptable.

We grew the TIP2;2-GFP/L*er*8-4 and the TIP2;2-GFP/*phyA*8-4 lines in the light and observed them by LSM. In both lines, the fluorescent intensity of TIP2;2-GFP in the outer sliced images containing the epidermis and cortex of the roots was similar (Figure 5a,b). However, in the inner slices containing endodermis of the roots in light-grown plants, the fluorescent intensity of *phyA*8-4 was slightly higher than that of L*er*8-4 (Figure 5a,c). The TIP2;2-GFP fluorescence of the roots of light-grown L*er*8-4 plants tended to be weak in the endodermis, but strong in the epidermis and cortex (Figure 5a). Based on the results, we estimate that the differences in TIP2;2-GFP expression levels between L*er* and *phyA* mutant backgrounds under light conditions (Figures 2–4) is largely derived from the differences in the insertion sites of the *TIP2;2-GFP* genes. In the light-grown plants, differences in fluorescent intensity between L*er* and *phyA* are thought to be limited to the inner region of roots (Figure 5).

2.2. Dark-Induced Increase in TIP2;2-GFP

Next, we examined the effects of dark adaptation on TIP2;2-GFP expression. In the TIP2;2-GFP/L*er* lines, the amount of TIP2;2-GFP protein increased during dark adaptation (Figure 3). The increase in amounts of TIP2;2-GFP is consistent with the fluorescent intensity of TIP2;2-GFP in the L*er* and Col lines (Figure 4). An increase in protein levels during dark adaptation was also observed in the TIP2;2-GFP/*phyA* lines, although the increase was low in TIP2;2-GFP/*phyA*8-4 (Figure 3). This tendency of each *phyA* line roughly matched the results from image analysis, but the TIP2;2-GFP fluorescence in the center region of *phyA*8-4 roots hardly changed during dark adaptation (Figure 4c). TIP1 also slightly increased during dark adaptation in L*er* but not *phyA* (Figure 3c).

Because the increase in TIP2;2-GFP fluorescence during dark adaptation differed among independent *phyA* lines, we analyzed the TIP2;2-GFP fluorescence of *phyA*8-4 and its progeny L*er*8-4 in detail. In both lines, fluorescent intensity of TIP2;2-GFP in the outer sliced images containing the epidermis and cortex of the roots increased during dark adaptation (Figure 5a,b). The increase in L*er*8-4 was higher than in *phyA*8-4. In the inner slices containing the endodermis of the roots, the fluorescent intensity increased during dark adaptation of L*er*8-4, but not *phyA*8-4 (Figure 5c). Strong TIP2;2-GFP fluorescence in the endodermal cells of dark-adapted L*er*8-4 plants was more frequently observed than in light-grown plants (Figure 5a). Furthermore, in the TIP2;2-GFP/Col lines, stronger fluorescence was observed in the endodermal cells of dark-adapted roots (Figure 4a).

Figure 5. Comparison of TIP2;2-GFP protein levels in epidermal, cortical and endodermal cells between L*er* and *phyA* mutant during dark adaptation. Seedlings grown for three days in the light were then grown continuously in the light or the dark for another three days. The insertion sites of *TIP2;2-GFP* transgenes in the genomes of both L*er*8-4 and *phyA*8-4 lines were identical. (a) Fluorescent images of TIP2;2-GFP in outer (epidermis-cortex) and inner (epidermis-endodermis) optical sections of the roots of light- and dark-grown *TIP2;2-GFP* transgenic plants with a L*er* or *phyA* background. Transverse sections (yz slices) were digitally reconstituted using multiple longitudinal optical sections (xy slices). The red and yellow lines in each yz image indicate the positions of the two xy longitudinal images beside it. The main roots were scanned at approximately 5 mm from the root tip. Bar, 50 μm; (b,c) Comparison of TIP2;2-GFP signals in the roots of light- and dark-grown TIP2;2-GFP transgenic plants with a L*er* or *phyA* background. The brightness of the GFP fluorescence of each image of epidermis-cortex (b) and epidermis-endodermis (c) slices was quantified. Results are means ± SD of three independent plants. Different letters indicate statistical differences found using Scheffe's F test ($p < 0.05$).

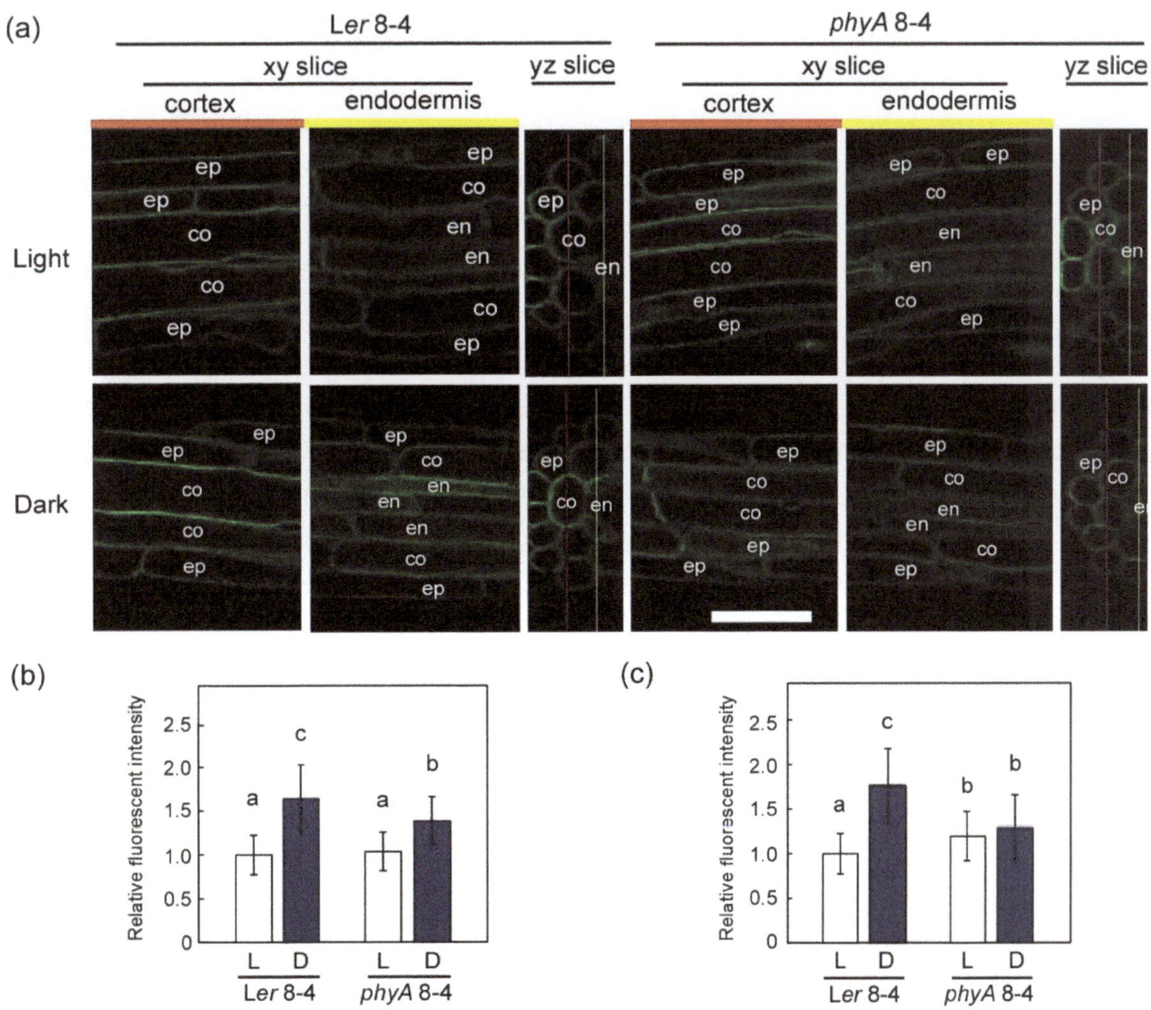

Lack of a dark-adapted increase in TIP2;2-GFP in the endodermal cells of the *phyA* mutant suggests that phyA signaling is involved in the mechanism leading to this change. Although it is not clear why an increase in TIP2;2-GFP in endodermal cells is more phyA-dependent than in the cortex and epidermis, this light response in the endodermal cells might be important for regulating water and solute transport in the roots because the endodermal tissues generally act as an efficient barrier to ions moving through the apoplast [53] and the site where transport of solutes is regulated [54,55].

In the dark-adapted plants, weak GFP fluorescence was also detected in the lumen of the vacuole (Figure 5a). This is consistent with low levels of proteins that had the same size as a GFP monomer (approximately 27 kD) also being detected in the dark-adapted plants (Figure 3c). Conversely, in the light, this presumed GFP monomer was not detected. Vacuolar GFP is rapidly degraded in the light [56], suggesting that tonoplast aquaporins taken up into vacuoles are also degraded at a constant speed in the light, and their degradation is slowed down in the dark. This behavior may lead to a dark-induced increase in the amount of aquaporins.

2.3. PhyA Regulation of Aquaporins

PhyA is the sole photoreceptor to respond to continuous FR light in *Arabidopsis* [57]. To confirm phyA regulation of the TIP2;2 protein, we compared the changes in protein levels in the roots of L*er* and a *phyA* mutant under FR illumination. In the roots of the L*er* lines, the amount of TIP2;2-GFP protein decreased 1 h after FR illumination, and further decreased by 6 h (Figure 3c). In *phyA* roots, the amount of TIP2;2-GFP slightly decreased at 1 h after FR illumination, and stayed at relatively high levels at 2 and 6 h after FR illumination. These different patterns of protein levels in L*er* and *phyA* as well as the lack of a dark-adapted increase in TIP2;2-GFP in the *phyA* mutant suggest that phyA signaling is partly involved in the reduction in TIP2;2-GFP in the light. Further studies are required to clarify the direct and indirect roles of phyA in regulation of the TIP2;2 abundance.

Modification of the rates of transcription [31] and translation, as well as degradation, of the protein might be involved in this rapid drop in TIP2;2-GFP 1 h after FR illumination (Figure 3c). In the *phyA* mutant, the observed slower rate of decrease suggests the involvement of phyA in the early phases of FR illumination. In later phases, TIP2;2-GFP levels were similar to those in the light, suggesting that other factors are also involved in this process. The involvement of other factors are also indicated by fluorescent analysis of epidermal and cortical cells, in which the *phyA* mutant showed a dark-induced increase in TIP2;2-GFP (Figure 5b) and by immunoblots using crude membrane proteins from whole roots (Figure 3c). The results indicate that a light-responsive mechanism to decrease the amount of TIP2;2 functions normally in wild-type plants, but this function is weakened in the *phyA* mutant. In the vacuolar lumen, all proteins seem to be more rapidly degraded in the light than in the dark [56]. It is not known how tonoplast proteins are degraded, or whether the degradation mechanisms of both the proteins in the vacuolar lumen and in the tonoplast are similar. Further analysis of the mechanisms behind the sudden drop in TIP2;2-GFP in the early phase of FR exposure may provide insight into tonoplast protein degradation systems in both the dark and light.

2.4. Why TIP2;2 Increases during Dark Adaptation

One of the mechanisms responsible for the increase in amount of TIP2;2 protein in the dark is transcriptional regulation of the gene. In addition to numerous *cis*-elements in light regulation, multiple potential sugar-repressive elements [50] are included in the *TIP2;2* promoter (Supplementary Table S1). Furthermore, the *TIP2;2* gene is induced by decreasing the sucrose concentration in the medium [31]. Thus, decreasing the amount of sucrose or photosynthetic products during a long dark period could cause an increase in aquaporin gene expression. Aquaporins can transport small neutral molecules as well as water [6]. It is thought that limited energy in the dark and degradation of various macromolecules [58] brings about the necessity of transporting toxic ammonia/ammonium either to the vacuole for isolation or to other subcellular compartments for production of other dark-required compounds. With respect to the pKa of 9.24 for the deprotonation of ammonium to ammonia and the cytosolic pH of 7.0 to 7.5, approximately 1% of the total cytoplasmic ammonia/ammonium is present in an uncharged form [16]. The uncharged form of ammonia can be passively transported by TIP2 subfamily of aquaporins, TIP2;1 and TIP2;3, from the cytosol to the vacuole [16]. The ammonia diffusing across the tonoplast subsequently binds a proton to form ammonium, which is trapped in the acidic vacuolar lumen.

TIP2;2 is quite similar to TIP2;1 and TIP2;3, which transport water, urea, and ammonia [15,16,18], thus suggesting that TIP2;2 also transports these molecules. The increase in the amount of TIP2;2-GFP during dark adaptation may compensate for the lowering transport efficiency of aquaporins. Increased aquaporins might facilitate efficient water and solute transport across the membranes, even if the driving forces such as osmotic and hydrostatic pressure gradients are lowered. The passive transport of ammonia by aquaporins is dependent on the pH gradient between the cytosol and the vacuolar lumen, which is mainly generated by two proton pumps [4]. The long dark condition without photosynthesis could lower the activities of proton pumps, and thus lower the pH-dependent transport of ammonia by aquaporins as well as other transporters whose transport is dependent on a chemical and/or electrical gradient between both sides of the tonoplast. Water transport by aquaporins is also correlated with solute transport because the osmotic pressure gradient is one of the driving forces of water movement [59].

Vacuoles can be classified by the type of TIP found in the tonoplast and TIP2;2 is a type of δ-TIP, which is generally in the tonoplast of storage vacuoles [4]. Because TIP2;2 is also one of the major aquaporins in *Arabidopsis* roots [46], it probably plays an important role in light-grown plants. The increase in TIP2;2 during dark adaptation suggests that vacuolar function is modified by light and that the physiological function of TIP2;2-containing vacuoles may be altered during long dark periods.

Alternatively, TIP2;2, in concert with other aquaporins, may facilitate water flow from the epidermis to the xylem in the central cylinder of roots. Intriguingly, TIP2;2-YFP and TIP2;3-YFP are present in the rows of pericycle cells that form the xylem poles [23]. We did not examine the fluorescent intensity of pericycle cells or the location of fluorescent cells, although several pericycle cells displayed clear fluorescence (Figures 4 and 5). The increase in TIP2;2 in the endodermal cells during dark adaptation is also associated with the important role of transport of water and solutes into the central cylinder. Further analysis of the mechanism of light regulation of aquaporins should lead to a better understanding of the tissue- and cell-specific roles of TIP2;2 in both the light and dark.

3. Experimental

3.1. Plant Material and Growth Conditions

Arabidopsis thaliana Ler (NW20) and *phyA-201* (N6219) lines were obtained from the NASC (Nottingham Arabidopsis Stock Centre, Loughborough, UK), and Col-0 (J58) was obtained from the SASSC (Sendai Arabidopsis Seed Stock Center, Sendai, Japan). Seeds were surface-sterilized and sown on the surface of Murashige and Skoog (MS) medium containing 1% agar and 1% sucrose in square styrene plastic cases (90 × 70 × 23 mm). Approximately 20 seeds (for fluorescent intensity analysis) or 300 seeds (for protein extraction) per case were sown in two rows. After vernalization at 4 °C for two days, seeds were cultured vertically at 22 °C under 12 or 16 h illumination (12L12D or 16L8D photoperiods) with cool white fluorescent light of 15 W/m^2. FR light was provided by a light-emitting diode (LED; 730 nm, 6.2 W/m^2; MIL-IF18; Sanyo, Tokyo, Japan) with MIL-U200 frames and a MIL-C1000T controller system. Fluence rates were measured with a model LI-250 radiometer (Li-Cor, Lincoln, NE, USA). Roots of the plants under dark adaptation were harvested using a glove box in total darkness, as described previously [31]. Sampling tubes containing roots were wrapped in aluminum foil before withdrawing them from the glove box, and were immediately frozen in liquid nitrogen. Frozen roots were stored at −80 °C.

3.2. Fluorescent Intensity Analysis

GFP fluorescence intensity was measured using a C1siR confocal LSM (Nikon Instruments, Tokyo, Japan). Fresh roots excised from seedlings were immediately immersed in *Arabidopsis* hydroponic medium [60] on slides and were observed by LSM. To image GFP, the 488 nm line was used for excitation, and emission was detected at 515/30 nm. For semiquantitative measurement of fluorescence intensity, the laser, pinhole and gain settings of the LSM were kept identical among treatments. The fluorescence intensity of each digital image was analyzed using EZ-C1 software [61] as shown in Supplementary Figure S2. Images were assembled using Photoshop software [62].

3.3. Protein Preparation

Frozen roots (100–200 mg) were crushed with a Multi-beads Shocker cell disruptor (Yasui Kikai, Osaka, Japan), after which 7 $\mu L\ mg^{-1}$ root of cold extraction buffer (100 mM Tris-HCl (pH 8.0), 300 mM NaCl, 20 mM EDTA, 20% (*w*/*v*) glycerol, 5 mM dithiothreitol (DTT) and 1× Complete Protease Inhibitor Cocktail (Roche Applied Science, Tokyo, Japan)) was immediately added, and the mixture was homogenized with the Multi-beads Shocker at 4 °C. The homogenate was strained through a filtration apparatus and centrifuged at 10,000 rpm for 10 min at 4 °C. The supernatant was recovered and centrifuged again at 100,000× *g* for 30 min at 4 °C. The pellet was used as the microsomal membrane fraction. The pellet was resuspended in 0.2 $\mu L\ mg^{-1}$ root extraction buffer. After protein concentrations were determined, each microsomal membrane fraction was frozen in liquid nitrogen and stored at −80 °C. Protein concentrations of the soluble and membrane protein samples were determined using a Bio-Rad protein assay kit (Bio-Rad, Tokyo, Japan) with γ-globulin as a standard.

3.4. Immunoblotting

Each microsomal membrane fraction was mixed with an equal volume of 2× SDS-PAGE sample buffer and incubated for 10 min at 70 °C. SDS-PAGE in 9% and 12% (*w*/*v*) polyacrylamide gels containing 0.1% SDS was carried out by the method of Laemmli (1970). For immunoblotting, proteins in polyacrylamide gels were transferred to a Clear Blot P polyvinylidene difluoride membrane (ATTO, Tokyo, Japan) with a semidry blotting apparatus using standard procedures. The membrane was blocked with 2% (*w*/*v*) ECL Advance Blocking Agent (GE Healthcare, Tokyo, Japan) in TBS-T buffer (20 mM Tris-HCl (pH 7.6), 0.14 M NaCl, 0.1% (*v*/*v*) Tween 20) overnight at 4 °C. The membrane was incubated with the appropriate primary antibody for 1 h at 20 °C. Concentrations of antibodies were: 1:500 dilution of anti-GFP antibody (Clontech A.v. peptide antibody, polyclonal; Takara-Bio, Shiga, Japan) and 1:10,000 dilution of anti-TIP1 antibody (γVM23, [63]). After washing for 15 min and another three times for 5 min with TBS-T, blots were incubated with secondary antibody, a 1:100,000 dilution of ECL anti-rabbit IgG, horseradish peroxidase-linked species specific whole antibody (from donkey, NA934V, GE Healthcare, Tokyo, Japan) for 1 h at 20 °C. After washing as described above, antibody conjugates were detected using an ECL Advance chemiluminescent reagent (GE Healthcare, Tokyo, Japan) according to the manufacturer's instructions. Chemiluminescence images were acquired using a LAS-3000 luminescent imaging system (Fujifilm, Tokyo, Japan), and were quantitated using MultiGauge software [64]. After detection, each immunoblot was stripped of probe according to Kaufmann and Kellner [65], and reprobed using an anti-TIP1 antibody. All protein samples used for immunoblotting were electrophoresed on other gels and gels were stained by standard procedures with silver or Coomassie brilliant blue R-250.

3.5. Generation of TIP2;2-GFP Transgenic Lines

In order to generate a TIP2;2 promoter::TIP2;2::sGFP construct (TIP2;2-GFP), a 2.9-kb genomic region containing the *TIP2;2* gene and a 1.9 kb upstream region was PCR amplified from the genomic DNA of Col-0 using primers TIP2;2-1190F and TIP2;2-4035R (Supplementary Table S2). PCR products were cloned into vector ENTR/D-TOPO (Invitrogen, Tokyo, Japan) before recombination into the destination vector pGWB404 using Gateway technology [66]. Correct in-frame GFP insertion was verified by DNA sequencing. The construct was introduced into *Agrobacterium tumefaciens* GV3101 before transformation of Arabidopsis Col-0, L*er*, and *phyA-201* plants by floral dipping [67]. Transgenic plants were either selected on kanamycin or identified by GFP fluorescence of roots under an M165 FC fluorescence stereomicroscope (Leica Microsystems, Tokyo, Japan). From 15,000 seeds of Col and 5000 seeds each of L*er* and *phyA*, 119 Col, 24 L*er* and 30 *phyA* kanamycin-resistant T_1 plants were selected. More than 10 independent T_2 plants were further selected. The transgenic lines showing 100% kanamycin resistance were observed under a confocal LSM. F_1, F_2, and F_3 progeny of TIP2;2-GFP/*phyA* 8-4 crossed to L*er* were selected by the hypocotyl phenotype of the seedlings grown under FR illumination for five days [68] in addition to GFP fluorescence and kanamycin resistance.

3.6. Protoplast Preparation

Roots of six-day-old seedlings (Col-0, TIP2;2-GFP-46-5 and TIP2;2-GFP-40) were cut longitudinally with a razor while immersed in a few drops of Arabidopsis hydroponic medium (0.015 M NaH_2PO_4, 2.6 mM Na_2HPO_4, 1.5 mM $MgSO_4$, 0.03 M KNO_3, 0.14 mM Fe(III) EDTA, 0.02 M $Ca(NO_3)_2$, 0.1 mM $MnSO_4$, 0.3 mM H_3BO_3, 0.01 mM $ZnSO_4$, 9.6 μM $CuSO_4$, 0.24 μM $(NH_4)_6Mo_7O_{24}$, 1.2 μM $CoCl_2$ [60]) on a slide. Root tissues were transferred to a sample tube, and were incubated in 0.2 mL protoplast enzyme solution [2% (*w*/*v*) cellulase Onozuka RS (Yakult Pharmaceutical, Tokyo, Japan), 0.1% (*w*/*v*) macerozyme R-10 (Yakult Pharmaceutical, Tokyo, Japan), 0.03% (*w*/*v*) pectolyase Y23 (Kikkoman, Tokyo, Japan), 20 mM MES-KOH (pH 5.5), 2 mM DTT, 0.39 M sorbitol] at 22 °C for 60 to 90 min [69]. Enzyme solution was then gently removed and protoplasts were washed twice in 0.5 mL washing solution (0.4 M sorbitol in Arabidopsis hydroponic medium). Gravitated protoplasts were gently suspended in 20 μL of washing solution and were observed under a confocal LSM.

3.7. RT-PCR

Roots of two-month-old plants (Col-0, TIP2;2-GFP-46-5 and TIP2;2-GFP-40) were excised and immediately immersed in Ambion RNAlater reagent (Applied Biosystems, Tokyo, Japan). The immersed roots were kept at 4 °C overnight. Fixed tissues were removed from the reagent, immediately frozen in liquid nitrogen, and stored at −80 °C. Frozen samples were crushed using the Multi-beads Shocker and total RNA was isolated using an RNeasy Plant Mini Kit (Qiagen, Tokyo, Japan). After treatment with RNase-free DNase I (Invitrogen, Tokyo, Japan), a 1 μg aliquot of total RNA was reverse transcribed using 200 U PrimeScript II RTase (Takara Bio, Otsu, Japan). cDNA was amplified by PCR using Ex Taq polymerase (Takara Bio) and primers TIP2;2cDNA-1F and TIP2;2-sGFP-976R (Supplementary Table S2). TIP2;2-sGFP-1266R was also used. PCR of TIP2;2-GFP was performed as follows: 94 °C for 9 min; 32 cycles at 94 °C for 30 s, 57.3 °C for 30 s, and 72 °C for 1 min; then 72 °C for 10 min. Expression levels of Actin-8 (At1g49240) were also examined as an internal control. PCR of Actin-8 was performed under the same conditions as for TIP2;2-GFP except for annealing temperature (55 °C), extension time (30 s), and primers (ACT8F and ACT8R, Supplementary Table S2).

4. Conclusions

In this paper, we presented evidence for the involvement of a phyA signaling pathway in light regulation of aquaporin TIP2;2. We also proposed that phyA signaling directly or indirectly influences the tissue-specific localization of TIP2;2-GFP and/or functional modification of the tonoplast. These modifications to the amount and localization of TIP2;2 during dark adaptation and FR illumination could change the permeability to water and some solutes via the tonoplast of root cells. During a long dark period, various slow-acting cellular processes allow survival. TIP2;2 may play a role in such processes in the roots of dark-adapted plants. The findings in this paper provide a basis to clarify the mechanisms regulating water and solute transport based on a plant's light environment.

Acknowledgments

We would like to thank R. Nagai and M. Tango, NWU, for technical assistance with this study. We are grateful to A. Sakai and M. Araki, NWU, for helpful advice and use of equipment for analyzing aquaporin expression. We thank T. Nakagawa, Shimane Univ., for providing a Gateway binary vector, and M. Murai, NARCT, for helpful advice on protoplast preparation. This work was supported by a Nara Women's University Intramural Grant for Project Research to K.S.-N. and a Grant-in-Aid for Challenging Exploratory Research (KAKENHI 24657032) to K.S.-N.

Author Contributions

Yumi Uenishi and Ayako Tsuchihira generated a TIP2;2-GFP construct and transgenic lines with Col background. Kayo Hashimoto generated transgenic lines with L*er* and *phyA* backgrounds. Yumi Uenishi carried out fluorescent observation with Col, and Yukari Nakabayashi did with L*er* and *phyA* lines. Yumi Uenishi, Mari Takusagawa and Kumi Sato-Nara carried out immunoblot analysis and data analysis. Masayoshi Maeshima promoted this work by exactly the right instructions, especially, for immunoblot analysis of microsomal membrane fraction. Kumi Sato-Nara wrote the paper with Yumi Uenishi, Ayako Tsuchihira and Masayoshi Maeshima, and supervised all work.

Conflicts of Interest

The authors declare no conflict of interest.

References

1. Taiz, L. The plant vacuole. *J. Exp. Biol.* **1992**, *172*, 113–122.
2. Marty, F. Plant vacuoles. *Plant Cell* **1999**, *11*, 587–600.
3. Frigerio, L.; Hinz, G.; Robinson, D.G. Multiple vacuoles in plant cells: Rule or exception? *Traffic* **2008**, *9*, 1564–1570.
4. Martinoia, E.; Maeshima, M.; Neuhaus, H.E. Vacuolar transporters and their essential role in plant metabolism. *J. Exp. Bot.* **2007**, *58*, 83–102.
5. Maurel, C. Plant aquaporins: Novel functions and regulation properties. *FEBS Lett.* **2007**, *581*, 2227–2236.
6. Katsuhara, M.; Hanba, Y.T.; Shiratake, K.; Maeshima, M. Review: Expanding roles of plant aquaporins in plasma membranes and cell organelles. *Funct. Plant Biol.* **2008**, *35*, 1–14.
7. Maurel, C.; Santoni, V.; Luu, D.T.; Wudick, M.M.; Verdoucq, L. The cellular dynamics of plant aquaporin expression and functions. *Curr. Opin. Plant Biol.* **2009**, *12*, 690–698.
8. Johnson, K.D.; Herman, E.M.; Chrispeels, M.J. An abundant, highly conserved tonoplast protein in seeds. *Plant Physiol.* **1989**, *91*, 1006–1013.
9. Johanson, U.; Karlsson, M.; Johansson, I.; Gustavsson, S.; Sjövall, S.; Fraysse, L.; Weig, A.R.; Kjellbom, P. The complete set of genes encoding major intrinsic proteins in *Arabidopsis* provides a framework for a new nomenclature for major intrinsic proteins in plants. *Plant Physiol.* **2001**, *126*, 1358–1369.

10. Maurel, C.; Reizer, J.; Schroeder, J.I.; Chrispeels, M.J. The vacuolar membrane protein γ-TIP creates water specific channels in *Xenopus* oocytes. *EMBO J.* **1993**, *12*, 2241–2247.
11. Kaldenhoff, R.; Fischer, M. Functional aquaporin diversity in plants. *Biochim. Biophys. Acta* **2006**, *1758*, 1134–1141.
12. Maurel, C.; Kado, R.T.; Guern, J.; Chrispeels, M.J. Phosphorylation regulates the water channel activity of the seed-specific aquaporin α-TIP. *EMBO J.* **1995**, *14*, 3028–3035.
13. Leitão, L.; Prista, C.; Moura, T.F.; Loureiro-Dias, M.C.; Soveral, G. Grapevine aquaporins: Gating of a tonoplast intrinsic protein (TIP2;1) by cytosolic pH. *PLoS One* **2012**, *7*, e33219.
14. Vera-Estrella, R.; Barkla, B.J.; Bohnert, H.J.; Pantoja, O. Novel regulation of aquaporins during osmotic stress. *Plant Physiol.* **2004**, *135*, 2318–2329.
15. Daniels, M.J.; Chaumont, F.; Mirkov, T.E.; Chrispeels, M.J. Characterization of a new vacuolar membrane aquaporin sensitive to mercury at a unique site. *Plant Cell* **1996**, *8*, 587–599.
16. Loqué, D.; Ludewig, U.; Yuan, L.; von Wirén, N. Tonoplast intrinsic proteins AtTIP2;1 and AtTIP2;3 facilitate NH_3 transport into the vacuole. *Plant Physiol.* **2005**, *137*, 671–680.
17. Jahn, T.P.; Møller, A.L.; Zeuthen, T.; Holm, L.M.; Klaerke, D.A.; Mohsin, B.; Kühlbrandt, W.; Schjoerring, J.K. Aquaporin homologues in plants and mammals transport ammonia. *FEBS Lett.* **2004**, *574*, 31–36.
18. Liu, L.H.; Ludewig, U.; Gassert, B.; Frommer, W.B.; von Wirén, N. Urea transport by nitrogen-regulated tonoplast intrinsic proteins in *Arabidopsis*. *Plant Physiol.* **2003**, *133*, 1220–1228.
19. Bienert, G.P.; Moller, A.L.; Kristiansen, K.A.; Schulz, A.; Moller, I.M.; Schjoerring, J.K.; Jahn, T.P. Specific aquaporins facilitate the diffusion of hydrogen peroxide across membranes. *J. Biol. Chem.* **2007**, *282*, 1183–1192.
20. Gerbeau, P.; Güçlü, J.; Ripoche, P.; Maurel, C. Aquaporin Nt-TIPa can account for the high permeability of tobacco cell vacuolar membrane to small neutral solutes. *Plant J.* **1999**, *18*, 577–587.
21. Jauh, G.Y.; Phillips, T.E.; Rogers, J.C. Tonoplast intrinsic protein isoforms as markers for vacuolar functions. *Plant Cell* **1999**, *11*, 1867–1882.
22. Maeshima, M.; Hara-Nishimura, I.; Takeuchi, Y.; Nishimura, M. Accumulation of vacuolar H^+-pyrophosphatase and H^+-ATPase during reformation of the central vacuole in germinating pumpkin seeds. *Plant Physiol.* **1994**, *106*, 61–69.
23. Gattolin, S.; Sorieul, M.; Hunter, P.R.; Khonsari, R.H.; Frigerio, L. *In vivo* imaging of the tonoplast intrinsic protein family in *Arabidopsis* roots. *BMC Plant Biol.* **2009**, *9*, doi:10.1186/1471-2229-9-133.
24. Gattolin, S.; Sorieul, M.; Frigerio, L. Mapping of tonoplast intrinsic proteins in maturing and germinating *Arabidopsis* seeds reveals dual localization of embryonic TIPs to the tonoplast and plasma membrane. *Mol. Plant* **2011**, *4*, 180–189.
25. Hunter, P.R.; Craddock, C.P.; di Benedetto, S.; Roberts, L.M.; Frigerio, L. Fluorescent reporter proteins for the tonoplast and the vacuolar lumen identify a single vacuolar compartment in *Arabidopsis* cells. *Plant Physiol.* **2007**, *145*, 1371–1382.
26. Wudick, M.M.; Luu, D.T.; Maurel, C. A look inside: Localization patterns and functions of intracellular plant aquaporins. *New Phytol.* **2009**, *184*, 289–302.
27. Suga, S.; Komatsu, S.; Maeshima, M. Aquaporin isoforms responsive to salt and water stresses and phytohormones in radish seedlings. *Plant Cell Physiol.* **2002**, *43*, 1229–1237.

28. Maurel, C.; Javot, H.; Lauvergeat, V.; Gerbeau, P.; Tournaire, C.; Santoni, V.; Heyes, J. Molecular physiology of aquaporins in plants. *Int. Rev. Cytol.* **2002**, *215*, 105–148.
29. Kim, H.S.; Yu, Y.; Snesrud, E.C.; Moy, L.P.; Linford, L.D.; Haas, B.J.; Nierman, W.C.; Quackenbush, J. Transcriptional divergence of the duplicated oxidative stress-responsive genes in the *Arabidopsis* genome. *Plant J.* **2005**, *41*, 212–220.
30. Liu, F.; Vantoai, T.; Moy, L.P.; Bock, G.; Linford, L.D.; Quackenbush, J. Global transcription profiling reveals comprehensive insights into hypoxic response in *Arabidopsis*. *Plant Physiol.* **2005**, *137*, 1115–1129.
31. Sato-Nara, K.; Nagasaka, A.; Yamashita, H.; Ishida, J.; Enju, A.; Seki, M.; Shinozaki, K.; Suzuki, H. Identification of genes regulated by dark adaptation and far-red light illumination in roots of *Arabidopsis thaliana*. *Plant Cell Environ.* **2004**, *27*, 1387–1394.
32. Culianez-Macia, F.A.; Martin, C. DIP: A member of the MIP family of membrane proteins that is expressed in mature seeds and dark-grown seedlings of *Antirrhinum majus*. *Plant J.* **1993**, *4*, 717–725.
33. Sade, N.; Vinocur, B.J.; Diber, A.; Shatil, A.; Ronen, G.; Nissan, H.; Wallach, R.; Karchi, H.; Moshelion, M. Improving plant stress tolerance and yield production: Is the tonoplast aquaporin SlTIP2;2 a key to isohydric to anisohydric conversion? *New Phytol.* **2009**, *181*, 651–661.
34. Lin, W.L.; Peng, Y.H.; Li, G.W.; Arora, R.; Tang, Z.C.; Su, W.A.; Cai, W.M. Isolation and functional characterization of PgTIP1 a hormone-autotrophic cells-specific tonoplast aquaporin in ginseng. *J. Exp. Bot.* **2007**, *58*, 947–956.
35. Okubo-Kurihara, E.; Sano, T.; Higaki, T.; Kutsuna, N.; Hasezawa, S. Acceleration of vacuolar regeneration and cell growth by overexpression of an aquaporin NtTIP1;1 in tobacco BY-2 cells. *Plant Cell Physiol.* **2009**, *50*, 151–160.
36. Whitelam, G.C.; Halliday, K.J. *Light and Plant Development*; Blackwell Publisher Ltd.: Oxford, UK, 2007.
37. Van Buskirk, E.K.; Decker, P.V.; Chen, M. Photobodies in light signaling. *Plant Physiol.* **2012**, *158*, 52–60.
38. Clack, T.; Mathews, S.; Sharrock, R.A. The phytochrome apoprotein family in *Arabidopsis* is encoded by five genes: The sequences and expression of *PHYD* and *PHYE*. *Plant Mol. Biol.* **1994**, *25*, 413–427.
39. Kami, C.; Lorrain, S.; Hornitschek, P.; Fankhauser, C. Light-regulated plant growth and development. *Curr. Top. Dev. Biol.* **2010**, *91*, 29–66.
40. Franklin, K.A.; Quail, P.H. Phytochrome functions in *Arabidopsis* development. *J. Exp. Bot.* **2010**, *61*, 11–24.
41. Botto, J.F.; Sánchez, R.A.; Whitelam, G.C.; Casal, J.J. Phytochrome A mediates the promotion of seed germination by very low fluences of light and canopy shade light in *Arabidopsis*. *Plant Physiol.* **1996**, *110*, 439–444.
42. Hennig, L.; Büche, C.; Schäfer, E. Degradation of phytochrome A and the high irradiance response in *Arabidopsis*: A kinetic analysis. *Plant Cell Environ.* **2000**, *23*, 727–734.
43. Correll, M.J.; Kiss, J.Z. The roles of phytochromes in elongation and gravitropism of roots. *Plant Cell Physiol.* **2005**, *46*, 317–323.
44. De Simone, S.; Oka, Y.; Inoue, Y. Effect of light on root hair formation in *Arabidopsis thaliana* phytochrome-deficient mutants. *J. Plant Res.* **2000**, *113*, 63–69.

45. Kiss, J.Z.; Mullen, J.L.; Correll, M.J.; Hangarter, R.P. Phytochromes A and B mediate red-light-induced positive phototropism in roots. *Plant Physiol.* **2003**, *131*, 1411–1417.
46. Boursiac, Y.; Chen, S.; Luu, D.T.; Sorieul, M.; van den Dries, N.; Maurel, C. Early effects of salinity on water transport in *Arabidopsis* roots. Molecular and cellular features of aquaporin expression. *Plant Physiol.* **2005**, *139*, 790–805.
47. Higo, K.; Ugawa, Y.; Iwamoto, M.; Korenaga, T. Plant *cis*-acting regulatory DNA elements (PLACE) database: 1999. *Nucleic Acids Res.* **1999**, *27*, 297–300.
48. Castillon, A.; Shen, H.; Huq, E. Phytochrome Interacting Factors: Central players in phytochrome-mediated light signaling networks. *Trends Plant Sci.* **2007**, *12*, 514–521.
49. Terzaghi, W.B.; Cashmore, A.R. Light-regulated transcription. *Annu. Rev. Plant Physiol. Plant Mol. Biol.* **1995**, *46*, 445–474.
50. Tatematsu, K.; Ward, S.; Leyser, O.; Kamiya, Y.; Nambara, E. Identification of *cis*-elements that regulate gene expression during initiation of axillary bud outgrowth in *Arabidopsis*. *Plant Physiol.* **2005**, *138*, 757–766.
51. Choi, H.; Hong, J.; Ha, J.; Kang, J.; Kim, S.Y. ABFs, a family of ABA-responsive element binding factors. *J. Biol. Chem.* **2000**, *275*, 1723–1730.
52. Ulmasov, T.; Hagen, G.; Guilfoyle, T.J. Dimerization and DNA binding of auxin response factors. *Plant J.* **1999**, *19*, 309–319.
53. Steudle, E. Water uptake by roots: Effects of water deficit. *J. Exp. Bot.* **2000**, *51*, 1531–1542.
54. Neuhäuser, B.; Dynowski, M.; Mayer, M.; Ludewig, U. Regulation of NH_4^+ transport by essential cross talk between AMT monomers through the carboxyl tails. *Plant Physiol.* **2007**, *143*, 1651–1659.
55. Vitart, V.; Baxter, I.; Doerner, P.; Harper, J.F. Evidence for a role in growth and salt resistance of a plasma membrane H^+-ATPase in the root endodermis. *Plant J.* **2001**, *27*, 191–201.
56. Tamura, K.; Shimada, T.; Ono, E.; Tanaka, Y.; Nagatani, A.; Higashi, S.I.; Watanabe, M.; Nishimura, M.; Hara-Nishimura, I. Why green fluorescent fusion proteins have not been observed in the vacuoles of higher plants. *Plant J.* **2003**, *35*, 545–555.
57. Deng, X.W.; Quail, P.H. Signalling in light-controlled development. *Semin. Cell Dev. Biol.* **1999**, *10*, 121–129.
58. Buchanan-Wollaston, V.; Page, T.; Harrison, E.; Breeze, E.; Lim, P.O.; Nam, H.G.; Lin, J.F.; Wu, S.H.; Swidzinski, J.; Ishizaki, K.; *et al.* Comparative transcriptome analysis reveals significant differences in gene expression and signalling pathways between developmental and dark/starvation-induced senescence in *Arabidopsis*. *Plant J.* **2005**, *42*, 567–585.
59. Maurel, C. Aquaporins and water permeability of plant membranes. *Annu. Rev. Plant Physiol. Plant Mol. Biol.* **1997**, *48*, 399–429.
60. Naito, S.; Hirai, M.Y.; Chino, M.; Komeda, Y. Expression of a soybean (*Glycine max* [L.] Merr.) seed storage protein gene in transgenic *Arabidopsis thaliana* and its response to nutritional stress and to abscisic acid mutations. *Plant Physiol.* **1994**, *104*, 497–503.
61. EZ-C1 software, version 3.80. Nikon: Tokyo, Japan, 2007.
62. Photoshop CS5 extended software, version 12.1. Adobe Systems: San Jose, CA, USA, 2011.
63. Maeshima, M. Characterization of the major integral protein of vacuolar membrane. *Plant Physiol.* **1992**, *98*, 1248–1254.
64. MultiGauge software, version 3.0. Fujifilm: Tokyo, Japan, 2004.

65. Kaufmann, S.H.; Kellner, U. Erasure of western blots after autoradiographic or chemiluminescent detection. In *Immunochemical Protocols: Methods in Molecular Biology*; Pound, J.D., Ed.; Humanna Press: Totowa, NJ, USA, 1998; Volume 80, pp. 223–235.
66. Nakagawa, T.; Kurose, T.; Hino, T.; Tanaka, K.; Kawamukai, M.; Niwa, Y.; Toyooka, K.; Matsuoka, K.; Jinbo, T.; Kimura, T. Development of series of gateway binary vectors, pGWBs, for realizing efficient construction of fusion genes for plant transformation. *J. Biosci. Bioeng.* **2007**, *104*, 34–41.
67. Clough, S.J.; Bent, A.F. Floral dip: A simplified method for *Agrobacterium*-mediated transformation of *Arabidopsis thaliana*. *Plant J.* **1998**, *16*, 735–743.
68. Nagatani, A.; Reed, J.W.; Chory, J. Isolation and initial characterization of *Arabidopsis* mutants that are deficient in phytochrome A. *Plant Physiol.* **1993**, *102*, 269–277.
69. Suga, S.; Murai, M.; Kuwagata, T.; Maeshima, M. Differences in aquaporin levels among cell types of radish and measurement of osmotic water permeability of individual protoplasts. *Plant Cell Physiol.* **2003**, *44*, 277–286.

4

Auxin and Cell Wall Invertase Related Signaling during Rice Grain Development

Sarah Russell French [1], Yousef Abu-Zaitoon [2], Md. Myn Uddin [3], Karina Bennett [1] and Heather M. Nonhebel [1,*]

[1] Molecular and Cellular Biology, University of New England, Armidale, New South Wales 2351, Australia; E-Mails: SKRussellFrench@dow.com (S.R.F.); kbennett@peracto.com (K.B.)

[2] Department of Biology, Faculty of Science, Islamic University, Madinah 170, Saudi Arabia; E-Mail: yousefaz@yahoo.com

[3] Department of Microbiology, University of Chittagong, Chittagong 4331, Bangladesh; E-Mail: mynuddincu@gmail.com

* Author to whom correspondence should be addressed; E-Mail: hnonheb2@une.edu.au

---

Abstract: Indole-3-acetic acid (IAA) synthesis is required for grain-fill in maize and appears to be regulated by cell-wall invertase (CWIN) activity. *OsYUC12* is one of three IAA biosynthesis genes we previously reported as expressed during early rice grain development, correlating with a large increase in IAA content of the grain. This work aimed to investigate further the role of *OsYUC12* and its relationship to CWIN activity and invertase inhibitors (INVINH). The analysis shows a brief peak of *OsYUC12* expression early in endosperm development. Meta-analysis of microarray data, confirmed by quantitative expression analysis, revealed that *OsYUC12* is coexpressed with *OsIAA29*, which encodes an unusual AUX/IAA transcription factor previously reported as poorly expressed. Maximum expression of *OsYUC12* and *OsIAA29* coincided with maximum CWIN activity, but also with a peak in *INVINH* expression. Unlike *ZmYUC1*, *OsYUC12* expression is not reduced in the rice CWIN mutant, *gif1*. Several reports have investigated *CWIN* expression in rice grains but none has reported on expression of *INVINH* in this species. We show that rice has 54 genes encoding putative invertase/pectin methylesterase inhibitors, seven of which are expressed exclusively during grain development. Our results suggest a more complex relationship between IAA, CWIN, and INVINH than previously proposed.

Keywords: grain-fill; rice; indole-3-acetic acid; cell wall invertase; invertase inhibitor; OsIAA29; OsHAP3D; ZmEBE-like; lipid transfer protein; defensin-like

---

1. Introduction

Several recent publications indicate key intersecting signaling roles for indole-3-acetic acid (IAA) and cell wall invertases (CWIN) during cereal grain development. The maize *defective endosperm18* (*de18*) phenotype was recently shown to result from loss of expression of the IAA biosynthesis gene *ZmYUC1* and related low levels of IAA in the developing grain [1]. The cell wall invertase *miniature1* (*mn1*) mutant also shows poor grain fill, the low levels of cell wall invertase resulting in a defective basal endosperm transfer layer (BETL) with poorly developed wall in-growths [2]. Maize *mn1* mutants have low levels of IAA and low expression of *ZmYUC1*; glucose was able to increase *ZmYUC1* transcript levels in cultured kernels [3]. Forestan *et al.* [4] showed that IAA accumulates in the BETL, aleurone and embryo surrounding region (ESR) just before the endosperm starts to accumulate starch. In addition, both BETL and ESR showed a high level of auxin transporter *ZmPIN1* transcript and protein. The rice ortholog of *Mn1*, *GIF1/OsCIN2*, also appears to be important for grain development, with *gif1* mutants (that have much lower CWIN activity in developing grains) showing poor grain-fill; expression of *GIF1* during grain development is localized to the vascular trace [5]. The relationship between IAA and invertase in rice has not been investigated.

In a previous paper on IAA synthesis in developing rice grains we showed that IAA accumulates more than 50-fold in developing kernels during endosperm cellularisation and early starch deposition [6]. The rise in IAA content was correlated with a large increase in expression of IAA biosynthesis genes *OsYUC9*, *OsYUC11*, and *OsTAR1*. We also obtained some rather inconsistent data on a third endosperm-specific YUCCA gene, *OsYUC12*. Preliminary results suggested that *OsYUC12* was expressed only briefly during grain development and may have a distinct role from the more highly expressed *OsYUC9* and *OsYUC11*. The aim of the work reported here was to clarify the expression of *OsYUC12* in developing grains and to investigate its relationship to CWIN, as well as the unexpected observation that *OsYUC12* is coexpressed with a putative invertase inhibitor (INVINH).

A number of reports, e.g., Jin *et al.* [7], indicate that invertases in dicots are regulated by INVINH. In addition, ZM-INVINH1 inhibited maize CWIN activity *in vitro*, bound to a glycoprotein fraction including CWIN, and was localized to the ESR of young developing maize kernels [8]. Several papers report on expression of *CWIN* in rice, e.g., [9], however the role of invertase inhibitors in rice has been neglected. Furthermore, extraction of invertase for assay of its activity has normally been carried out at pH 7.5, e.g., [10], conditions under which any bound inhibitor would be expected to dissociate from the enzyme [11]. In this work, we carried out a comprehensive phylogenetic analysis of invertase inhibitor homologs in rice; we investigated the expression of Os04g49720, a co-ortholog of *ZM-INVINH1* previously designated as *OsINVINH3* [8]; we extracted and assayed CWIN activity under conditions that have been shown to preserve an enzyme/inhibitor complex [8]. Our data on expression of *OsYUC12* and *OsINVINH3*, as well as CWIN activity suggest that the relationship between IAA and CWIN is more complex than previously proposed.

In addition, preliminary work suggested a second auxin related gene, *OsIAA29*, encoding a putative AUX/IAA transcription factor is coexpressed with *OsYUC12* and *OsINVINH3*. Coexpression analysis has been used productively to predict networks of genes that may act together in a common pathway, e.g., [12]. To explore other genes that may act in the same signaling pathway as *OsYUC12*, we carried out a comprehensive coexpression meta-analysis using *OsYUC12*, *OsIAA29*, and *OsINVINH3* as bait genes. The search for coexpressed genes revealed a number of sequences encoding uncharacterized proteins, some of which are homologs of proteins with an identified signaling role earlier in plant reproductive development. In addition, a number of the coexpressed genes encode homologs of molecules specifically found in endosperm transfer cells of other cereals. We suggest that the coexpressed genes may encode novel signaling factors.

2. Results

2.1. OsYUC12 and Other Endosperm YUCCA Genes Are Conserved in Cereal Species

In preparation for investigating further the role of *OsYUC12* in grain-fill we asked the question whether this gene had conserved orthologs in other cereal species. A phylogenetic analysis of *OsYUC9*, *OsYUC11*, and *OsYUC12* in maize, sorghum, and rice (Supplemental Figure S1) showed that each of these YUCCA genes has clear orthologs in each species. Secondly, we explored the expression of *GRMZM2G107761_T01*, the maize ortholog of *OsYUC12*, using microarray data from Sekhon *et al.* [13] accessed via PLEXdb [14] as experiment ZM37. This revealed that *GRMZM2G107761_T01* was also expressed primarily during grain development with a peak at eight days after pollination (DAP) (Supplemental Figure S2).

2.2. Expression of OsYUCCA Genes in the gifl CWIN Mutant

In light of the low expression and activity of ZmYUC1 in the maize CWIN mutant *mn1*, we investigated expression of *OsYUC9*, *OsYUC11*, and *OsYUC12* in the equivalent *gif1* mutant of rice using data from Wang *et al.* [5] "Global gene expression profiles of *Oryza sativa* wild type Zhonghua11 and mutant *gif*1 in filling stage" accessible as accession OS17 in PLEXdb [14]. The data shown in Figure 1 indicates that expression of *OsYUC9* only is reduced in the *gif1* mutant.

Figure 1. Comparison of expression of *OsYUC9*, *OsYUC11* and *OsYUC12* in developing grains (7 DAP) from the *gif1* mutant and wild type using RMA normalized microarray data from Wang *et al.* [5] "Global gene expression profiles of *Oryza sativa* wild type Zhonghua11 and mutant *gif*1 in filling stage" accessible as accession OS17 in PLEXdb [14]. Data represent the mean of three samples ± standard error.

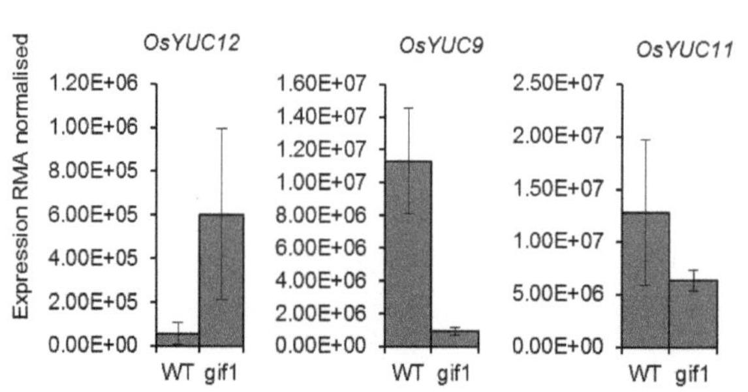

2.3. *Invertase Inhibitor Phylogeny*

A BlastP search of the rice proteome using query sequences ZM-INVINH1, ZM-INVINH2, and ZM-INVINH3 initially identified 20 homologous proteins. Additional rice homologs, as well as sequences from sorghum, were obtained using the homologous protein function in Phytozome. The phylogenetic tree shown in Figure 2 resulted from analysis of 54 protein sequences from rice, 34 sequences from sorghum, as well as experimentally characterized pectin methylesterase inhibitors (PMEIs) from wheat [15] and *Arabidopsis* [16], INVINHs from tomato [7], potato [17], and Arabidopsis [16] and F2DCS9 HORVD, an invertase inhibitor homolog from barley identified as up-regulated under heat stress by Mangelsen *et al.* [18]. The comprehensive set of homologs from the sorghum proteome was included to identify rice sequences that were conserved in more than one cereal species.

A number of bootstrap values for larger clades are low due to the highly diverse nature of the sequences. However, an examination of the multiple sequence alignment indicated that the four conserved cysteine residues previously shown to form disulfide bridges in PMEIs and INVINHs [19] were present in most sequences and were correctly aligned (Supplemental Figure S3). Branches with high bootstrap values indicate reliable prediction of phylogenetic relationships. These show that sorghum has unambiguous orthologs of ZM-INVINH1, ZM-INVINH2, and ZM-INVINH3. Rice also has clear orthologs of both ZM-INVINH1 (Os04g49730) and ZM-INVINH3 (Os01g20970). Further analysis of sequences most similar to ZM-INVINH1 indicated that the tandem repeats Os04g49730 and Os04g49720, as well as Os02g46360 are co-orthologous to ZM-INVINH1, the only characterized invertase inhibitor from a cereal [8]. Interestingly, characterized invertase inhibitors from dicot species are in separate clades from ZM-INVINH1. Arabidopsis invertase inhibitor, C/VIF1, as well as potato and tomato invertase inhibitors do not have any unambiguous cereal orthologs. On the other hand, C/VIF2 falls into a well-supported clade containing ZM-INVINH2, two rice sequences and four sorghum sequences. ZM-INVINH3 and Os01g20970 are orthologs of TaPMEI and are, therefore, likely to be PMEIs rather than invertase inhibitors. The Arabidopsis PMEIs appear in a separate branch from cereal PMEIs. The tree contains many orthologous pairs from rice and sorghum; however rice has many more sequences which appear to be due to recent gene duplication as evidenced by groups of tandem repeats, particularly on chromosome 8. The barley protein F2DCS9_HORVD, which has been given the name INVINH1 [18], is only distantly related to other characterized invertase inhibitors but has unambiguous orthologs in both rice and sorghum.

2.4. *Analysis of OsCIN2/GIF1 and Putative INVINH Expression in Developing Caryopses Using Published Microarray Data*

The extracellular invertase GIF1/OsCIN2 has been shown to be responsible for most of the CWIN activity in developing rice grains [5]. We mined microarray data for information on expression of *GIF1* as well as the 54 invertase inhibitor homologs during grain development. Data was accessed via PLEXdb [14] from experiment accessions OS5 "Expression data for reproductive development in rice" [20], OS8 "Expression data from rice embryo, endosperm, root, leaf, and seedling" [21], OS16

"Genome-wide gene expression profiling of rice stigma" [22], OS44 "Rice expression atlas (3): Early embryogenesis" [23] and OS89 "Expression data from rice embryo and endosperm development" [24].

Figure 2. Phylogenetic tree showing relationships between INVINH and PMEI homologs from rice and Sorghum; protein sequences for ZM-INVINH1, ZM-INVINH2, ZM-INVINH3, and experimentally characterized PMEIs and INVINH from wheat, potato, tomato, and *Arabidopsis* are also included for comparison. The tree was produced in MEGA5.2 [25] using the Neighbor-Joining method [26]. MUSCLE [27] was used for multiple sequence alignment. The bootstrap consensus tree was inferred from 500 replicates [28]. Evolutionary distances were computed using the Poisson correction method [29]. All ambiguous positions were removed for each sequence pair. Highlighted sequences are encoded by genes expressed in endosperm during early caryopsis development.

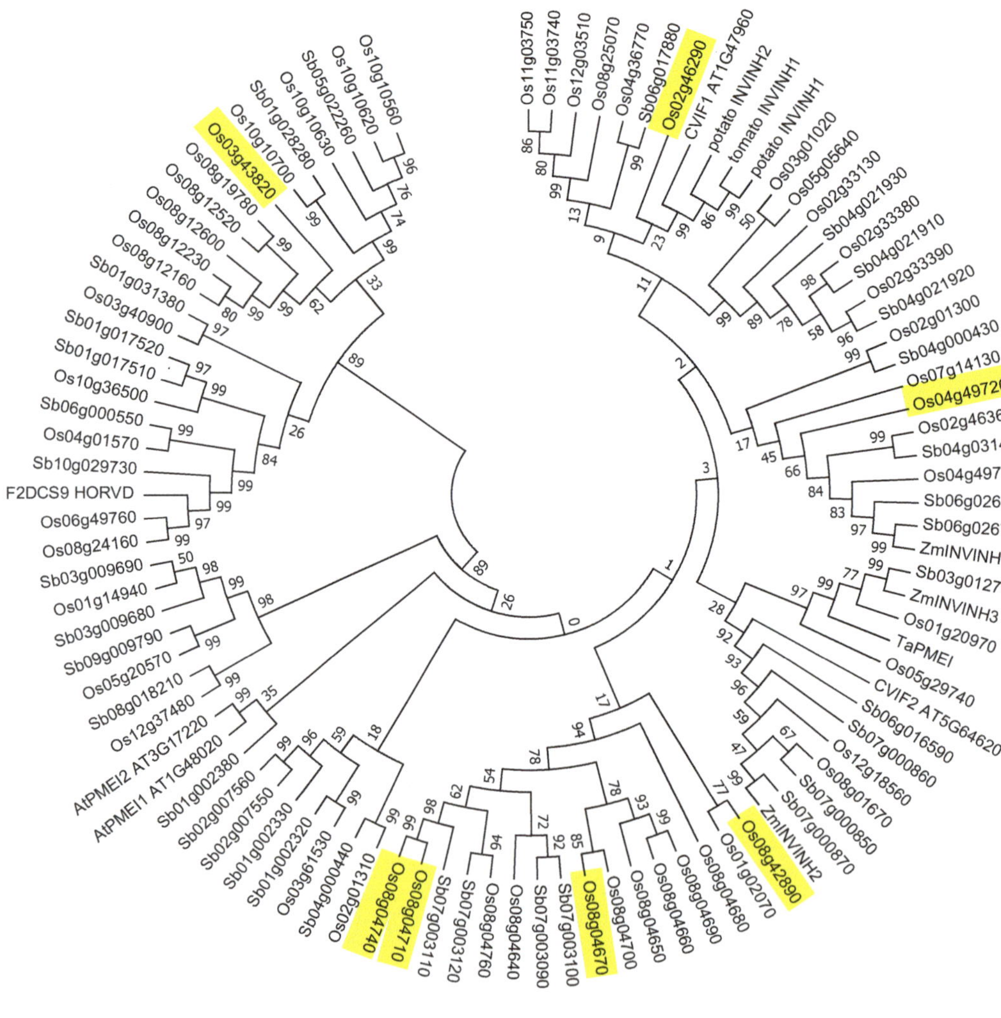

Seven *INVINH/PMEI* homologs (Table 1) were found to be expressed exclusively during caryopsis development with expression highest in samples three to four days after pollination (DAP); expression appeared specific to the endosperm. Caryopsis-expressed INVINH/PMEI homologs included Os04g49720 which is co-orthologous with Os04g49730 and Os02g46360 to ZM-INVINH1. Os04g49730 and Os02g46360 appeared to be only weakly expressed; Os04g49720 has previously been designated as OsINVINH3 [8]. In contrast to the large changes in expression of *OsINVINH3*, there was little change in the expression of *GIF1* from prior to anthesis until 20 DAP (Figure 3). Interestingly examination of endosperm-specific genes in Figure 2 showed them to be present in diverse clades. We also noted that the expression profiles of genes in Table 1 matched that of IAA biosynthesis gene *OsYUC12* and the gene encoding putative AUX/IAA transcriptional regulator OsIAA29.

Table 1. Invertase inhibitor/pectin methylesterase inhibitor (INVINH/PMEI) homologs from rice identified as expressed during caryopsis development using microarray data accessed via PLEXdb [14] from experiment accessions OS5 "Expression data for reproductive development in rice" [26], OS8 "Expression data from rice embryo, endosperm, root, leaf, and seedling" [21], OS16 "Genome-wide gene expression profiling of rice stigma" [27], OS44 "Rice expression atlas (3): Early embryogenesis" [28], and OS89 "Expression data from rice embryo and endosperm development" [29].

Locus ID	Expression
Os02g46290.1	2 to 4 DAP, endosperm
Os03g43820.1	3-4 DAP, 5-10 DAP, endosperm
Os04g49720.1	3-4 DAP, 5-10 DAP, endosperm
Os08g04670.1	3-4 DAP, endosperm
Os08g04710.1	3-4 DAP, endosperm
Os08g04740.1	3-4 DAP, 5-10 DAP, endosperm
Os08g42890.1	3-4 DAP, 5-10 DAP, endosperm

Figure 3. Comparison of expression of cell wall invertase *GIF1* (Os04g33740) and *OsINVINH3* (Os04g49720) during grain development using RMA normalized data from Expression data for reproductive development in rice [26] accessed as accession OS5 from PLEXdb [14]. Data represent the mean of three samples ± standard error. Data on expression of other cell wall invertases during grain development from the same source is provided as Supplemental Figure S4.

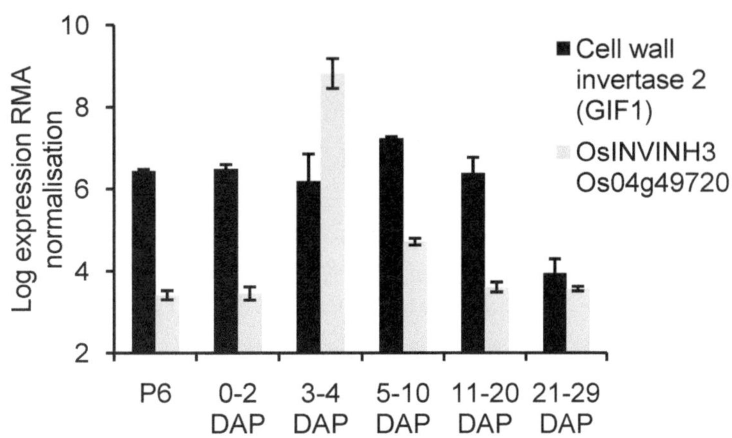

2.5. Coexpression Meta-Analysis Using Bait Genes OsIAA29, OsYUC12 and OsINVINH3

Several on-line platforms are available for coexpression analysis including RiceNet [30], GeneCAT [31], and Rice Oligonucleotide Array Database [32]. Initial investigations indicated that the limited data set used in RiceNet led to the identification of some genes that had poorly correlated expression with our genes of interest in data sets, such as OS44 [28], which includes samples collected daily during the early stages after pollination. Rice Oligonucleotide Array Database contains more data sets but as most of these did not contain caryopsis tissue and the site does not allow for condition dependent selection of data sets, results were similarly unsatisfactory. We therefore downloaded RMA normalized datasets via PLEXdb for previously mentioned accessions OS5, OS8, OS16, OS44, and OS89 and calculated pairwise mutual coexpression ranks [33] for three bait genes: *OsIAA29*, *OsYUC12*, and *OsINVINH3*. Examination of the expression profiles of genes with a mutual co-expression rank <50 in individual experiments indicated very close correspondence with expression of bait genes. All appear to be expressed exclusively in the endosperm of developing grains with maximal expression at 3–4 DAP.

Genes with a mutual co-expression rank less than 20 (MR < 20) are shown in Table 2, with a larger set (MR < 50) shown in Supplemental Table S1. All database annotations have been manually checked against the literature; where the annotation differs from the literature, the published protein designation has been used. Table 2 includes two lipid transfer protein-like genes (*LTPL*) and four defensin/defensin like genes (*DEF/DEFL*). This group of genes encoding cysteine-rich peptides (CRPs) was also highly represented in the larger MR < 50 group, with a total of 13 genes present. Also present was a *ZmEBE-like* (embryo sac/basal endosperm transfer layer/embryo surrounding region) gene; a further four homologs of this gene are found in the MR <50 group. A third group of homologous proteins found in the MR <50 group included two additional INVINH/PMEI homologs. Table 2 also includes genes encoding OsHAP3D/OsLEC1A (Leafy cotyledon1) [34] and a leucine-rich repeat receptor-like kinase. Genes with a likely signaling role in the MR <50 group also include OsRR33, a type-B response regulator [35] an ortholog of OBERON1/2 [36] and at least two other proteins with a predicted nuclear location.

Table 2. Genes with a pairwise mutual coexpression rank < 20 with each of the bait genes, *INVINH3*, *OsIAA29*, and *OsYUC12*. Genes listed in order of increasing mutual co-expression rank.

Transcript ID	Description
Os12g13960.1	Lipid transfer protein-like LTPL33 [37]
Os02g07628.1	Defensin DEF5 [37]
Os11g45360.1	Defensin-like DEFL15 [37]
Os01g70680.1	Defensin DEF1 [37]
Os02g07624.1	Defensin DEF4 [37]
Os11g18140.1	homologous to ZmEBE 1 (embryo sac/basal endosperm transfer layer/embryo surrounding region [38]
Os06g17480.1	Nuclear factor YC OsHAP3D /OsLEC1A [34]
Os05g07850.1	Leucine-rich repeat receptor-like kinase [39]
Os11g14880.1	Lipid transfer protein-like LTPL32 [37]

2.6. Quantitative Expression Analysis of Genes Coexpressed with OsIAA29, OsYUC12, and OsINVINH3

To confirm the expression profile of bait genes *OsYUC12, OsIAA29*, and *INVINH3* along with coexpressed genes *LTPL32, DEFL13, ZmEBE-like Os11g18140*, and *OsHAP3D* in plants of the same variety and grown under the same conditions as our previously published IAA measurements, we carried out quantitative RT-PCR analysis. *LTPL32, DEFL13, ZmEBE-like Os11g18140* were chosen as representatives from the LTPL, DEFL, and ZmEBE-like gene families, as these genes contained introns allowing for primers to be designed across exon boundaries. Expression data are presented in Figure 4. All genes showed a short-lived increase in expression during early grain development; under our plant growth conditions an increase in expression was seen at 4 DAP (apart from *OsINVINH3*) with highest expression apparent in 7 DAP samples. Large error bars for some 4 DAP samples resulted from one of the biological replicates in which the major increase in expression had occurred earlier than the other two samples. The early increase in expression was seen in most genes when using the same RNA template.

Figure 4. Quantitative RT–PCR results for *OsIAA12*, *OsIAA29*, *OsINVINH3*, *OsLTPL32*, *OsDEFL13*, *ZmEBE-like Os11g18140*, and *OsHAP3D*. Total RNA was extracted from developing rice grains 1–15 DAP using a Bioline ISOLATE Plant RNA Mini Kit. Each tube contained 50 ng RNA per 20 μL reaction and a final primer concentration of 0.5 μM. Amplification was carried out using a Bioline SensiFAST™SYBR No-ROX RT-PCR kit in a Rotor-Gene Q (QIAGEN) thermocycler fitted with a 100-tube rotor for 45 cycles. Controls (no reverse transcriptase) were included for *OsINVINH3* and *OsHAP3D.* Expression was calculated relative to reference gene UBC using ROTORGENE software (QIAGEN). Melt curve analysis confirmed that a single product was obtained for each gene amplified. Results represent the means from three biological replicates ± standard error of the means.

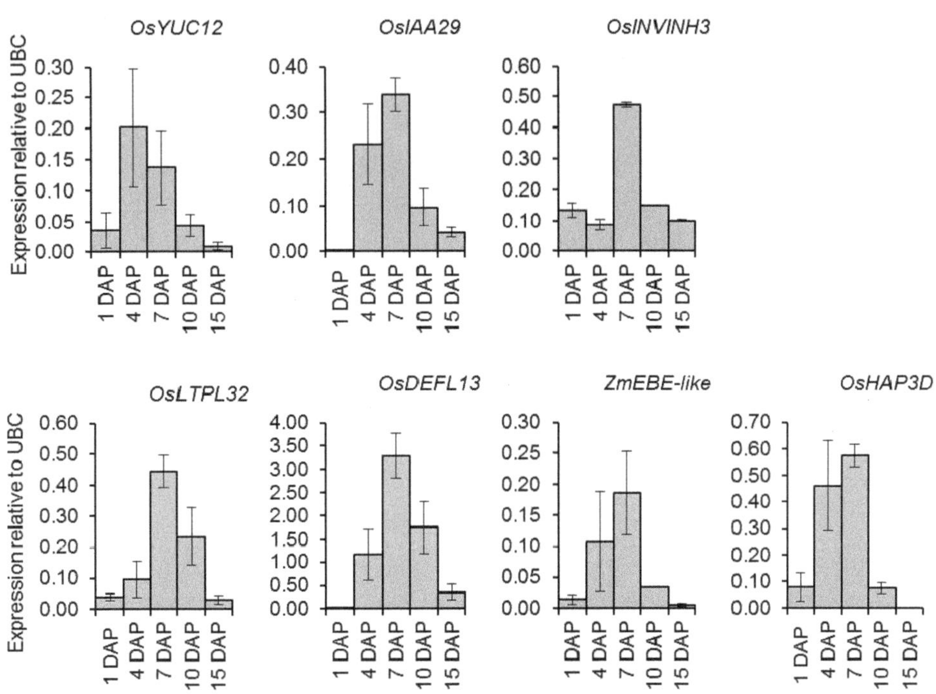

2.7. Invertase Activity in Developing Rice Grains

Cell wall invertase activity during grain development is shown in Figure 5. Total cell wall invertase shown on a per grain basis increased to a maximum at 7 DAP; when calculated on a per g fresh weight basis, no significant difference was seen between samples from 1–7 DAP with a substantial decrease at 10 DAP.

Figure 5. Changes in activity of cell wall invertase expressed per grain extracted from developing rice grains. Values represent the mean of three biological replicates ± standard error.

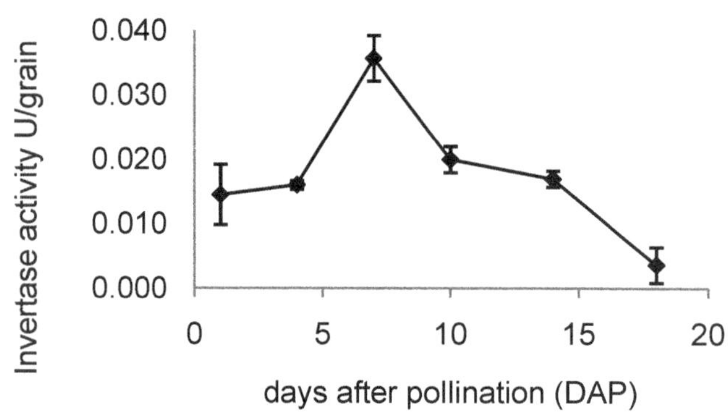

3. Discussion

3.1. Expression of IAA Related Genes, OsYUC12 and OsIAA29 Coincides with Endosperm Cellularisation

Previous research using maize and rice has suggested a key role for IAA and CWIN during early grain-fill however the exact roles of IAA and CWIN remain to be established. Our results indicated that the relationship between IAA and CWIN is complex, with differences in expression of three IAA biosynthesis genes *OsYUC9*, *OsYUC11*, and *OsYUC12* during grain development. In this paper we explore the role of one IAA biosynthesis gene, *OsYUC12*. We chose to investigate this gene as to our knowledge there have been no other studies of it or its orthologs. We have shown that *OsYUC12* has conserved orthologs in maize and sorghum, it is specifically expressed in endosperm for a short time during early grain development and this expression pattern also appears in maize. We, therefore, suggest that this gene has a specific and conserved signaling role during grain development. The aim of this work was to identify other possible components of the signaling network that could be investigated.

An examination of on-line accessible microarray data indicated that AUX/IAA gene *OsIAA29* has a very similar expression pattern to *OsYUC12*; this was confirmed by qRT-PCR. Our quantitative expression data confirm that expression of *OsYUC12* and *OsIAA29* is limited to a short period during early grain-fill. The increase in expression coincided with our previously reported large increase in the IAA content of developing grains at 7 DAP in plants grown under the same conditions [6]. Examination of material under the light microscope indicated that, at 4 DAP, the endosperm syncytium contained many nuclei and a few small irregularly shaped cells. At day seven there were clearly defined endosperm cells containing amyloplasts, as well as differentiated outer cellular layers. The coexpression of *OsYUC12* and *OsIAA29* in the endosperm over a very limited period during grain development, suggests a specific role for IAA during this period. However, OsIAA29 is an unusual

AUX/IAA protein that lacks domains I and II [40]. As domain II is responsible for interacting with the F-box protein, IAA co-receptor [41] it is unclear how OsIAA29 may interact with IAA. Dimerization domains III and IV, which mediate interactions between AUX/IAA proteins and ARFs are still present in OsIAA29 [40]. In work with seedling tissue, Jain *et al.* [40] reported very low expression of *OsIAA29* and no expression of its tandem repeat *OsIAA28*. Our observations showing high expression at a very specific stage of development suggest that *OsIAA29* may in fact play a key but as yet unidentified role. This is further supported by the existence of *OsIAA29* orthologs in both sorghum and maize and our recent unpublished observation that the sorghum ortholog is also expressed during grain-fill.

3.2. Expression of OsINVINH3 Coincides with Maximum Extracellular Invertase Activity

LeClere *et al.* [42] have provided evidence that *ZmYUC1* expression may be regulated by glucose levels produced as a result of CWIN activity. We therefore investigated the expression of *GIF1* during grain development. Examination of microarray data indicated *GIF1* expression appeared to change little from prior to pollination until 20 DAP. However, we were surprised to find that expression of Os04g49720 (*OsINVINH3*), a co-ortholog of ZM-INVINH1 [8] changed dramatically during early grain-development. Invertase inhibitor ZM-INVINH1 has been reported to bind CWIN during grain development in maize [8]; both ZM-INVINH1 and OsINVINH3 have predicted signal peptides. Quantitative expression analysis confirmed that Os04g49720 showed a large, short-lived increase in expression during early grain development. In plants grown under our conditions maximum expression was found in 7-DAP samples, also coinciding with the peak in *OsYUC12* and *OsIAA29* expression and rapidly increasing IAA levels [6]. We were surprised to also observe maximum invertase activity on a per grain basis at 7 DAP. The unexpected result was confirmed by repetition of the experiment by a second person in the laboratory. Clearly, further work is required to clarify the relationship between CWIN and OsINVINH3, however, our data support the suggestion by Wang *et al.* [5] that CWIN in developing grains may be tightly regulated.

3.3. Rice Has 54 INVINH Homologs Several of Which Are Expressed Specifically in Developing Endosperm

Invertase inhibitors have similar structure and are homologous to pectin methylesterase inhibitors [19]. A comprehensive search for members of this family revealed 54 diverse homologous proteins in rice. Sorghum homologs were also included to distinguish between species-specific proteins and those that have been conserved. Although the high sequence diversity resulted in some ambiguous relationships, both sorghum and rice have clear orthologs of ZM-INVINH1. In addition, a small well-supported clade containing the experimentally characterized wheat PMEI suggested that ZM-INVINH3, Sb03g012790, Os01g20970 and Os05g29740 are likely to be PMEIs. *OsINVINH3* is not the only member from the family to be expressed exclusively during grain development. Os02g46290, Os03g43820, Os08g42890, Os08g04670, Os08g04710, and Os08g04740 are also grain specific and expressed maximally at 3–4 DAP. Exploration of data from OS8 [21], OS16 [27], and OS89 [29] indicated that expression is restricted to the endosperm. The seven endosperm-specific proteins are found in four clades; Os03g43820 is present in a large clade quite separate from the others, but including the barley protein F2DCS9 HORVD. Os08g04670, Os08g04710, and Os08g04740 are part of a cluster of nine genes found in tandem; Os08g42890 appears to be part of the same clade though this is not strongly supported. Os02g46290 is a possible ortholog of C/VIF1, an experimentally characterized vacuolar invertase inhibitor from Arabidopsis. The high diversity of these proteins in separate clades of a tree that is known to contain proteins of different biological functions raises the possibility that not all are invertase inhibitors or PMEIs. We suggest that some members of the INVINH/PMEI protein family may regulate other carbohydrate metabolizing enzymes. An investigation of the predicted sub-cellular locations of the gene products, using CELLO [43] and TargetP [44], suggests that all proteins have a signal peptide, however an examination of extracellular and vacuolar invertase inhibitors from potato [17] shows that domain prediction programs cannot reliably distinguish between these two sub-cellular destinations.

3.4. Putative Signaling Proteins Coexpressed with OsYUC12, OsIAA29 and OsINVINH3

The observation that *OsYUC12*, *OsIAA29*, and *OsINVINH3* are expressed exclusively in endosperm samples from developing grains 3–10 DAP suggest they may be co-regulated and may be part of a common signaling pathway. A meta-analysis of microarray data indicated genes from three families, *CRPs*, *ZmEBE-like*, and *INVINH/PMEI-like*, all encoding putative extracellular proteins, are prominent among genes coexpressed with *OsYUC12*, *OsIAA29*, and *OsINVINH1*. The CRPs come from two distinct groups; the so-called lipid transfer protein-like group and the defensin/defensin-like group. We note that despite their reported antimicrobial activity, defensins have recently been shown to have a signaling role earlier in maize reproductive development [45]. In addition, the CRP MEG1 is essential for the differentiation of ETCs in maize [46]. *OsLTPL29/OsPR602* and *DEFL OsPR9a*, as well as *ZmEBE1*, *ZmEBE2*, and *ZM-INVINH1*, are also expressed primarily in ETCs and ESR [38,47]. Differentiating transfer cells in maize kernels accumulate high levels of auxin transporter *OsPIN1* transcripts [4]; the BETL and ESR also show high levels of PIN proteins as well as an accumulation of IAA itself. The specific localization of rice proteins found in Table 2 should be investigated to determine whether they are also restricted to the ETCs and/or ESR.

Genes with a mutual co-expression rank of <20 included also *OsHAP3D* [34] and Os05g07850, a putative leucine-rich repeat receptor-like kinase. *OsHAP3D*, also referred to as *OsLEC1a*, encodes a nuclear factor YC protein, homologous to Arabidopsis LEC1/L1L. These Arabidopsis proteins have a crucial role in embryogenesis [48] including the up-regulation of *YUCCA10* [49]. Rice has two homologs of LEC1, OsHAP3D, and OsHAP3E. An examination of their expression profile via PLEXdb indicates that both are expressed in developing grains with a peak at 3–4 DAP. However, whereas *OsHAP3D* is expressed in endosperm, *OsHAP3E* expression appears to be restricted to the embryo. Os05g07850 encodes a leucine-rich repeat receptor-like kinase which has a predicted trans-membrane domain and serine/threonine kinase domain. It is homologous but not orthologous to the brassinolide receptor BRI1 and aligns in the same clade as Arabidopsis NSP interacting kinases. A partially homologous receptor-like kinase has been shown to form part of a signaling network that also includes a DEFL during fertilization in maize [50]. The identification of these coexpressed genes provides a framework for investigating protein-protein interactions that could form part of a signaling network.

4. Experimental

4.1. Bioinformatic Analysis

Rice protein sequences were downloaded via Phytozome 8.0 [51] from MSU Release 7.0 of the Rice Genome Annotation; protein sequences for ZM-INVINH1, ZM-INVINH2, ZM-INVINH3, and experimentally characterized pectin methyl esterase inhibitors (PMEI) and invertase inhibitors (INVINH) from wheat, potato, tomato, and Arabidopsis were downloaded from Uniprot. The phylogenetic analysis of protein sequences was carried out in MEGA5.2 [25] using the Neighbor-Joining method [26]. MUSCLE [27] was used for multiple sequence alignment. The bootstrap consensus tree was inferred from 500 replicates [28]. Evolutionary distances were computed using the Poisson correction method [24]. All ambiguous positions were removed for each sequence pair. Normalized microarray data from accessions OS5 [20], OS8 [21], OS16 [22], OS17 [5], OS44 [23], and OS89 [24] was downloaded via PLEXdb [14] and analyzed in Excel. Pairwise mutual coexpression ranks were calculated as per Obayashi *et al.* [33].

4.2. Plant Material

Rice grains (ssp. Japonica cv. Jarrah) were direct sown into flooded soil-filled cylindrical plastic pots and thinned out to three plants per pot when the seedlings had reached the 2–3 leaf growth stage. Plants were grown in a glasshouse under a natural light with day/night temperatures 28 °C/18 °C. Plants were fertilized fortnightly using Aquasol® (8 g/5 L). When spikelets in the top half of the panicle had reached anthesis, panicles were tagged to record the date; panicles were harvested at 1, 4, 7, 10, and 15 days after pollination (DAP). Grain samples were placed in microfuge tubes, the weight and number of grains recorded before freezing in liquid nitrogen and storage at −80 °C.

4.3. Quantitative Reverse Transcriptase PCR

Total RNA was extracted from 60–80 mg grain samples following instructions for the Bioline ISOLATE Plant RNA Mini Kit. RNA concentration and 260/280 ratio were determined using a NanoDrop ND-1000 Spectrophotometer (NanoDrop Technologies, Inc., Wilmington, DE, USA). Only samples with A_{260}/A_{280} of 1.8–2.0 were used for further analysis. The RNA quality was also checked using agarose gel electrophoresis for two clear bands of 18s and 28s rRNA [52].

Gene-specific primers (See Supplemental Table S2) were designed with melting points in the range 59–61 °C and product sizes between 100–150 bp. Primers were tested for amplification of a single product of the expected size using a QIAGEN OneStep RT-PCR kit (QIAGEN, Limburg, The Netherlands), with agarose gel analysis of products. To avoid the possibility of amplifying genomic DNA, either left or right primers were designed to span an exon-exon boundary. Controls (no reverse transcriptase) were included for *INVINH1* and *OsHAP3D* genes, which had no introns; amplification was not detected in the controls. Quantitative RT-PCR was carried out following manufacturer's instructions using 50 ng RNA per 20 µL reaction, a final primer concentration of 0.5 µM and reagents from a Bioline SensiFAST™SYBR No-ROX RT-PCR kit (Bioline, London, UK). Reaction tubes were prepared by using a QIAgility robotic liquid handling system (QIAGEN). Reactions were monitored in a Rotor-Gene Q (QIAGEN) thermocycler fitted with a 100-tube gene disc rotor. The amplification program included 48 °C min, 95 °C 2 min and 45 cycles of 95 °C 5 s, 58 °C 10 s, and 72 °C 5 s. Expression was calculated relative to reference gene for ubiquitin-conjugating enzyme E2 (*OsUBC*) [53] using ROTORGENE software (QIAGEN). Melt curve analysis confirmed that a single product was obtained for each gene amplified.

4.4. Invertase Assay

Rice grain samples were extracted using a procedure modified from Tomlinson *et al.* [54]. Samples (200 mg) were ground to a fine powder in liquid N_2 in a mortar and pestle then transferred to a close fitting glass homogenizer and ground further with 1 mL of extraction buffer (pH 4.8, 100 mM acetic acetate buffer, containing 8 mM $MgCl_2$, 2 mM EDTA and 12.5% (w/v) glycerol, 10 mM dithiothreitol and 1 mM phenylmethanesulfonyl fluoride). The homogenate plus a 400 µL wash were transferred to a 2 mL microfuge tube and centrifuged at 16,000 ×*g* for 10 min. The pellet was washed and re-centrifuged twice with 500 µL of extraction buffer before resuspending in two volumes of salt extraction buffer (pH 4.8 100 mM acetate buffer, 1 M NaCl). Samples were extracted for 1 h on ice with gentle agitation, as well as 3× sonication for 60 s. Following salt extraction, samples were centrifuged at 16,000 ×*g* for 10 min and the supernatant retained for assay of CWIN activity.

Reaction mixtures containing 80 µL of enzyme preparation, 10 mM sucrose and pH 4.8, 100 mM acetate buffer to a total volume of 440 µL were incubated at 37 °C for 30 min with gentle shaking. Three 40 µL aliquots were taken for glucose determination at 0 and 30 min. Glucose in the reaction mixture was measured using the Somogyi-Nelson method [55]; and calculated in comparison with a set of standards containing 5 to 100 µg glucose.

5. Conclusions

In conclusion, we have shown that the endosperm-specific IAA synthesis gene *OsYUC12* and the AUX/IAA gene *OsIAA29* are coexpressed, with expression limited to a short period during endosperm cellularisation. Extractable CWIN activity, as well as expression of putative invertase inhibitor *OsINVINH3*, also reached a maximum at 7 DAP, coinciding with maximal expression of *OsYUC12* and *OsIAA29*. The possible regulation of CWIN activity by OsINVINH3 needs to be considered as part of any investigation of the role of CWIN during grain development. We have shown that rice has at least 54 genes encoding diverse proteins homologous to invertase inhibitors and PMEIs. In addition to *OsINVINH3*, a rice ortholog of *ZM-INVINH1*, six other *INVINH/PMEI* genes from widely divergent clades were specifically expressed in endosperm during early grain-fill. We explored other genes co-expressed with *OsYUC12*, *OsIAA29*, and *OsINVINH3*. Seven out of the nine genes with MR<20 appear to encode extracellular proteins homologous to molecules expressed exclusively in ETCs/ESR of other plants; these include six cysteine-rich proteins, some with similarity to BETL1 from maize. In addition, the presence of an endosperm-specific *LEC1* ortholog, *OsHADP3D* and a receptor–like kinase gene suggests that these genes and their encoded proteins are a priority for experimental investigation in regard to a signaling network regulating differentiation of ETCs in rice. The expression of *OsYUC12* is distinct from *OsYUC9* and *OsYUC11*. It has orthologs in sorghum and maize; the maize ortholog is also expressed for a short time during grain development, suggesting conserved subfunctionalization. Expression of *OsYUC9* but not *OsYUC12* is reduced in the *gif1* CWIN mutant. We suggest that further investigation of the specific roles of three IAA biosynthesis genes will be necessary to reveal the full extent of the relationship between CWIN and IAA.

Acknowledgements

This work was supported by funding from the University of New England.

Author Contributions

Sarah Russell French carried out quantitative gene expression analysis, Yousef Abu Zaitoon completed the coexpression analysis, Myn Uddin carried out the search for invertase inhibitor/PMEI sequences, invertase enzyme assays were done by Karina Bennett. Heather Nonhebel supervised all work, carried out data analysis and wrote the paper.

Conflicts of Interest

The authors declare no conflict of interest.

References

1. Bernardi, J.; Lanubile, A.; Li, Q.-B.; Kumar, D.; Kladnik, A.; Cook, S.D.; Ross, J.J.; Marocco, A.; Chourey, P.S. Impaired auxin biosynthesis in the defective endosperm18 mutant is due to mutational loss of expression in the ZmYUC1 gene encoding endosperm-specific YUCCA1 protein in maize. *Plant Physiol.* **2012**, *160*, 1318–1328.
2. Kang, B.-H.; Xiong, Y.; Williams, D.S.; Pozueta-Romero, D.; Chourey, P.S. Miniature1-encoded cell wall invertase is essential for assembly and function of wall-in-growth in the maize endosperm transfer cell. *Plant Physiol.* **2009**, *151*, 1366–1376.
3. LeClere, S.; Schmelz, E.A.; Chourey, P.S. Sugar levels regulate tryptophan-dependent auxin biosynthesis in developing maize kernels. *Plant Physiol.* **2010**, *153*, 306–318.
4. Forestan, C.; Meda, S.; Varotto, S. ZmPIN1-mediated auxin transport is related to cellular differentiation during maize embryogenesis and endosperm development. *Plant Physiol.* **2010**, *152*, 1373–1390.
5. Wang, E.; Wang, J.; Zhu, X.D.; Hao, W.; Wang, L.Y.; Li, Q.; Zhang, L.X.; He, W.; Lu, B.R.; Lin, H.X.; *et al.* Control of rice grain-filling and yield by a gene with a potential signature of domestication. *Nat. Genet.* **2008**, *40*, 1370–1374.
6. Abu-Zaitoon, Y.M.; Bennett, K.; Normanly, J.; Nonhebel, H.M. A large increase in IAA during development of rice grains correlates with the expression of tryptophan aminotransferase OsTAR1 and a grain-specific yucca. *Physiol. Plant.* **2012**, *146*, 487–499.
7. Jin, Y.; Ni, D.-A.; Ruan, Y.-L. Posttranslational elevation of cell wall invertase activity by silencing its inhibitor in tomato delays leaf senescence and increases seed weight and fruit hexose level. *Plant Cell Online* **2009**, *21*, 2072–2089.

8. Bate, N.J.; Niu, X.P.; Wang, Y.W.; Reimann, K.S.; Helentjaris, T.G. An invertase inhibitor from maize localizes to the embryo surrounding region during early kernel development. *Plant Physiol.* **2004**, *134*, 246–254.
9. Cho, J.I.; Lee, S.K.; Ko, S.H.; Kim, H.K.; Jun, S.H.; Lee, Y.H.; Bhoo, S.H.; Lee, K.W.; An, G.H.; Hahn, T.R.; *et al.* Molecular cloning and expression analysis of the cell-wall invertase gene family in rice (*Oryza sativa* L.). *Plant Cell Rep.* **2005**, *24*, 225–236.
10. Ishimaru, T.; Hirose, T.; Matsuda, T.; Goto, A.; Takahashi, K.; Sasaki, H.; Terao, T.; Ishii, R.; Ohsugi, R.; Yamagishi, T. Expression patterns of genes encoding carbohydrate-metabolizing enzymes and their relationship to grain filling in rice (*Oryza sativa* L.): Comparison of caryopses located at different positions in a panicle. *Plant Cell Physiol.* **2005**, *46*, 620–628.
11. Hothorn, M.; van den Ende, W.; Lammens, W.; Rybin, V.; Scheffzek, K. Structural insights into the pH-controlled targeting of plant cell-wall invertase by a specific inhibitor protein. *Proc. Natl. Acad. Sci. USA* **2010**, *107*, 17427–17432.
12. Fu, F.-F.; Xue, H.-W. Coexpression analysis identifies rice starch regulator1, a rice AP2/EREBP family transcription factor, as a novel rice starch biosynthesis regulator. *Plant Physiol.* **2010**, *154*, 927–938.
13. Sekhon, R.S.; Lin, H.N.; Childs, K.L.; Hansey, C.N.; Buell, C.R.; de Leon, N.; Kaeppler, S.M. Genome-wide atlas of transcription during maize development. *Plant J.* **2011**, *66*, 553–563.
14. Dash, S.; van Hemert, J.; Hong, L.; Wise, R.P.; Dickerson, J.A. PLEXdb: Gene expression resources for plants and plant pathogens. *Nucleic Acids Res.* **2011**, *40*, D1194–D1201.
15. Hong, M.J.; Kim, D.Y.; Lee, T.G.; Jeon, W.B.; Seo, Y.W. Functional characterization of pectin methylesterase inhibitor (PMEI) in wheat. *Genes Genet. Syst.* **2010**, *85*, 97–106.
16. Rausch, T.; Greiner, S. Plant protein inhibitors of invertases. *Biochim. Biophys. Acta* **2004**, *1696*, 253–261.
17. Liu, X.; Song, B.T.; Zhang, H.L.; Li, X.Q.; Xie, C.H.; Liu, J. Cloning and molecular characterization of putative invertase inhibitor genes and their possible contributions to cold-induced sweetening of potato tubers. *Mol. Genet. Genomics* **2010**, *284*, 147–159.
18. Mangelsen, E.; Kilian, J.; Harter, K.; Jansson, C.; Wanke, D.; Sundberg, E. Transcriptome analysis of high-temperature stress in developing barley caryopses: Early stress responses and effects on storage compound biosynthesis. *Mol. Plant* **2011**, *4*, 97–115.
19. Scognamiglio, M.A.; Ciardiello, M.A.; Tamburrini, M.; Carratore, V.; Rausch, T.; Camardella, L. The plant invertase inhibitor shares structural properties and disulfide bridges arrangement with the pectin methylesterase inhibitor. *J. Protein Chem.* **2003**, *22*, 363–369.
20. Jain, M.; Nijhawan, A.; Arora, R.; Agarwal, P.; Ray, S.; Sharma, P.; Kapoor, S.; Tyagi, A.K.; Khurana, J.P. F-box proteins in rice. Genome-wide analysis, classification, temporal and spatial gene expression during panicle and seed development, and regulation by light and abiotic stress. *Plant Physiol.* **2007**, *143*, 1467–1483.
21. Xue, H.W. Institute of Plant Physiology and Ecology,Shanghai Institutes of Biological Sciences, Shanghai, China. Unpublished data, 2009.
22. Li, M.N.; Xu, W.Y.; Yang, W.Q.; Kong, Z.S.; Xue, Y.B. Genome-wide gene expression profiling reveals conserved and novel molecular functions of the stigma in rice. *Plant Physiol.* **2007**, *144*, 1797–1812.

23. Fujita, M.; Horiuchi, Y.; Ueda, Y.; Mizuta, Y.; Kubo, T.; Yano, K.; Yamaki, S.; Tsuda, K.; Nagata, T.; Niihama, M.; *et al.* Rice expression atlas in reproductive development. *Plant Cell Physiol.* **2010**, *51*, 2060–2081.
24. Gao, L.-L.; Xue, H.-W. Global analysis of expression profiles of rice receptor-like kinase genes. *Mol. Plant* **2012**, *5*, 143–153.
25. Tamura, K.; Peterson, D.; Peterson, N.; Stecher, G.; Nei, M.; Kumar, S. Mega5: Molecular evolutionary genetics analysis using maximum likelihood, evolutionary distance, and maximum parsimony methods. *Mol. Biol. Evol.* **2011**, *28*, 2731–2739.
26. Saitou, N.; Nei, M. The neighbor-joining method—A new method for reconstructing phylogenetic trees. *Mol. Biol. Evol.* **1987**, *4*, 406–425.
27. Edgar, R.C. Muscle: Multiple sequence alignment with high accuracy and high throughput. *Nucleic Acids Res.* **2004**, *32*, 1792–1797.
28. Felsenstein, J. Confidence limits on phylogenies: An approach using the bootstrap. *Evolution* **1985**, *39*, 783–791.
29. Zuckerkandl, E.; Pauling, L. Evolutionary divergence and convergence in proteins. In *Evolving Genes and Proteins*; Bryson, V., Vogel, H.J., Eds.; Academic Press: New York, NY, USA, 1965; pp. 97–166.
30. Mutwil, M.; Klie, S.; Tohge, T.; Giorgi, F.M.; Wilkins, O.; Campbell, M.M.; Fernie, A.R.; Usadel, B.; Nikoloski, Z.; Persson, S. PlaNet: Combined sequence and expression comparisons across plant networks derived from seven species. *Plant Cell* **2011**, *23*, 895–910.
31. Mutwil, M.; Obro, J.; Willats, W.G.T.; Persson, S. GeneCAT—Novel webtools that combine blast and co-expression analyses. *Nucleic Acids Res.* **2008**, *36*, W320–W326.
32. Cao, P.; Jung, K.-H.; Choi, D.; Hwang, D.; Zhu, J.; Ronald, P. The rice oligonucleotide array database: An atlas of rice gene expression. *Rice* **2012**, *5*, doi:10.1186/1939-8433-5-17.
33. Obayashi, T.; Hayashi, S.; Saeki, M.; Ohta, H.; Kinoshita, K. ATTED-II provides coexpressed gene networks for arabidopsis. *Nucleic Acids Res.* **2009**, *37*, D987–D991.
34. Thirumurugan, T.; Ito, Y.; Kubo, T.; Serizawa, A.; Kurata, N. Identification, characterization and interaction of hap family genes in rice. *Mol. Genet. Genomics* **2008**, *279*, 279–289.
35. Tsai, Y.-C.; Weir, N.R.; Hill, K.; Zhang, W.; Kim, H.J.; Shiu, S.-H.; Schaller, G.E.; Kieber, J.J. Characterization of genes involved in cytokinin signaling and metabolism from rice. *Plant Physiol.* **2012**, *158*, 1666–1684.
36. Saiga, S.; Möller, B.; Watanabe-Taneda, A.; Abe, M.; Weijers, D.; Komeda, Y. Control of embryonic meristem initiation in arabidopsis by phd-finger protein complexes. *Development* **2012**, *139*, 1391–1398.
37. Silverstein, K.A.T.; Moskal, W.A.; Wu, H.C.; Underwood, B.A.; Graham, M.A.; Town, C.D.; VandenBosch, K.A. Small cysteine-rich peptides resembling antimicrobial peptides have been under-predicted in plants. *Plant J.* **2007**, *51*, 262–280.
38. Magnard, J.L.; Lehouque, G.; Massonneau, A.E.; Frangne, N.; Heckel, T.; Gutierrez-Marcos, J.F.; Perez, P.; Dumas, C.; Rogowsky, P.M. ZmEBE genes show a novel, continuous expression pattern in the central cell before fertilization and in specific domains of the resulting endosperm after fertilization. *Plant Mol. Biol.* **2003**, *53*, 821–836.

39. Hwang, S.G.; Kim, D.S.; Jang, C.S. Comparative analysis of evolutionary dynamics of genes encoding leucine-rich repeat receptor-like kinase between rice and arabidopsis. *Genetica* **2011**, *139*, 1023–1032.
40. Jain, M.; Kaur, N.; Garg, R.; Thakur, J.K.; Tyagi, A.K.; Khurana, J.P. Structure and expression analysis of early auxin-responsive AUX/IAA gene family in rice (*Oryza sativa*). *Funct. Integr. Genomics* **2006**, *6*, 47–59.
41. Gray, W.M.; Kepinski, S.; Rouse, D.; Leyser, O.; Estelle, M. Auxin regulates SCFTIR1-dependent degradation of AUX/IAA proteins. *Nature* **2001**, *414*, 271–276.
42. LeClere, S.; Schmelz, E.A.; Chourey, P.S. Cell wall invertase-deficient miniature1 kernels have altered phytohormone levels. *Phytochemistry* **2008**, *69*, 692–699.
43. Yu, C.S.; Chen, Y.C.; Lu, C.H.; Hwang, J.K. Prediction of protein subcellular localization. *Proteins* **2006**, *64*, 643–651.
44. Emanuelsson, O.; Nielsen, H.; Brunak, S.; von Heijne, G. Predicting subcellular localization of proteins based on their *N*-terminal amino acid sequence. *J. Mol. Biol.* **2000**, *300*, 1005–1016.
45. Amien, S.; Kliwer, I.; Marton, M.L.; Debener, T.; Geiger, D.; Becker, D.; Dresselhaus, T. Defensin-like ZmES4 mediates pollen tube burst in maize via opening of the potassium channel kzm1. *PLoS Biol.* **2010**, *8*, doi:10.1371/journal.pbio.1000388.
46. Costa, L.M.; Yuan, J.; Rouster, J.; Paul, W.; Dickinson, H.; Gutierrez-Marcos, J.F. Maternal control of nutrient allocation in plant seeds by genomic imprinting. *Curr. Biol.* **2012**, *22*, 160–165.
47. Li, M.; Singh, R.; Bazanova, N.; Milligan, A.S.; Shirley, N.; Langridge, P.; Lopato, S. Spatial and temporal expression of endosperm transfer cell-specific promoters in transgenic rice and barley. *Plant Biotechnol. J.* **2008**, *6*, 465–476.
48. Kwong, R.W.; Bui, A.Q.; Lee, H.; Kwong, L.W.; Fischer, R.L.; Goldberg, R.B.; Harada, J.J. Leafy cotyledon1-like defines a class of regulators essential for embryo development. *Plant Cell* **2003**, *15*, 5–18.
49. Junker, A.; Mönke, G.; Rutten, T.; Keilwagen, J.; Seifert, M.; Thi, T.M.N.; Renou, J.-P.; Balzergue, S.; Viehöver, P.; Hähnel, U.; *et al.* Elongation-related functions of leafy cotyledon1 during the development of *Arabidopsis thaliana*. *Plant J.* **2012**, *71*, 427–442.
50. Boisson-Dernier, A.; Kessler, S.A.; Grossniklaus, U. The walls have ears: The role of plant CrRLK1Ls in sensing and transducing extracellular signals. *J. Exp. Botany* **2011**, *62*, 1581–1591.
51. Goodstein, D.M.; Shu, S.; Howson, R.; Neupane, R.; Hayes, R.D.; Fazo, J.; Mitros, T.; Dirks, W.; Hellsten, U.; Putnam, N.; *et al.* Phytozome: A comparative platform for green plant genomics. *Nucleic Acids Res.* **2012**, *40*, D1178–D1186.
52. Nolan, T.; Hands, R.E.; Bustin, S.A. Quantification of mrna using real-time RT-PCR. *Nat. Protoc.* **2006**, *1*, 1559–1582.
53. Li, Q.F.; Sun, S.S.M.; Yuan, D.Y.; Yu, H.X.; Gu, M.H.; Liu, Q.Q. Validation of candidate reference genes for the accurate normalization of real-time quantitative RT-PCR data in rice during seed development. *Plant Mol. Biol. Rep.* **2010**, *28*, 49–57.
54. Tomlinson, K.L.; McHugh, S.; Labbe, H.; Grainger, J.L.; James, L.E.; Pomeroy, K.M.; Mullin, J.W.; Miller, S.S.; Dennis, D.T.; Miki, B.L.A. Evidence that the hexose-to-sucrose ratio does not control the switch to storage product accumulation in oilseeds: Analysis of tobacco seed development and effects of overexpressing apoplastic invertase. *J. Exp. Botany* **2004**, *55*, 2291–2303.

55. Fournier, E. Colorimetric quantification of carbohydrates. In *Current Protocols in Food Analytical Chemistry*; John Wiley & Sons, Inc.: Hoboken, NJ, USA, 2001.

5

Plant Responses to Simultaneous Biotic and Abiotic Stress: Molecular Mechanisms

Ines Ben Rejeb, Victoria Pastor and Brigitte Mauch-Mani *

Faculty of Sciences, Institute of Biology, University of Neuchâtel, Rue Emile Argand 11, 2000 Neuchâtel, Switzerland; E-Mails: ines.benrejeb@unine.ch (I.B.R.); victoria.pastor@unine.ch (V.P.)

* Author to whom correspondence should be addressed; E-Mail: brigitte.mauch@unine.ch

External Editors: Dr. Sylvie Renault and Dr. Germán Avila Sakar

---

Abstract: Plants are constantly confronted to both abiotic and biotic stresses that seriously reduce their productivity. Plant responses to these stresses are complex and involve numerous physiological, molecular, and cellular adaptations. Recent evidence shows that a combination of abiotic and biotic stress can have a positive effect on plant performance by reducing the susceptibility to biotic stress. Such an interaction between both types of stress points to a crosstalk between their respective signaling pathways. This crosstalk may be synergistic and/or antagonistic and include among others the involvement of phytohormones, transcription factors, kinase cascades, and reactive oxygen species (ROS). In certain cases, such crosstalk can lead to a cross-tolerance and enhancement of a plant's resistance against pathogens. This review aims at giving an insight into cross-tolerance between abiotic and biotic stress, focusing on the molecular level and regulatory pathways.

Keywords: cross-tolerance; biotic stress; abiotic stress; plant hormones

---

1. Introduction

Plants have to deal with various and complex types of interactions involving numerous environmental factors. In the course of evolution, they have evolved specific mechanisms allowing them to adapt and survive stressful events. Exposure of plants to biotic and abiotic stress induces a disruption in plant metabolism implying physiological costs [1–4], and thus leading to a reduction in fitness and ultimately in productivity [5]. Abiotic stress is one of the most important features of and has a huge impact on growth and, consequently, it is responsible for severe losses in the field. The resulting growth reductions can reach >50% in most plant species [6]. Moreover, biotic stress is an additional challenge inducing a strong pressure on plants and adding to the damage through pathogen or herbivore attack [7–11].

A crucial step in plant defense is the timely perception of the stress in order to respond in a rapid and efficient manner. After recognition, the plants' constitutive basal defense mechanisms [12] lead to an activation of complex signaling cascades of defense varying from one stress to another [13,14]. Following exposure to abiotic and/or biotic stress, specific ion channels and kinase cascades [15] are activated, reactive oxygen species (ROS) [16], phytohormones like abscisic acid (ABA), salicylic acid (SA), jasmonic acid (JA), and ethylene (ET) [17] accumulate, and a reprogramming of the genetic machinery results in adequate defense reactions and an increase in plant tolerance in order to minimize the biological damage caused by the stress [18].

In recent years, research has mainly concentrated on understanding plant responses to individual abiotic or biotic stresses [19–22], although the response to simultaneous stresses is bound to lead to a much more complex scenario [18]. From the perception of the stimulus (stress) to the final response in cells, plants use various signaling pathways depending on the challenge(s). It seems that plants respond in a specific manner when they have to face more than one stress simultaneously, and the response cannot be predicted based on the plant's response to the individual single stresses [23]. Research on multiple stresses has been trying to simulate natural conditions, but in the field, conditions are not controlled, and one stress can strongly influence the primary stress defense response of the plants [18]. Moreover, plants can show different degrees of sensitivity depending on the field condition and the developmental stage of the plant [24]. Additional factors that can influence an interaction are the intensity of the stress and the plant species. Various interactions can take place between the defenses induced after perception of the stresses. They depend on the specific combination of stresses and even on the degree of simultaneity [15,25,26]. It is not clear whether simultaneous stresses are rather antagonistic, synergistic or additive, inducing more or less susceptibility to a specific kind of stress [27,28]. Combination of two stressors can have a negative and additive effect on plants, the second stress being the one that leads to a greater damage [29]. On the other hand, the combination of stresses can also lead to antagonistic responses in the plants [30,31]. Common beans exposed to drought stress display more symptoms when infected by *Macrophomina phaseolina* [29] and treatment of detached tomato leaves with exogenously applied ABA increases the susceptibility of wild type plants to *Botrytis cinerea* [32].

Interestingly, one possible outcome of multiple stress exposure is that plants that are able to defend themselves facing one stress can become more resistant to other stresses [33]. This phenomenon is called cross-tolerance, showing that plants possess a powerful regulatory system that allows them to adapt quickly to a changing environment [33–35]. Wounding, for instance, increases salt tolerance in tomato plants [34]. Furthermore, in tomato plants again, localized infection by *Pseudomonas syringae* pv. tomato (*Pst*) induces systemic resistance to the herbivore insect *Helicoverpa zea* [36]. The association between abiotic and biotic stress is also possible [13], as demonstrated by the reduced infection of tomato by *Botrytis cinerea* and *Oidium neolycopersici* following the application of drought stress [37].

Ozone exposure can induce resistance to virulent *Pseudomonas syringae* strains in *Arabidopsis* [38]. Conversely, biotic stress can also interfere to increase the resistance to abiotic stress. This effect is visible when plants are under pathogen attack. Infection may cause stomatal closure to hinder pathogen entry and as a consequence water loss is reduced and leads to an enhanced plant resistance under abiotic stress [39]. Xu and colleagues [40] show that viral infection protects plants against drought stress. Verticillium infection in Arabidopsis plants induced the expression of the Vascular-Related No Apical meristem ATAF and Cup-Shaped Cotyledon (NAC) domain (VND) transcription factor *VND7*. VND7 induced *de novo* xylem formation ensuring the water storage capacity and as a consequence, increased plant drought tolerance [41]. Stress combination induces different signaling pathways, which share some components and common outputs [14–25]. This could help plants to minimize energy costs and create a flexible signaling network [42].

Resistance to both biotic and abiotic stress has been well documented in a variety of crops through priming of defenses. This component of induced resistance can be achieved through specific chemical stimuli like the resistance inducers BABA (beta-aminobutyric acid) or BTH (benzothiadiazole) [43,44], genetic manipulation of genes and proteins [45] or by previous contact with a pathogen [46]. Due to the complexity of interactions in defense, in the present review, we aim to focus on the cross-tolerance between abiotic and biotic stress as a part of induced resistance for defense.

2. Cross-Tolerance between Abiotic and Biotic Stress

Plants are able to manage simultaneous exposure to abiotic and biotic stress, and there is evidence for a link between the responses to these two stressful situations [23,47–49]. Usually, environmental pressure by abiotic and biotic stress can induce plant resistance. However, some plants confronted with each stress individually have also been reported to be more susceptible compared to a simultaneous exposure to two different stresses [50,51]. In addition, certain environmental stresses have the possibility to predispose the plant in order to allow it to respond faster and in a resistant manner to additional challenges. Therefore, cross-tolerance between environmental and biotic stress may induce a positive effect and enhanced resistance in plants and have significant agricultural implications. Interestingly, abiotic stress regulates the defense mechanisms at the site of pathogen infection as well as in systemic parts, thus ensuring an enhancement of the plant's innate immunity system [31]. Likewise, osmotic and proton stress are inducers of resistance in barley against powdery mildew. This induced resistance depends on the formation of callose-containing papillae capable of blocking fungal growth [48]. This kind of resistance is similar to the chemically induced resistance by BTH and INA (isonicotinic acid) [52]. Achuo *et al.* [37] demonstrated that drought stress increased the ABA content of tomato leaves, concomitantly with increasing the resistance against the necrotophic fungus *Botrytis cinerea* and that salt stress reduced susceptibility towards the biotrophic fungus *Odium neolycopersici* but not against *Botrytis cinerea*. This difference between drought and salt stress is in accordance with the observation that they both induce different gene expression patterns [53]. Additionally, the acclimation of *Nicotiana benthamiana* to moderate drought stress (60% of field capacity) reduced the growth of *P. syringae* pv. *tabaci* [26]. Recently, Atkinson and Urwin [23] reviewed the interaction of abiotic and biotic stress where they showed the common threads in pathways leading to a regulation of plant responses. Therefore, in order to prepare the plant for the battle, the activation of various detoxifying enzymes, control hormones, signaling pathways, and gene expression are indispensable [4,42,54].

The defense response of plants exposed to different stressors is expected to be complex including the interconnection of various signaling pathways regulating numerous metabolic networks [55].

3. Signaling Pathways Induced by Multiple Stress Responses

The interaction between abiotic and biotic stress induces complex responses to the different stressors. Under stress, the accumulation of certain metabolites positively affects a plant's response to both stresses and therefore protects it from multiple aggressors [25,47]. Callose accumulation, changes in ions fluxes, ROS, and phytohormones are the first responses induced to combat the stress and the resulting signal transduction triggers metabolic reprogramming towards defense [31,56].

3.1. Reactive Oxygen Species

A rapid generation of ROS is observed after stress sensing [57,58]. One of the major roles of ROS is to serve as signaling molecules in the cells [58–62]. The production of ROS is fine-modulated by the plant to avoid tissue damage [58,63–71]. ROS have long been known to be destructive and harmful compounds in stressed organisms. However, it has been shown that while high levels of ROS lead to cell death, lower levels are mostly responsible to regulate the plant's stress responses [67–69]. In biotic stress, ROS are mainly involved in signaling. This again might attenuate the oxidative stress caused by abiotic stress [70]. Furthermore, ROS could interfere in cross-tolerance [33]. ROS are involved in stress-induced tolerance in *Arabidopsis thaliana* after infection with the vascular pathogen *Verticillium* spp. by increasing drought tolerance due to *de novo* xylem formation and the resulting enhanced water flow [68]. Additionally, the production of ROS can help in cell-to-cell communication by amplifying the signal through the *Respiratory Burst Oxidase Homologue D* (*RBOHD*; [72]) and can act as a secondary messenger by modifying protein structures and activating defense genes [61,73]. ROS respond to abiotic and biotic stress, but differently from one stress to another [47]. Davletova *et al.* [74] showed that the transcription factor *Zat12* was involved in both abiotic and biotic stress and that *Zat12* could be a regulator in ROS scavenging. ROS may possibly be the central process mediating cross-tolerance between abiotic and biotic stress responsive networks [23]. In Arabidopsis, ROS production can be sensed by ROS-sensitive transcription factors [75,76] leading to the induction of genes participating in the stress responses. Gechev *et al.* [77] proposed that ROS were inducers of tolerance by activating stress response-related factors like mitogen-activated protein kinases (MAPKs), transcription factors, antioxidant enzymes, dehydrins, and low-temperature-induced-, heat shock-, and pathogenesis-related proteins.

Priming for stress tolerance induced after application of specific chemicals is responsible for certain modifications in ROS signaling [70–78]. Treatment of cucumber plants with brassinosteroids lead to a rise in H_2O_2 levels and primed the plants for both biotic and abiotic stress tolerance [68]. H_2O_2 priming for salt tolerance in citrus moderately increased the abundance of oxidized and *S*-nitrosylated proteins, and the level remained the same after stress application, however, non-treated plants were more sensitive to the stress [78].

3.2. Mitogen-Activated Protein Kinase (MAPK) Cascades

Following perception and recognition of stress stimuli, Mitogen-Activated Protein Kinase (MAPK) cascades are activated. They control the stress response pathways [79,80]. MAPKs are highly conserved in all eukaryotes and are responsible for the signal transduction of diverse cellular processes under various abiotic and biotic stress responses, and certain kinases are involved in both kind of stress [18,81,82]. Since MAPKs are involved in different stress responses, they could have a role in the combination of abiotic and biotic stress [83,84]. For instance, in cotton the kinase *GhMPK6a* negatively regulates both biotic and abiotic stress [85]. MAPK pathways activated by pathogen attack are mediated by SA, and the resulting expression of *PR* genes induces defense reactions [86]. The Arabidopsis protein VIP1 is translocated into the nucleus after phosphorylation by MPK3 and acts as an indirect inducer of *PR1* [87]. Chinchilla *et al.* [88] showed that pathogen associated molecular patterns (PAMPs) like flagellin trigger MAPK cascades in order to establish pathogen response signaling. In addition, MAPK such as MPK3, MPK4, and MPK6 also responded to various abiotic stresses [89,90]. MAPK cascades are important in controlling cross-tolerance between stress responses [12]. MPK3 and MPK6 are essential to show full primed defense responses [91], therefore, these two kinases could be important for mediating tolerance to further stresses. Over-expression of the *OsMPK5* gene and also kinase activity of OsMPK5 induced by ABA contributes to increased abiotic and biotic stress tolerance. *OsMPK5* seems to play a double role in the rice stress response, one as a positive regulator of resistance to the necrotrophic brown spot pathogen *Cochliobolus miyabeanus* and the second as a mediator of abiotic stress tolerance [81,92]. Tomato plants activate MPK1 and MPK2 against UV-B, wounding, and pathogens in order to enhance their defense reactions [93]. MAPK signaling also interacts with ROS and ABA signaling pathways leading to enhanced plant defense and induction of cross-acclimation to both abiotic and biotic stress [94–96].

3.3. Relevance of Hormone Signaling under Stress Interaction

The control of every kind of stress by specific hormones allows defense responses against defined environmental conditions. ABA is considered the primary hormone involved in the perception of many abiotic stresses [97]. Increases in ABA concentration modulate the abiotic stress-regulation network [98] while biotic stress responses are preferentially mediated by antagonism between other stress hormones such as SA and acid JA/ET [99]. In certain cases, ABA has been shown to accumulate after infection [18,27,100,101]. For instance, higher levels of ABA were observed after *Pst* DC 3000 infection [102], and this provoked a suppression of other defense responses [103]. However, recent findings show a positive effect of ABA on biotic stress resistance [30,104,105]. This dual effect makes ABA a controversial molecule that can switch from "good to bad" depending on the environmental conditions (type and timing of the stress; [105]). Moreover, under combination of abiotic and biotic stress, ABA mostly acts antagonistically with SA/JA/ethylene inducing a susceptibility of the plant against disease and herbivore attack [28,31,32,106,107]. However, since an increase of ABA under the effect of abiotic stress induces stomatal closure, as a "secondary effect", the entry of biotic assailants through these passive ports of the plant is prevented. Hence, under such circumstances, the plant is protected from abiotic as well as from biotic stress [108]. There are three different phases showing the influence of ABA on pathogen infection [23,30]. The first effect of ABA on the combination of both, abiotic and biotic stress is related only to abiotic stress because an infection takes more time to establish itself and the plants react therefore later to it [30–109]. At this moment, ABA induces stomatal closure [110], which allows a reduction in water loss and, as a consequence, the maintenance of a beneficial water potential. In this first phase, SA, JA and ethylene might not yet be activated and ABA can antagonize their induction. In this situation, future responses against potential pathogens are modified. The second phase concerns the post-infection reactions. Callose is an important inducible defense that can prevent pathogen invasion [111]. After infection, an intact ABA signaling pathway is required to increase callose accumulation in attacked plants [44,112], and the presence of ABA can induce or repress additional callose accumulation [98] depending on the environmental conditions. Therefore, ABA variation by a previous stress can influence the final output following biotic stress, such as strengthening the resistance phenotype through accumulation of callose or by inducing other defense pathways [96,108]. The third phase finally starts when PAMPs stimulate the accumulation of specific hormones that are SA, JA, and ethylene in order to regulate the defense reaction [27,96,113]. In summary, the exact role of ABA as a regulator of the dialogue between abiotic and biotic stress strongly depends on the timing of the stress perception: does the infection hit a plant that had already been exposed previously to abiotic stress or does an infected plant become additionally exposed to abiotic stress [30,97,114]?

The beneficial role of SA in the relationship between plants and pathogens has been extensively studied. What is known is that ABA and SA have an antagonistic role in plant defense against stressors [31]. However, Miura and Tada [88] have shown that in addition to ABA, SA seems to also be important in plant responses to drought stress. Furthermore, SA increased barley resistance against water deficit [115].

3.4. Transcription Factors and Molecular Responses in Cross-Tolerance

Changes in gene expression occur after detection of a given stress, and the reprogramming of the molecular machinery is regulated by the action of transcription factors. The altered expression of certain genes is a key event in helping plants to set up an effective defensive state, and there is convincing evidence that many genes are multifunctional and able induce tolerance in plants towards more than one stress [49–116]. The activity of such genes involved in defense is mediated by specific phytohormones like ABA, SA, JA, and Ethylene. For example, the activity of the *BOTRYTIS SUSCEPTIBLE1* (*BOS1*) gene is mediated by both ABA and JA and induces resistance against osmotic stress and necrotrophic pathogens [117], and *bos1* mutant plants are more susceptible to both stresses [117]. In Arabidopsis, the transcription factor *MYB96* plays an important role in plant protection under pathogen infection by mediating the molecular link between both ABA induced by drought stress and SA expressed following pathogen infection [118]. *SlAIM1* in tomato responds positively to the combination of abiotic stress and infection with *Botrytis cinerea* [13] and *OsMAPK5*, which has kinase activity, is a positive regulator of the rice response to drought, salt, and cold tolerance and disease resistance [86].

Interestingly, many *PR* genes are also induced upon exposure of a plant to abiotic stress ensuring disease resistance [118]. PR proteins are crucial for plant resistance against pathogens, and their expression is strongly up-regulated when plants are attacked [118]. Over-expression of certain transcription factors in plants confronted with cold stress and infection activates cold-responsive *PR* genes, thereby conferring protection against both stressors [119]. The up-regulation of some transcription factors after exposure to abiotic stress leads to an accumulation of PR proteins. The transcription factors C-repeat Binding Factors (*CBF*), Dehydration-Responsive Element-Binding proteins (*DREB*) and No Apical meristem ATAF and Cup-Shaped Cotyledon (*NAC*) have been extensively studied as players of the primary abiotic stress signaling pathways ensuring tolerance under stress [120–122]. *CBF* is induced under cold stress together with a group of PR proteins [123]. Transgenic Arabidopsis overproducing the *NAC* transcription factor *NTL6*, which is induced by cold stress, enhance their defense response against pathogen attack by promoting an up-regulation of the *PR1* gene [118–124]. Tsutsui *et al.* [125] showed that the transcription factor *DREB* could regulate the response of cross-tolerance between abiotic and biotic stress insuring the resistance of Arabidopsis response to cold and pathogen (Figure 1).

Figure 1. Abiotic stress can enhance the expression of specific transcription factors (TFs) like C-repeat Binding Factors (*CBF*), No Apical meristem ATAF and Cup-Shaped Cotyledon (*NAC*), *MYB* mediated by abscisic acid (ABA). Although the exact role of ABA in plant pathogen interactions is still a matter of debate, in some specific cases it has been shown to promote resistance against biotic stress following abiotic stress. This is attributed to the over-expression of TFs inducing the up-regulation of *PR* genes.

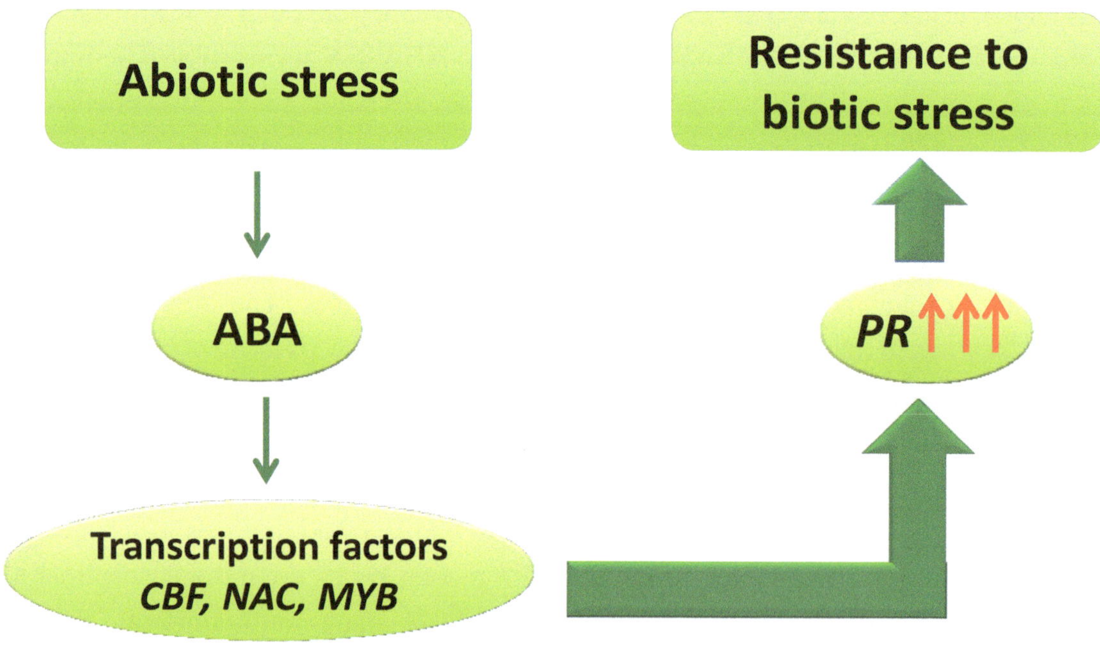

Recently, it has been proposed that the WHIRLY1 protein and *REDOX-RESPONSIVE TRANSCRIPTION FACTOR1* (*RRTF1*) could participate in the traffic of communication between plastids and the nucleus [126]. WHIRLY1 perceives the redox changes in the plastid and carries the information to the nucleus in an NPR1-independent manner. The authors propose this protein as an ideal component in retrograde signaling that will lead to acclimation and adaption to new stresses. In the same way, *RRTF1,* which is induced by biotic and abiotic stresses, could be priming distant leaves to defend themselves against further stresses.

4. Conclusions and Outlook

A plant's response following exposure to abiotic/biotic stress strongly depends on its developmental stage [127] and the environmental conditions to which it is subjected [99]. Many stress combinations lead to phenotypic damage and, as mentioned above, the expression of defense is affected according to the type of abiotic stress and the pathogens involved. Overall, the complex response of the plant stems from the interplay of specific signaling pathways involved in abiotic and biotic stress. The combination of both stress types leads to an increased accumulation of a large number of signaling compounds that, in an ideal case, will be expressed as cross-tolerance (Figure 2).

Plants perceive the information signal of each stress and consequently activate specific molecules. Only some of them, which are common to both stressors, will participate in the defense response to the specific stress combination and thus contribute to protect the plant and enhance its resistance.

Various novel approaches can help plants to resist under combinatorial stress. The "Omics" technology is one of these approaches. Transcriptomics, proteomics, and metabolomics have revealed plant responses under stress and their underlying mechanisms and point to potential target genes, proteins or metabolites for inducing tolerance and improve plant responses. Little is known about the "Omics" characterization of abiotic and biotic stress combinations, but recently, several reports have addressed this question [16,51,70,128,129]. Although complete genome sequences are available for an increasing number of crop and model plants, in comparison, protein and metabolite databases are still rather incomplete, hence complicating the task of integrating all observations. Additionally, different plant species or even cultivars may behave differently, plant responses are also often organ-dependent, and results obtained with whole plants may be misleading.

Another approach might consist of molecular engineering of specific genes and their introduction into crop plants. By modifying a gene coding for a small antimicrobial peptide and introducing it into potato, the resistance of potato to biotic and abiotic stress was increased [130].

The manipulation of common regulators is also a promising approach. Boosting the accumulation of flavonoid biosynthesis mitigates the negative effects of abiotic and biotic stress [131,132]. Polyamines are another example. These substances have long been known to mediate resistance to pathogens [133] but they are also involved in abiotic stress resistance [134]. Genetic manipulation of polyamine accumulation could lead to multi stress tolerance [135].

A further possibility to promote cross-tolerance is the exploitation of priming. Some chemicals have been shown to prime plants for both biotic and abiotic stresses under laboratory conditions [136], and their application might allow a better management of multiple stresses under field conditions. The ultimate goal in every case is to maintain or even enhance plant performance, yield, and productivity under adverse conditions.

Figure 2. Elements possibly involved in cross-tolerance between biotic and abiotic stress. Both biotic and abiotic stress have to be first sensed by the plant cell, and then the information is transduced to appropriate downstream-located pathway(s). Sensors as well as signal transducers might be shared by both types of stressors. Reactive oxygen species (ROS) and Ca^{++} are known among others to play a prominent role as transducers (messengers) and mitogen-activated protein kinases (MAPK) cascades have been shown to be used by both types of stresses. MAPKs are centrally positioned in Ca^{2+}-ROS crosstalk as well as in the signal output after exposure to a specific stress. The importance of ROS has repeatedly been described for both types of stresses too, and, therefore, ROS might represent crucial elements in the integration of both stresses during cross-tolerance. Plant hormone signaling is of utter importance for stress adaptation. While abscisic acid (ABA) is predominantly involved in abiotic stress adaptation, salicylic acid (SA) and jasmonate/ethylene (JA/ET) are more responsible for the plant's reaction to biotic stress. However, there is a tremendous amount of crosstalk taking place between the various hormonal pathways, and the exact nature of this crosstalk during simultaneous biotic and abiotic stress remains to be investigated. ABA signaling contributes positively to pre-invasion defense and is responsible for enhancing callose deposition. ABA presents a positive interaction with JA/ET signaling. The activation of SA signaling by pathogen challenge can attenuate ABA responses. ABA signaling negatively affects signals that trigger systemic acquired resistance, enhancing pathogen spread from the initial site of infection. The interaction of SA, JA, and ET signaling results in increased resistance to pathogens. Hormones, secondary metabolites, priming agents, and further chemicals located in the cytoplasm finally up-regulate transcription factors (TF), pathogenesis related (PR) and defense genes, heat shock protein (HSP) genes, and further genes involved in protection against stress and thus lead to the phenotypic expression known as cross-tolerance. Arrows: induction; flat-ended lines: repression.

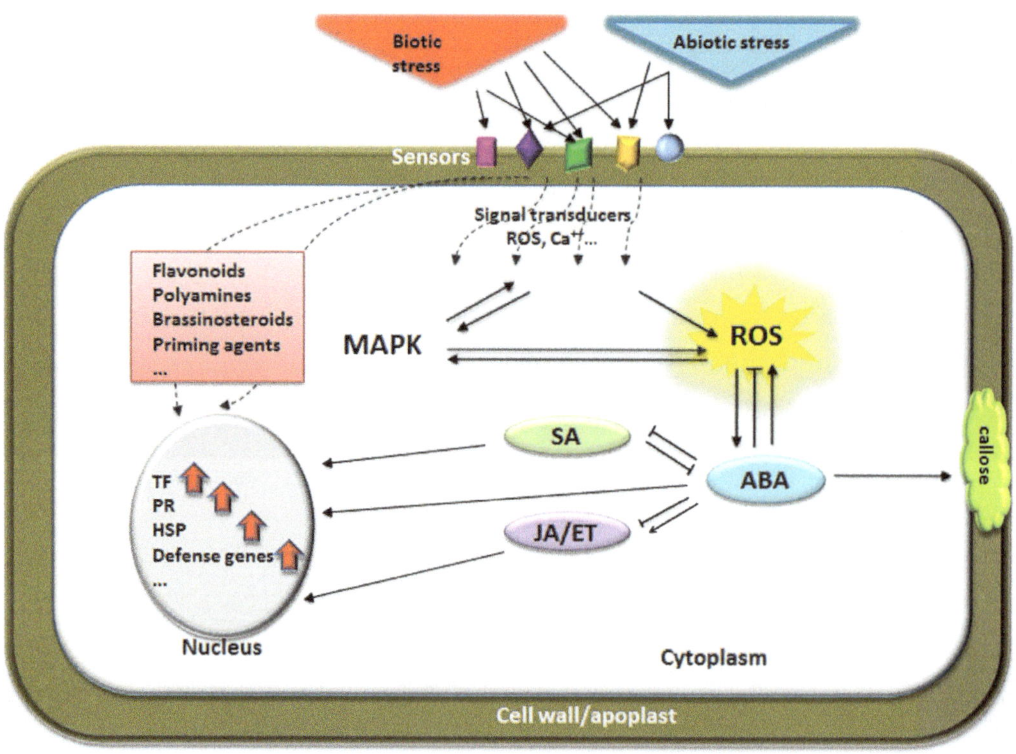

Acknowledgments

We apologize to all the colleagues whose work has not been cited in this short review. The financial support of the Swiss National Science Foundation, grant 31003A_140593, is gratefully acknowledged. We also thank the Swiss Confederation for attributing a Swiss Government Excellence Research Scholarship for foreign students to I. Ben Rejeb.

Author Contributions

All authors contributed equally to this review article.

Conflicts of Interest

The authors declare no conflicts of interest.

References

1. Heil, M.; Bostock, R.M. Induced systemic resistance (ISR) against pathogens in the context of induced plant defences. *Ann. Bot.* **2002**, *89*, 503–512.
2. Swarbrick, P.J.; Schulze-Lefert, P.; Scholes, J.D. Metabolic consequences of susceptibility and resistance in barley leaves challenged with powdery mildew. *Plant Cell Environ.* **2006**, *29*, 1061–1076.
3. Bolton, M.V. Primary metabolism and plant defense-Fuel for the fire. *Mol. Plant Microbe Interact.* **2009**, *22*, 487–497.
4. Massad, T.J.; Dyer, L.A.; Vega, C.G. Cost of defense and a test of the carbon-nutrient balance and growth-differentation balance hypotheses for two co-occurring classes of plant defense. *PLoS One* **2012**, *7*, e7554.
5. Shao, H.B.; Chu, L.Y.; Jaleel, C.A.; Zhao, C.X. Water-deficit stress—Induced anatomical changes in higher plants. *C. R. Biol.* **2008**, *331*, 215–225.
6. Wang, W.; Vinocur, B.; Altman, A. Plant responses to drought; salinity and extreme temperatures: Towards genetic engineering for stress tolerance. *Planta* **2003**, *218*, 1–14.
7. Mordecai, E.A. Pathogen impacts on plant communities: Unifying theory, concepts, and empirical work. *Ecol. Monogr.* **2011**, *81*, 429–441.
8. Maron, J.L.; Crone, E. Herbivory: Effects on plant abundance, distribution and population growth. *Proc. R. Soc. B* **2006**, *273*, 2575–2584.
9. Maron, J.L.; Kauffman, M. Habitat-specific consumer impacts on plant population dynamics. *Ecology* **2006**, *87*, 113–124.
10. Strauss, S.Y.; Zangerl, A.R. Plant-insect interactions in terrestrial ecosystems. In *Plant-Animal Interactions. An Evolutionary Approach*; Herrera, C.M., Pellmyr, O., Eds.; Blackwell Science: Oxford, UK, 2002; pp. 77–106.
11. Brown, J.K.M.; Hovmoller, M.S. Aerial dispersal of pathogens on the global and continental scales and its impact on plant disease. *Science* **2002**, *297*, 537–541.
12. Andreasson, E.; Ellis, B. Convergence and specificity in the *Arabidopsis* MAPK nexus. *Trends Plant Sci.* **2010**, *15*, 106–113.

13. Abou Qamar, S.; Luo, H.; Laluk, K.; Mickelbart, V.M.; Mengiste, T. Crosstalk between biotic and abiotic stress responses in tomato is mediated by AIM1 transcription factor. *Plant J.* **2009**, *58*, 1–13.
14. Chinnusamy, V.; Schumaker, K.; Zhu, J.K. Molecular genetics perspectives on cross-talk and specifcity in abiotic stress signalling in plants. *J. Exp. Bot.* **2004**, *55*, 225–236.
15. Fraire-Velázquez, S.; Rodríguez-Guerra, R.; Sánchez-Calderón, L. *Abiotic and Biotic Stress Response Crosstalk in Plants-Physiological, Biochemical and Genetic Perspectives*; Shanker, A., Ed.; InTech Open Access Company: Rijeka, Croatia, 2011; pp. 1–26.
16. Laloi, C.; Appel, K.; Danon, A. Reactive oxygen signalling: The latest news. *Curr. Opin. Plant Biol.* **2004**, *7*, 323–328.
17. Spoel, S.H.; Dong, X. Making sense of hormone crosstalk during plant immune response. *Cell Host Microbe* **2008**, *3*, 348–351.
18. Fujita, M.; Fijita, Y.; Noutoshi, Y.; Takahashi, F.; Narusaka, Y.; Yamaguchi-Shinozaki, K.; Shinozaki, K. Crosstalk between abiotic and biotic stress responses: A current view from the points of convergence in the stress signaling networks. *Curr. Opin. Plant Biol.* **2006**, *9*, 436–442.
19. Todaka, D.; Nakashima, K.; Shinozaki, K.; Yamaguchi-Shinozaki, K. Toward understanding transcriptional regulatory networks in abiotic stress responses and tolerance in rice. *Rice J.* **2012**, *5*, 1–9.
20. Stotz, H.U.; Mitrousia, G.K.; de Wit, P.J.G.M.; Fitt, B.D.L. Effector-triggered defence against apoplastic fungal pathogens. *Trends Plant Sci.* **2014**, *19*, 491–500.
21. Thakur, M.; Sohal, B.S. Role of elicitors in inducing resistance in plants against pathogen infection: A review. *ISRN Biochem.* **2013**, doi:10.1155/2013/762412.
22. Qin, F.; Shinozaki, K.; Yamaguchi-Shinozaki, K. Achievements and challenges in understanding plant abiotic stress responses and tolerance. *Plant Cell Physiol.* **2011**, *52*, 1569–1582.
23. Atkinson, N.; Urwin, P.-E. The interaction of plant biotic and abiotic stresses: From genes to the field. *J. Exp. Bot.* **2012**, *63*, 3523–3544.
24. Mittler, R.; Blumwald, E. Genetic engineering for modern agriculture: Challenges and perspectives. *Ann. Rev. Plant Biol.* **2010**, *61*, 443–462.
25. Rasmussen, S.; Barah, P.; Suarez-Rodriguez, M.C.; Bressendorff, S.; Friis, P.; Costantino, P.; Bones, A.M.; Nielsen, H.B.; Mundy, J. Transcriptome responses to combinations of stresses on Arabidopsis. *Plant Physiol.* **2013**, *161*, 1783–1794.
26. Ramegowda, V.; Senthil-Kumar, M.; Ishiga, Y.; Kaundal, A.; Udayakumar, M.; Mysore, K.S. Drought stress acclimation impacts tolerance to *Sclerotina sclerotiorum* and *Pseudomonas syringae* in *Nicotiana benthamiana. Int. J. Mol. Sci.* **2013**, *14*, 9497–9513.
27. Anderson, J.P.; Badruzsaufari, E.; Schenk, P.M.; Manners, J.; Desmond, O.J.; Ehlert, C.; Maclean, D.J.; Ebert, P.R.; Kazan, K. Antagonistic interaction between abscisic acid and jasmonate-ethylene signaling pathways modulates defense gene expression and disease resistance in Arabidopsis. *Plant Cell* **2004**, *16*, 3460–3479.
28. Asselbergh, B.; Achuo, A.E.; Hofte, M.; van Gijegem, F. Abscisic acid deficiency leads to rapid activation of tomato defence responses upon infection with *Erwinia chrysanthemi. Mol. Plant Pathol.* **2008**, *9*, 11–24.

29. Suleman, P.; Al-Musallam, A.; Menezes, C.A. The effect of solute potential and water stress on black scorch caused by *Chalara paradoxa* and *Chalara radicicola* on date palms. *Plant Dis.* **2001**, *85*, 80–83.
30. Ton, J.; Ent, V.D.S.; Hulten, V.M.; Pozo, M.; Oosten, V.V.; Loon, L.C.V; Mauch-Mani, B.; Turlings, T.C.J.; Pieterse, C.M.J. Priming as a mechanism behind induced resistance against pathogens; insects and abiotic stress. *IOBC/wprs Bull.* **2009**, *44*, 3–13.
31. Yasuda, M.; Ishikawa, A.; Jikumaru, Y.; Seki, M.; Umezawa, T.; Asami, T.; Maruyama-Nakashita, A.; Kudo, T.; Shinozaki, K.; Yoshida, S.; *et al.* Antagonistic interaction between systemic acquired resistance and the abscisic acid-mediated abiotic stress response in Arabidopsis. *Plant Cell* **2008**, *20*, 1678–1692.
32. Audenaert, K.; de Meyer, G.B.; Hofte, M. Abscisic acid determines basal susceptibility of tomato to *Botrytis cinerea* and suppresses salicylic acid-dependent signaling mechanisms. *Plant Physiol.* **2002**, *128*, 491–501.
33. Bowler, C.; Fluhr, R. The role of calcium and activated oxygens as signals for controlling cross-tolerance. *Trends Plant Sci.* **2000**, *5*, 241–246.
34. Capiati, D.A.; Pais, S.M.; Tellez-Iñon, M.T. Wounding increases salt tolerance in tomato plants: Evidence on the participation of calmodulin-like activities in cross-tolerance signaling. *J. Exp. Bot.* **2006**, *57*, 2391–2400.
35. Suzuki, N.; Koussevitzky, S.; Mittler, R.; Miller, G. ROS and redox signalling in the response of plants to abiotic stress. *Plant Cell Environ.* **2012**, *35*, 259–270.
36. Stout, M.J.; Fidantsef, A.L.; Duffey, S.S.; Bostock, R.M. Signal interactions in pathogen and insect attack: Systemic plant-mediated interactions between pathogens and herbivores of the tomato. *Lycopersicum esculentum. Physiol. Mol. Plant Pathol.* **1999**, *54*, 115–130.
37. Achuo, E.A.; Prinsen, E.; Hofle, M. Influence of drought; salt stress and abscisic acid on the resistance of tomato to *Botrytis cinerea* and *Oidium neolycopersici*. *Plant Pathol.* **2006**, *55*, 178–186.
38. Sharma, Y.; Leon, J.; Raskin, I.; Davis, K.R. Ozone-induced responses in *Arabidopsis thaliana*: The role of salicylic acid in the accumulation of defense-related transcripts and induced resistance. *Plant Biol.* **1996**, *93*, 5099–5104.
39. Goel, A.K.; Lundberg, D.; Torres, M.A.; Matthews, R.; Akimoto-Tomiyama, C.; Farmer, L.; Dangl, J.L.; Grant, S.R. The *Pseudomonas syringae* type III effector HopAM1 enhances virulence on water-stressed plants. *Mol. Plant Microbe Interact.* **2008**, *21*, 361–370.
40. Xu, P.; Chen, F.; Mannas, J.P.; Feldman, T.; Sumner, L.W.; Roossinck, M.J. Virus infection improves drought tolerance. *New Phytol.* **2008**, *180*, 911–921.
41. Reusche, M.; Thole, K.; Janz, D.; Truskina, J.; Rindfleisch, S.; Drübert, C.; Polle, A.; Lipka, V.; Teichmann, T. Verticillium infection triggers VASCULAR-RELATED NAC DOMAIN7-dependent *de novo* xylem formation and enhances drought tolerance in arabidopsis. *Plant Cell* **2012**, *24*, 3823–3837.
42. Koornneef, A.; Pieterse, C.M.J. Cross talk in defense signaling. *Plant Physiol.* **2014**, *146*, 839–844.
43. Jakab, G.; Ton, J.; Flors, V; Zimmerli, L; Métraux, J.P.; Mauch-Mani, B. Enhancing *Arabidopsis* salt and drought stress tolerance by chemical priming for its abscisic acid responses. *Plant Physiol.* **2005**, *139*, 267–274.

44. Ton, J.; Mauch-Mani, B. Beta-amino-butyric acid-induced resistance against necrotrophic pathogens is based on ABA-dependent priming for callose. *Plant J.* **2004**, *38*, 119–130.
45. Umezawa, T.; Fujita, M.; Fujita, Y.; Yamaguchi-Shinozaki, K.; Shinozaki, K. Engineering drought tolerance in plants: Discovering and tailoring genes to unlock the future. *Curr. Opin. Biotechnol.* **2006**, *17*, 113–122.
46. Chalfoun, N.R.; Castagnaro, A.P.; Díaz Ricci, J.C. Induced resistance activated by a culture filtrate derived from an avirulent pathogen as a mechanism of biological control of anthracnose in strawberry. *Biol. Control* **2001**, *58*, 319–329.
47. Pastori, G.M.; Foyer, C.H. Common components, networks, and oathways of cress-tolerance to stress. The central role of "redox" and abscisic acid-mediated controls. *Plant Physiol.* **2002**, *129*, 460–468.
48. Weise, J.; Kranz, T.; Schbert, S. Induction of pathogen resistance in barley by abiotic stress. *Plant Biol.* **2004**, *6*, 529–536.
49. Ramirez, V.; Coego, A.; Lopez, A.; Agorio, A.; Flors, V.; Vera, P. Drought tolerance in Arabidopsis is controlled by the *OCP3* disease resistance regulator. *Plant J.* **2009**, *58*, 578–591.
50. Tippmann, H.F.; Schluter, U.; Collinge, D.B. Common themes in Biotic and abiotic stress signaling in plants. *Floric. Ornam. Plant Biotechnol.* **2006**, *3*, 52–67.
51. Suzuki, N.; Rivero, R.M.; Shulaev, V.; Blumwald, E.; Mittler, R. Abiotic and biotic stress combinations. *New Phytol.* **2014**, *203*, 32–43.
52. Besser, K.; Jarosch, B.; Langen, G.; Kogel, K.-H. Expression analysis of genes induced in barley after chemical activation reveals distinct disease resistance pathways. *Mol. Plant Pathol.* **2000**, *5*, 277–286.
53. Takahashi, S.; Seki, M.; Ishida, G.; Satou, M.; Sakurai, T.; Narusaka, M.; Kamiya, A.; Nakajima, M.; Enju, A.; Akiyama, K.; *et al.* Monitoring the expression profiles of genes induced by hyperosmotic; high salinity and oxidative stress and abscisic acid treatment in *Arabidopsis* cell culture using a full-length cDNA microarray. *Mol. Biol.* **2004**, *56*, 29–55.
54. Mittler, R. Abiotic stress, the field environment and stress combination. *Trends Plant Sci.* **2006**, *11*, 15–19.
55. Nakashima, K.; Ito, Y.; Yamaguchi-Shinozaki, K. Transcriptional regulatory networks in response to abiotic stresses in arabidopsis and grasses. *Plant Physiol.* **2009**, *149*, 88–95.
56. Bartoli, C.G.; Casalongué, C.A.; Simontacchi, M.; Marquez-Garcia, B.; Foyer, C.H. Interactions between hormone and redox signaling pathways in the control of growth and cross-tolerance to stress. *Environ. Exp. Bot.* **2013**, *94*, 73–88.
57. Wojtaszek, P. Oxidative burst: An early plant response to pathogen infection. *Biochem. J.* **1997**, *322*, 681–692.
58. Foyer, C.; Noctor, G. Redox homeostasis and antioxidant signaling: A metabolic interface between stress perception and physiological responses. *Plant Cell* **2005**, *17*, 1866–1875.
59. Apel, K.; Hirt, H. Reactive oxygen species: Metabolism; oxidative stress; and signal transduction. *Ann. Rev. Plant Biol.* **2004**, *55*, 373–399.
60. Hancock, J.; Desikan, R.; Harrison, J.; Bright, J.; Hooley, R.; Neill, S. Doing the unexpected: Proteins involved in hydrogen peroxide perception. *J. Exp. Bot.* **2006**, *57*, 1711–1718.

61. Spoel, S.H.; Loake, G.J. Redox-based protein modifications: The missing link in plant immune signalling. *Curr. Opin. Plant Biol.* **2011**, *14*, 358–364.
62. Meng, L.; Wong, J.H.; Feldman, L.J.; Lemaux, P.G.; Buchanan, B.B. A membrane-associated thioredoxin required for plant growth moves from cell to cell; suggestive of a role in intercellular communication. *Proc. Natl. Acad. Sci. USA* **2010**, *107*, 3900–3905.
63. Mittler, R.; Vanderauwera, S.; Suzuki, N.N.; Miller, G.; Tognetti, V.B.; Vandepoele, K.; Gollery, M.; Shulaev, V.; van Breusegem, F. ROS signaling: The new wave? *Trends Plant Sci.* **2011**, *16*, 1360–1385.
64. Vinocur, B.; Altman, A. Recent advances in engineering plant tolerance to abiotic stress: Achievement and limitations. *Curr. Opin. Biotech.* **2005**, *16*, 123–132.
65. Bhattacharjee, S. The language of reactive oxygen species signaling in plants. *J. Bot.* **2012**, *2012*, 1–22.
66. Foyer, C.; Noctor, G. Ascorbate and glutathione: The heart of the redox hub. *Plant Physiol.* **2011**, *155*, 1–18.
67. Choudhury, S.; Panda, P.; Sahoo, L.; Panda, S.K. Reactive oxygen species signaling in plants under abiotic stress. *Plant Signal. Behav.* **2013**, *8*, e23681.
68. Xia, X.-J.; Wang, Y.-J.; Zhou, Y.-H.; Tao, Y.; Mao, W.-H.; Shi, K.; Asami, T.; Chen, Z.; Yu, J.-Q. Reactive oxygen species are involved in brassinosteroid-induced stress tolerance in cucumber. *Plant Physiol.* **2012**, *158*, 1034–1045.
69. Mori, I.C.; Schroeder, J.I. Reactive oxygen species activation of plant Ca^{2+} channels. A signaling mechanism in polar growth, hormone transduction, stress signaling, and hypothetically mechanotransduction. *Plant Physiol.* **2004**, *135*, 702–708.
70. Kissoudis, C.; van de Wiel, C.; Visser, R.G.F.; van der Linden, G. Enhancing crop resilience to combined abiotic and biotic stress through the dissection of physiological and molecular crosstalk. *Front. Plant Sci.* **2014**, *5*, e207.
71. Pastor, V.; Luna, E.; Ton, J.; Cerezo, M.; García-Agustín, P.; Flors, V. Fine tunning of reactive oxygen species homeostasis regulates primed immune responses in *Arabidopsis*. *Mol. Plant Microbe Interact.* **2013**, *11*, 1334–1344.
72. Miller, G.; Suzuki, N.; Ciftci-Yilmaz, S.; Mittler, R. Reactive oxygen species homeostasis and signalling during drought and salinity stresses. *Plant Cell Environ.* **2010**, *33*, 453–467.
73. Spoel, S.H.; Tad, Y.; Loake, G.J. Post-translational protein modification as a tool for transcription reprogramming. *New Phytol.* **2010**, *186*, 333–339.
74. Davletova, S.; Schlauch, K.; Coutu, J.; Mittler, R. The Zinc-Finger protein Zat12 plays a central role in reactive oxygen and abiotic stress signaling in Arabidopsis. *Plant Physiol.* **2005**, *139*, 847–856.
75. Desikan, R.; Mackerness, S.A.H.; Hancock, J.T.; Neill, S. Regulation of the Arabidopsis transcriptome by oxidative stress. *Plant Physiol.* **2001**, *127*, 159–172.
76. Miller, G.; Shulaev, V.; Mittler, R. Reactive oxygen signaling and abiotic stress. *Physiol. Plant.* **2008**, *133*, 481–489.
77. Gechev, T.S.; van Breusegem, F.; Stone, J.M.; Denev, I.; Laloi, C. Reactive oxygen species as signals that modulate plant stress responses and programmed cell death. *Bioessays* **2006**, *28*, 1091–1101.

78. Tanou, G.; Molassiotis, A.; Diamantidis, G. Hydrogen peroxide and nitric oxide-induced systemic antioxidant prime-like activity under NaCl-stress and stress-free conditions in citrus plants. *J. Plant Physiol.* **2009**, *166*, 1904–1913.
79. Zhang, D.; Martyniuk, C.J.; Trudeau, V.L. SANTA domain: A novel conserved protein module in Eukaryote with potential involvement in chromatin regulation. *Bioinformatics* **2006**, *22*, 2459–2462.
80. Wurzinger, B.; Mair, A.; Pfister, B.; Teige, M. Cross-talk of calcium-dependent protein kinase and MAP kinase signaling. *Plant Signal. Behav.* **2011**, *6*, 8–12.
81. Teige, M.; Scheikl, E.; Eulgem, T.; Roczi, F.; Ichimura, K.; Shinozaki, K.; Dangl, J.L.; Hirt, H. The MKK2 pathway mediates cold and salt stress signaling in *Arabidopsis. Mol. Cell* **2004**, *15*, 141–152.
82. Brader, G.; Djamei, A.; Teige, M.; Palva, E.T.; Hirt, H. The MAP kinase kinase MKK2 affects disease resistance in *Arabidopsis. Mol. Plant Microbe Interact.* **2007**, *20*, 589–596.
83. Zhang, S.; Liu, Y. Activation of salicylic acid-induced protein kinase; a mitogen-activated protein kinase; induces multiple defense responses in tobacco. *Plant Cell* **2001**, *13*, 1877–1889.
84. Šamajová, O.; Plíhal, O.; Al-Yousif, M.; Hirt, H.; Šamaj, J. Improvement of stress tolerance in plants by genetic manipulation of mitogen-activated protein kinases. *Biotechnol. Adv.* **2013**, *31*, 118–128.
85. Li, Y.; Zhang, L.; Wang, X.; Hao, L.; Chu, X.; Guo, X. Cotton GhMPK6a negatively regulates osmotis tolerance and bacterial infection in transgenic *Nicotiana benthaminana* and plays a pivotal role in development. *FEBS J.* **2013**, *280*, 5128–5144.
86. Xiong, L.; Yang, Y. Disease resistance and abiotic stress tolerance in rice are inversely modulated by an abscisic acid-inducible mitogen-activated protein kinase. *Plant Cell* **2003**, *15*, 745–759.
87. Pitzschke, A.; Djamel, A.; Teige, M.; Hirt, H. VIP1 response elements mediate mitogen-activated protein kinase 3-induced stress gene expression. *Proc. Natl. Acad. Sci. USA* **2009**, *106*, 18414–18419.
88. Chinchilla, D.; Zipfel, C.; Robatzek, S.; Kemmerling, B.; Nurnberger, T.; Jones, J.D.G.; Felix, G.; Boller, T. A flagellin-induced complex of the receptor FLS2 and BAK1 initiates plant defence. *Nature* **2007**, *448*, 497–500.
89. Ichimura, K.; Mizoguchi, T.; Yoshida, R.; Yuasa, T.; Shinozaki, K. Various abiotic stresses rapidly activate Arabidopsis MAP kinases ATMPK4 and ATMPK6. *Plant J.* **2000**, *24*, 655–665.
90. Gudesblat, G.E.; Iusem, N.D.; Morris, P.C. Guard cell-specific inhibition of *Arabidopsis* MPK3 expression causes abnormal stomatal responses to abscisic acid and hydrogen peroxide. *New Phytol.* **2007**, *173*, 713–721.
91. Beckers, G.J.M.; Jaskiewicz, M.; Conrath, U. Mitogen-Activates Protein Kinases 3 and 6 are required for full priming of stress responses in *Arabidopsis thaliana. Plant Cell* **2009**, *21*, 944–953.
92. Sharma, R.; de Vleesschauwer, D.; Sharma, M.K.; Ronald, P.C. Recent advances in dissecting stress-regulatory crosstalk in rice. *Mol. Plant* **2013**, *6*, 250–260.

93. Holley, S.R.; Yalamanchili, R.D.; Moura, D.S.; Ryan, C.A.; Stratmann, J.W. Convergence of signaling pathways induced by systemin; oligosaccharide elicitors; and Ultraviolet-B radiation at the level of mitogen-activated Protein Kinases in *Lycopersicon peruvianum* suspension-cultured cells. *Plant Physiol.* **2003**, *132*, 1728–1738.
94. Lu, C.; Han, M.-H.; Guevara-Garcia, A.; Fedoroff, N.V. Mitogen-activated protein kinase signaling in postgermination arrest of development by abscisic acid. *Proc. Natl. Acad. Sci. USA* **2002**, *99*, 15812–15817.
95. Miura, K.; Tada, Y. Regulation of water; salinity and cold stress responses by salicylic acid. *Front. Plant Sci.* **2014**, *5*, e4.
96. Zhou, J.; Xia, X.-J.; Zhou, Y.-H.; Shi, K.; Chen, Z.; Yu, J.-Q. RBOH1-dependent H_2O_2 production and subsequent activation of MPK1/2 play an important role in acclimation-induced cross-tolerance in tomato. *J. Exp. Bot.* **2014**, *65*, 595–607.
97. Cramer, G.R.; Urano, K.; Delrot, S.; Pezzotti, M.; Shinozaki, K. Effects of abiotic stress on plants: A systems biology perspective. *BMC Plant Biol.* **2011**, *11*, 163–177.
98. Xiong, L.; Schumaker, K.S.; Zhu, J.K. Cell signaling during cold; drought; and salt stress. *Plant Cell.* **2002**, *14*, 165–183.
99. Liu, C.; Ruan, Y.; Lin, Z.; Wei, R.; Peng, Q.; Guan, C.; Ishii, H. Antagonism between acibenzolar-*S*-methyl-induced systemic acquired resistance and jasmonic acid-induced systemic acquired susceptibility to *Colletotrichum orbiculare* infection in cucumber. *Physiol. Mol. Plant Pathol.* **2008**, *72*, 141–145.
100. Flors, V.; Ton, J.; van Doorn, R.; Jakab, G.; García -Agustín, P.; Mauch-Mani, B. Interplay between JA, SA and ABA signalling during basal and induced resistance against *Pseudomonas syringae* and *Alternaria brassicicola*. *Plant J.* **2008**, *54*, 81–92.
101. Ton, J.; Flors, V.; Mauch-Mani, B. The multifaceted role of ABA in disease resistance. *Trends Plant Sci.* **2009**, *14*, 310–317.
102. Truman, W.; Torres de Zabala, M.; Grant, M. Type III effectors orchestrate a complex interplay between transcriptional networks to modify basal defence responses during pathogenesis and resistance. *Plant J.* **2006**, *46*, 14–33.
103. De Torres-Zabala, M.; Truman, W.; Bennett, M.H.; Lafforgue, G.; Mansfield, J.W.; Egea, P.R.; Bogre, L.; Grant, M. *Pseudomonas syringae* pv. *Tomato* hijicks the Arabidopsis abscisic acid signaling pathway to cause disease. *EMBO J.* **2007**, *26*, 1434–1443.
104. García-Andrade, J.; Ramirez, V.; Flors, V.; Vera, P. Arabidopsis ocp3 mutant reveals a mechanism linking ABA and JA to pathogen-induced callose deposition. *Plant J.* **2011**, *67*, 783–794.
105. Luna, E.; Pastor, V.; Robert, J.; Flors, V.; Mauch-Mani, B.; Ton, J. Callose deposition: A multifaceted plant defense response. *Mol. Plant Microbe Interact.* **2011**, *24*, 183–193.
106. Mauch-Mani, B.; Mauch, F. The role of abscisic acid in plant-pathogen interactions. *Curr. Opin. Plant Biol.* **2005**, *8*, 409–414.
107. Robert-Seilaniantz, A.; Navarro, L.; Bari, R.; Lones, J.D.G. Pathological hormone imbalances. *Curr. Opin. Plant Biol.* **2007**, *10*, 372–379.
108. Melotto, M.; Underwood, W.; Koczan, J.; Nomura, K.; He, S.Y. Plant stomata function in innate immunity against bacterial invasion. *Cell* **2006**, *126*, 969–980.

109. Chul Lee, S.; Luan, S. ABA signal transduction at the crossroad of biotic and abiotic stress responses. *Plant Cell Environ.* **2012**, *35*, 53–60.
110. Kim, T-H.; Hauser, F.; Ha, T.; Xue, S.; Bohmer, M.; Nishimura, N.; Munemasa, S.; Hubbard, K.; Peine, N.; Lee, B.; *et al.* Chemical genetics reveals negative regulation of abscisic acid signaling by a plant immune response pathway. *Curr. Biol.* **2011**, *21*, 990–997.
111. Garcia-Andrade, J.; Ramirez, V.; Lopez, A.; Vera, P. Mediated plastid RNA editing in plant immunity. *PLoS Pathog.* **2013**, *9*, e1003713.
112. Ton, J.; Jakab, G.; Toquin, V.; Flors, V.; Iavicoli, A.; Maeder, M.N.; Metraux, J.P.; Mauch-Mani, B. Dissecting the β-aminobutyric acid induced priming phenomenon in Arabidopsis. *Plant Cell* **2005**, *17*, 987–999.
113. Nishimura, M.T.; Stein, M.; Hou, B.H.; Vogel, J.P.; Edwards, H.; Somerville, S.C. Loss of a callose synthase results in salicylic acid-dependent disease resistance. *Science* **2003**, *301*, 969–972.
114. Jensen, M.K.; Hagedorn, P.H.; de Torres-Zabala, M.; Grant, M.R.; Rung, J.H.; Collinge, D.B.; Lyngkjaer, M.F. Transcriptional regulation by an NAC (NAM-ATAF1;2-CUC2) transcription factor attenuates ABA signalling for efficient basal defence towards *Blumeria graminis* f. sp. *hordei* in *Arabidopsis. Plant J.* **2008**, *56*, 867–880.
115. Bandurska, H.; Stroiński, A. The effect of salicylic acid on barley response to water deficit. *Acta Physiol. Plant.* **2005**, *27*, 379–386.
116. Swindell, W.R. The association among gene expression responses to nine abiotic stress treatments in *Arabidopsis thaliana. Genet. Soc. Am.* **2006**, *174*, 1811–1824.
117. Mengiste, T.; Chen, X.; Salmeron, J.; Dietrich, R. The *BORTYTIS SUSCEPTIBLE1* gene encodes an *R2R3MYB* transcription factor protrin that is required for biotic and abiotic stress responses in Arabidopsis. *Plant Cell* **2003**, *15*, 2551–2565.
118. Seo, P.J.; Kim, M.J.; Park, J.Y.; Kim, S.Y.; Jeon, J.; Lee, Y.H.; Kim, J.; Park, C.M. Cold activation of a plasma membrane-tethered NAC transcription factor induces a pathogen resistance response in Arabidopsis. *Plant J.* **2010**, *61*, 661–671.
119. Seo, P.J.; Park, C.-M. *MYB96*-mediated abscisic acid signals induce pathogen resistance response by promoting salicylic acid biosynthesis in *Arabidopsis. New Phythol.* **2010**, *186*, 471–483.
120. Oh, S.J.; Song, S.I.; Kim, Y.S.; Jang, H.J.; Kim, S.Y.; Kim, M.; Kim, Y.K.; Nahm, B.H.; Kim J.K. Arabidopsis CBF3/DREB1A and ABF3 in transgenic rice increased tolerance to abiotic stress without stunting growth. *Plant Physiol.* **2005**, *138*, 341–351.
121. Tran, L.S.P.; Nishiyama, R.; Yamaguchi-Shinozaki, K.; Shinozaki, K. Potential utilization of NAC transcription factors to enhance abiotic stress tolerance in plants by biotechnological approach. *GM Crops* **2010**, *1*, 32–39.
122. Shinozaki, K.; Yamaguchi-Shinozaki, K. Gene networks involved in drought stress response and tolerance. *J. Exp. Bot.* **2007**, *58*, 221–227.
123. Snider, C.S.; Hsiang, T.; Zhao, G.; Griffith, M. Role of ice nucleation and antifreeze activities in pathogenesis and growth of snow molds. *Phytopathology* **2000**, *90*, 354–361.
124. Kim, S.-Y.; Kim, S.-G.; Kim, Y.-S.; Seo, P.-J.; Bae, M.; Yoon, H.-K.; Park, C.-M. Exploring membrane-associated NAC transcription factors in Arabidopsis: Implications for membrane biology in genome regulation. *Nucleic Acids Res.* **2007**, *35*, 203–213.

125. Tsutsui, T.; Kato, W.; Asada, Y.; Sako, K.; Sato, T.; Sonoda, Y.; Kidokoro, S.; Yamaguchi-Shinozaki, K.; Tamaoki, M.; Arakawa, K.; *et al.* DEAR1; a transcriptional repressor of DREB protein that mediates plant defense and freezing stress responses in Arabidopsis. *J. Plant Res.* **2009**, *122*, 633–643.
126. Foyer, C.H.; Karpinska, B.; Krupinska, K. The functions of WHIRLY1 and REDOX-RESPONSIVE TRANSCRIPTION FACTOR 1 in crosstolerance responses in plants: A hypothesis. *Philos. Trans. R. Soc. Lond. B Biol. Sci.* **2014**, *369*, 20130226.
127. Zhang, Y.; Butelli, E.; de Stefano, R.; Schoonbeek, H.J.; Magusin, A.; Pagliarani, C.; Wellner, N.; Hill, L.; Orzaez, D.; Granell, A.; *et al.* Anthocyanins double the shelf life of tomatoes by delaying over ripening and reducing susceptibility to graymold. *Curr. Biol.* **2013**, *23*, 1094–1100.
128. Marco, F.; Altabella, T.; Alcázar, R.; Cuevas, J.C.; Bortolotti, C.; González, M.E.; Carrasco, P. Transcriptome analysis of polyamine overproducers reveals activation of plant stress responses and related signalling pathways tolerance in plants. *Omics Plant Abiotic Stress Toler.* **2011**, *1*, 82–90.
129. Atkinson, N.J.; Lilley, C.J.; Urwin, P.E. Identification of genes involved in the response of *Arabidopsis* to simultaneous biotic and abiotic stresses. *Plant Physiol.* **2013**, *162*, 2028–2041.
130. Goyal, R.K.; Hancock, R.E.W.; Mattoo, A.K.; Misra, S. Expression of an engineered heterologous antimicrobial peptide in potato alters plant development and mitigates normal abiotic and biotic responses. *PLoS One* **2013**, *8*, e77505.
131. Treutter, D. Significance of flavonoids in plant resistance: A review. *Environ. Chem. Lett.* **2006**, *4*, 147–157.
132. Nakabayashi, R.; Yonekura-Sakakibara, K.; Urano, K.; Suzuki, M.; Yamada, Y.; Nishizawa, T.; Matsuda, F.; Kojima, M.; Sakakibara, H.; Shinozaki, K.; *et al.* Enhancement of oxidative and drought tolerance in *Arabidopsis* by over accumulation of antioxidant flavonoids. *Plant J.* **2014**, *77*, 367–379.
133. Walters, D. Resistance to plant pathogens: Possible roles for free polyamines and polyamine catabolism. *New Phytol.* **2003**, *159*, 109–115.
134. Bitrián, M.; Zarza, X.; Altabella, T.; Tiburcio, A.F.; Alcázar, R. Polyamines under abiotic stress: Metabolic crossroads and hormonal crosstalks in plants. *Metabolites* **2012**, *2*, 516–528.
135. Hussain, S.S.; Ali, M.; Ahmad, M.; Siddique, K.H. Polyamines: Natural and engineered abiotic and biotic stress tolerance in plants. *Biotechnol. Adv.* **2011**, *29*, 300–311.
136. Prime-A-Plant Group; Conrath, U.; Beckers, G.J.; Flors, V.; García-Agustín, P.; Jakab, G.; Mauch, F.; Newman, M.A.; Pieterse, C.M.; Poinssot, B.; *et al.* Priming: Getting ready for battle. *Mol. Plant Microbe Interact.* **2006**, *19*, 1062–1071.

6

Characterization of Four Bifunctional Plant IAM/PAM-Amidohydrolases Capable of Contributing to Auxin Biosynthesis

Beatriz Sánchez-Parra [1,†], Henning Frerigmann [2,†,‡], Marta-Marina Pérez Alonso [1], Víctor Carrasco Loba [1], Ricarda Jost [3], Mathias Hentrich [2] and Stephan Pollmann [1,*]

[1] Center for Plant Biotechnology and Genomics (U.P.M.-I.N.I.A.), Technical University Madrid, Montegancedo Campus, Crta. M-40, km 38, 28223 Pozuelo de Alarcón (Madrid), Spain; E-Mails: beatriz.sanchez@upm.es (B.S.-P.); martamarina.perez@upm.es (M.-M.P.A.); victorcarrascoloba@gmail.com (V.C.L.)

[2] Department of Plant Physiology, Faculty of Biology and Biotechnology, Ruhr-University Bochum, Universitätsstraße 150, 44801 Bochum, Germany; E-Mails: henning.frerigmann@uni-koeln.de (H.F.); mathias.hentrich@rub.de (M.H.)

[3] School of Plant Biology, University of Western Australia, 35 Stirling Highway, Crawley, WA 6009, Australia; E-Mail: ricarda.jost@uwa.edu.au

† These authors contributed equally to this work.

‡ Present address: Institute of Botany Chair II, University of Cologne, Zülpicher Straße 47b, 50674 Cologne, Germany.

* Author to whom correspondence should be addressed; E-Mail: stephan.pollmann@upm.es

---

Abstract: Amidases [EC 3.5.1.4] capable of converting indole-3-acetamide (IAM) into the major plant growth hormone indole-3-acetic acid (IAA) are assumed to be involved in auxin *de novo* biosynthesis. With the emerging amount of genomics data, it was possible to identify over forty proteins with substantial homology to the already characterized amidases from *Arabidopsis* and tobacco. The observed high conservation of amidase-like proteins throughout the plant kingdom may suggest an important role of theses enzymes in plant development. Here, we report cloning and functional analysis of four, thus far, uncharacterized plant amidases from *Oryza sativa*, *Sorghum bicolor*, *Medicago truncatula*, and *Populus trichocarpa*. Intriguingly, we were able to demonstrate that the examined amidases are also capable of converting phenyl-2-acetamide (PAM) into phenyl-2-acetic acid (PAA), an auxin endogenous to several plant species including *Arabidopsis*. Furthermore, we compared the subcellular localization of the enzymes to that of *Arabidopsis* AMI1, providing further evidence for similar enzymatic functions. Our results point to the presence of a presumably conserved pathway of auxin biosynthesis via IAM, as amidases, both of monocot, and dicot origins, were analyzed.

Keywords: amidase; auxin; indole-3-acetic acid; indole-3-acetamide; phenyl-2-acetic acid; phenyl-2-acetamide

1. Introduction

The collective term auxin refers to a class of compounds sharing similar physiological functions in plants. Auxins are well characterized as growth promoting phytohormones, occurring at low concentrations across plant species. They are known to be essential for the regulation of a wealth of different processes, such as cell elongation, initiation of lateral and adventitious root growth, flower and fruit development, and fruit ripening [1,2]. Albeit indole-3-acetic acid (IAA) is considered the major auxin in plants, a number of additional naturally occurring compounds that exert auxin effects have been reported to date (e.g., 4-chloroindole-3-acetic acid [3], and phenyl-2-acetic acid (PAA) [4,5]). Moreover, a variety of IAA precursors such as indole-3-acetamide (IAM), indole-3-acetonitrile, indole-3-pyruvic acid (IPyA) [1,6], and indole-3-butyric acid [7,8] exhibit auxin-like effects, most likely due to their conversion to IAA.

Although much is known about the physiological functions and effects of auxins, the biosynthesis of substances of this compound class still remains partially elusive (Figure 1).

The main portion of free IAA in plants is seemingly produced via a two-step pathway involving tryptophan aminotransferases (TAA1 and TARs) and flavin-containing monooxygenases of the YUCCA family (YUC1-11) [9,10]. However, there is additional evidence that the biosynthesis of IAA, at least to a certain extent, may also proceed via a number of other metabolic routes, which are classified in terms of their intermediates. These routes are thought to act either in parallel or in a developmentally regulated manner [11–14]. One of these biosynthetic pathways is supposed to lead from L-tryptophan via IAM to IAA and appears to be one of the two major pathways commonly used in bacteria to produce IAA [15–17]. Previous work provided growing evidence for the existence of a comparable IAM-pathway to be operative *in planta* as well, suggested by the following observations: (i) in a number of independent studies IAM has been identified as a natural constituent of several plant species including *Citrus unshiu* [18], *Prunus jamasakura* [19], *Curcubita maxima* [20], as well as in *Arabidopsis thaliana*, *Oryza sativa*, *Zea mays*, and *Nicotiana tabacum* [21,22]; (ii) IAM hydrolyzing activities have been reported from *Triticum aestivum* and *Pisum sativum* (whole plant extracts) [6], *O. sativa* (callus extracts) [23,24], as well as from *Poncirus trifoliate* (fruit extracts) [25]; and (iii) amidases from *A. thaliana* and *N. tabacum* capable of converting IAM to IAA *in vitro* have been identified [26,27].

Figure 1. Proposed pathways of L-tryptophan-dependent IAA biosynthesis in plants. The IAOx-pathway that is seemingly restricted to indole glucosinolate-producing plant species is given in the grey box. In yellow, the IPyA-pathway is shown; a possible side-branch (IAAld-pathway) is added in green to the IPyA-pathway. In the middle, the IAM-pathway is highlighted in blue. Dashed lines indicate assumed reaction steps for which the corresponding enzymes have yet to be identified.

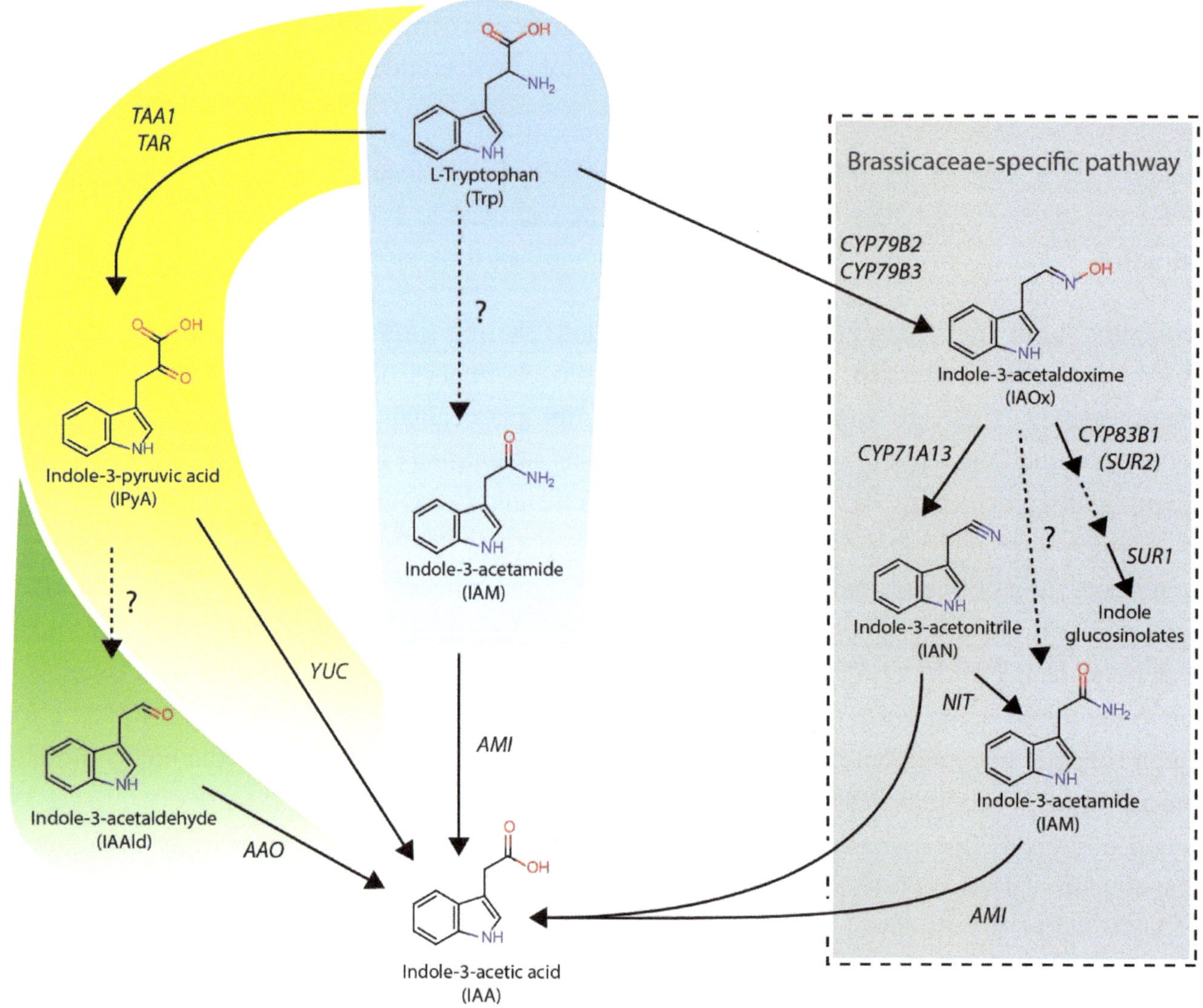

Enzymes are abbreviated as follows: AAO, ARABIDOPSIS ALDEHYDE OXIDASE 1; AMI, AMIDASE; CYP71A13, CYTOCHROME P450 MONOOXYGENASE 71A13; CYP79B2/B3, CYTOCHROME P450 MONOOXYGENASE 79B2/B3; CYP83B1, CYTOCHROME P450 MONOOXYGENASE 83B1; NIT, NITRILASE; SUR1, SUPERROOT 1 (S-ALKYL-THIOHYDROXIMATE LYASE); TAA1, TRYPTOPHAN AMINOTRANSFERASE OF ARABIDOPSIS 1; TAR2, TRYPTOPHAN AMINOTRANSFERASE RELATED 2; YUC, YUCCA.

Like the bacterial IAM-hydrolases from *Agrobacterium tumefaciens* [28] and *Pseudomonas syringae* [29], *At*AMI1 is a member of the amidase-signature superfamily, characterized by a glycine and serine rich stretch of approximately 50 amino acids, which contains most of the catalytically relevant amino acid residues [30]. *At*AMI1 is located in both the cytoplasm and nucleoplasm, acts as an obligate monomer of approximately 46 kDa, and is mainly expressed in tissues with high meristematic activity, for instance young leaves and floral buds [31,32]. Moreover, *At*AMI1 specifically converts IAM to IAA *in vitro*, while refusing most other naturally occurring amides as substrates. On this basis, a tentative involvement of *At*AMI1 in auxin biosynthesis in *A. thaliana* has been suggested.

Due to these lines of evidence, implying a possibly broader importance of IAA biosynthesis through the IAM-pathway in plants, it appeared intriguing to us to investigate whether IAM-amidohydrolases are restricted to only very few plant families or if this enzyme class shows a broader distribution in the plant kingdom. The identification and functional analysis of comparable AMI1-like enzymes from various plant species would likewise highlight a broader presence of the IAM-dependent IAA production and point towards a general concept in auxin biosynthesis.

Thoroughly analyzing the wealth of emerging genome data from plants, we were able to identify a reasonably high number of AMI1-like proteins. Phylogenetic comparison revealed substantial sequence homology between those proteins and *Arabidopsis* AMI1. To gain further insight into the function of these enzymes, we chose four amidases from *O. sativa*, *Sorghum bicolor*, *Medicago truncatula*, and *Populus trichocarpa* in this study and compared them to *At*AMI1 [26,31]. With respect to enzymatic properties and subcellular localization, we observed considerable similarity between the four candidate enzymes and *At*AMI1, although the selected amidases displayed lower specific activities towards IAM when compared to the properties of *At*AMI1. Furthermore, we were capable of demonstrating that all examined amidases are additionally able to convert phenyl-2-acetamide (PAM) to the naturally occurring auxin, PAA. We also show that PAA is endogenous to *Arabidopsis*, and that it exerts an auxin-like effect in *Arabidopsis* root growth bioassays.

2. Results and Discussion

2.1. Identification of AMI1-like Proteins in Plant Genomes

Homologs of the *Arabidopsis YUC* gene family have been identified and characterized from several plant species including petunia, rice, tomato, maize, and pea [33–37], suggesting a widespread occurrence of the YUC-dependent IAA biosynthetic pathway in the plant kingdom. In contrast, CYP79B2/B3 and NIT1-3 enzymes seem to be restricted to the *Brassicaceae* [22,38,39], which contradicts a general importance of these enzymes in auxin biosynthesis. To investigate the distribution of AMI1-like proteins in the plant kingdom, the *Arabidopsis* IAM-hydrolase (*At*AMI1, At1g08980) nucleic acid sequence, as well as the translated primary amino acid sequence, were used to query publicly available databases (e.g., Phytozome, Plant Gene Duplication Database) for candidate IAM-hydrolase orthologs. The *in silico* analyses provided indication for 47 *At*AMI1-like proteins from 38 plant species (Figure 2).

It is noteworthy to mention that, at least to the best of our current knowledge, all plant genomes published thus far contain one or more *At*AMI1-like sequences, which may suggest an important function of this enzyme class in plant development, as the gene is conserved across a large number of plant species. The notion of a wider distribution of AMI1-like proteins in higher plants is further strengthened by the previously mentioned identification of IAM and detection of IAM-hydrolase activities in various other plant species, respectively. In the first place, however, the obtained results indicate a broader distribution of *At*AMI1-like proteins in the plant kingdom.

Figure 2. Cladogram of 51 AMI1-like proteins from plants and bacteria.

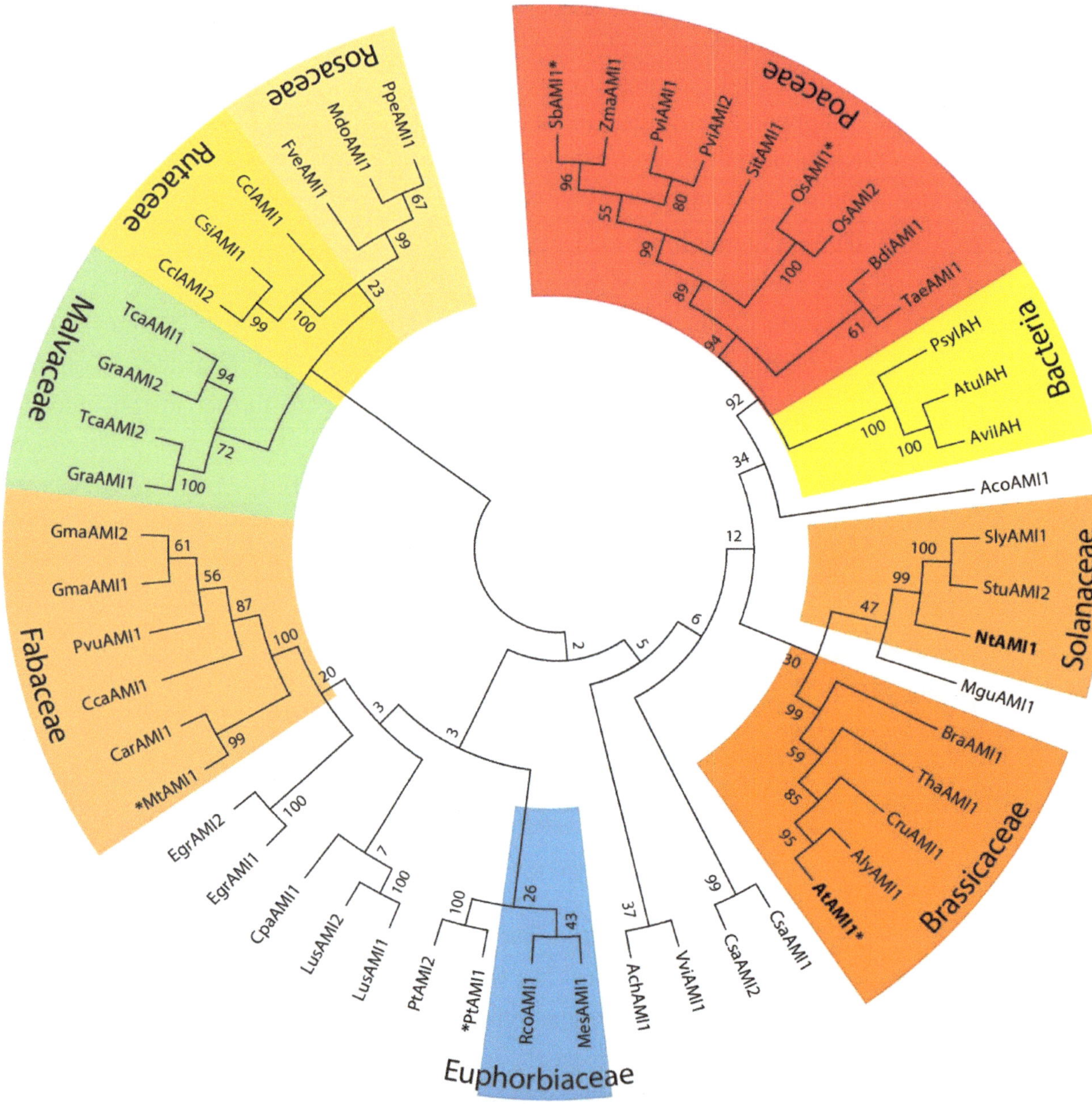

A bootstrap consensus tree inferred from 500 replicates was generated to represent the phylogenetic relationships of the proteins analyzed. Proteins from the plant species highlighted in bold letters have already been enzymatically characterized in our group [26,30,31] or by other laboratories [27]. The amidases further studied in this work are labeled with asterisks. An amino acid sequence alignment of the selected enzymes can be taken from Appendix Figure A1. Plant species that belong to the same family are given in colored boxes. Species names are abbreviated as follows with gene/protein identifiers given in parenthesis: Ach: *Actinidia chinensis* (ACHN083681); Aco: *Aquilegia coerulea* (Aquca_05400135.1); Aly: *Arabidopsis lyrata* (16043473); At: *Arabidopsis thaliana* (At1g08980); Atu: *Agrobacterium tumefaciens* (YP_001967410); Avi: *Agrobacterium vitis* (P25016); Bdi: *Brachypodium distachyon* (Bradi5g27490); Bra: *Brassica rapa* (Bra031652); Car: *Cicer arietinum* (XP_004495736); Cca: *Cajanus cajan* (C.cajan06830); Ccl: *Citrus clementine* (AMI1: Ciclev10001239m, AMI2: Ciclev10001213m); Cpa: *Carica papaya* (evm.model.supercontig_48.34); Cru: *Capsella rubella* (Crubv10006860m); Csa: *Cucumis sativus* (AMI1: Cucsa.11590.1, AMI2: Cucsa.182140.1); Csi: *Citrus sinensis* (orange1.1g044294m); Egr: *Eucalyptus grandis* (AMI1: Eucgr.B01994.1, AMI2: Eucgr.B01996.1); Fve: *Fragaria vesca* (Mrna22572.1-v1.0-hybrid); Gma: *Glycine max* (AMI1: Glyma10g29640.1, AMI2: Glyma20g37660.1); Gra: *Gossypium raimondii* (AMI1: Gorai.007G295700.1,

AMI2: Gorai.007G295900.1); Lus: *Linum usitatissimum* (AMI1: Lus10029941, AMI2: Lus10004466); Mdo: *Malus x domestica* (MDP0000302761); Mes: *Manihot esculentum* (cassava4.1_010118m); Mgu: *Mimulus guttatus* (mgv1a006860m); Mt: *Medicago truncatula* (Medtr1g082750.1); Nt: *Nicotiana tabacum* (AB457638); Os: *Oryza sativa* (AMI1: Os04g02780.1, AMI2: Os04g02754.1); Psy: *Pseudomonas syringae* (P52831); Ppe: *Prunus persica* (ppa005982m); Pt: *Populus trichocarpa* (AMI1: Potri.013g024100.1, AMI2: Potri.013g024200.1); Pvi: *Panicum virgatum* (AMI1: Pavirv00043873m, AMI2: Pavirv00038995m); Pvu: *Phaseolus vulgaris* (Phvul.007G180900.1); Rco: *Ricinus communis* (30128.m008759); Sb: *Sorghum bicolor* (Sb02g039510.1); Sit: *Setaria italica* (Si001473m); Sly: *Solanum lycopersicum* (Solyc10g086170.1.1); Stu: *Solanum tuberosum* (PGSC0003DMP400065555); Tae: *Triticum aestivum* (AK330138); Tca: *Theobroma cacao* (AMI1: Thecc1EG046817t1, AMI2: Thecc1EG046818t1); Tha: *Thellungiella halophile* (Thhalv10007679m); Vvi: *Vitis vinifera* (GSVIVT01000036001); Zma: *Zea mays* (GRMZM2G169087_T01).

2.2. Subcellular Localization of Selected Plant Amidases

The high degree of sequence homology between the *At*AMI1-like proteins compared in Figure 2 made it tempting to speculate that the identified enzymes also share similar functions. In order to gain deeper insight into their functional properties, we selected four candidate amidases both from monocot and dicot species, namely *O. sativa Os*AMI1 (Os04g02780), *S. bicolor Sb*AMI1 (Sb02g039510), *M. truncatula Mt*AMI1 (Mt1g099370), and *P. trichocarpa Pt*AMI1 (Pt13g02200), for further investigation. On a protein level, the selected candidates show an overall amino acid identity of 57% (*Os*AMI1), 58% (*Sb*AMI1), 66% (*Mt*AMI1), and 68% (*Pt*AMI1), respectively, relative to *At*AMI1. As shown in Appendix Figure A1, in particular the region of the amidase signature, containing most of the catalytically relevant amino acid residues [30], shows a high degree of amino acid identity. Consequently, there is reason to suspect that they share similar function.

In a first set of experiments, the subcellular localization of chimeric GFP fusion proteins was investigated in transient assays. In each experiment, 5 μg of plasmid DNA, facilitating the *35S*-promoter driven expression of the corresponding fusion construct, were transferred into leaf epidermal cells of *A. thaliana* by particle bombardment. Like the AMI1 enzyme from *A. thaliana*, the amidases from *O. sativa*, *S. bicolor*, *M. truncatula*, and *P. trichocarpa*, appear located to the cytoplasm of epidermal cells when transiently expressed as GFP fusion constructs. Moreover, the microscopic assessment of the intracellular protein localization shown in Figure 3 provided further evidence for a diffusion of all studied amidase-constructs into the nucleus and an accumulation in the nuclear pocket, thus, delivering results similar to those obtained for the *Arabidopsis* AMI1:GFP fusion construct that has been shown to be able to diffuse through nuclear pores. However, as previously shown, the nuclear localization of *At*AMI1 is most likely not by reason of a specific nuclear import, but rather due to the compact conical shape of the protein [30,31].

The obtained results are consistent with predictions of the TargetP [40] and PredictNLS server [41], detecting neither nuclear localization signals or signals for the secretory pathway nor mitochondrial and plastidial import signals in the primary amino acid sequence of the analyzed enzymes. In any case, the similar cellular localization pattern of the tested amidases adds an additional cue highlighting a possibly comparable function of the polypeptides.

Figure 3. Localization of amidase GFP fusion proteins in epidermal cells of *Arabidopsis thaliana*. The images show typical results of confocal laser scanning microscopic studies, using differential emission light filter sets. The upper row provides transmission images, followed by the GFP channel (GFP-fluorescence), showing the fluorescence of the GFP-fluorophore between 500 and 530 nm. Next, the chlorophyll-autofluorescence is shown in a separate channel (650 to 798 nm). In the bottom row, an overlay of the fluorescence channels is depicted. (**A**) Transformation with an empty GFP-vector (pSP-EGFP), cytoplasmic control; (**B**) Transformation with an *AMI1*:*GFP* construct from *A. thaliana* [31] (positive control); (**C**) Transformation of an epidermis cell with the *Os*AMI1:GFP construct; (**D**) Transformation with the chimeric GFP:*Mt*AMI construct; (**E**) Transformation with the *Pt*AMI:GFP vector; (**F**) Transformation of the *Sb*AMI1:GFP construct. Scale bars for each set of pictures are included in the figure.

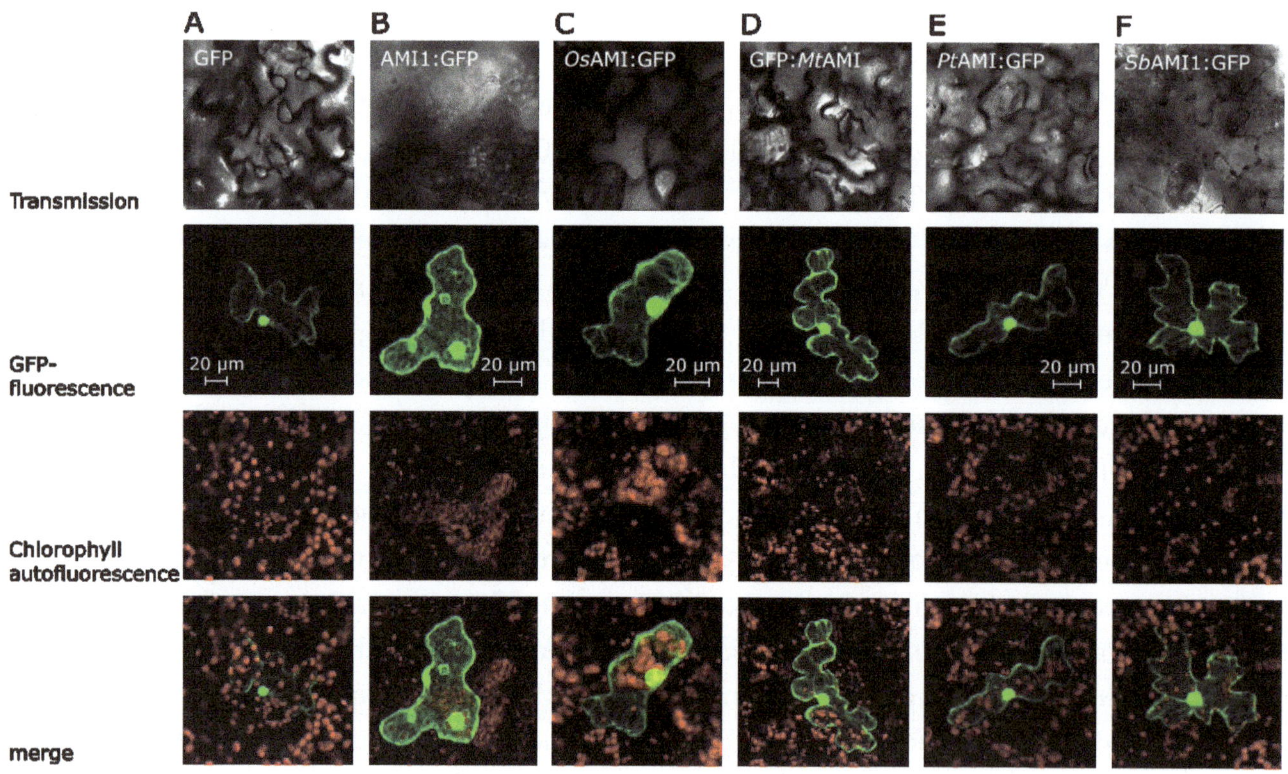

2.3. Functional Analysis of the Selected Plant Amidases

To assess whether the proteins possess IAM-amidohydrolase activity *in vitro*, the respective cDNAs from *O. sativa*, *S. bicolor*, *M. truncatula*, and *P. trichocarpa* were amplified by RT-PCR and cloned into the prokaryotic expression vector pTrcHis2 (Invitrogen) for in-frame expression with a c-myc/hexahistidine double tag in *Escherichia coli*, strain BL21-AI (Invitrogen). Recombinant proteins were isolated from bacterial lysates by Ni^{2+}-nitrilotriacetate affinity chromatography. As displayed in Figure 4, due to general difficulties in expressing amidases in bacteria, the amount as well as the homogeneity of the purified protein fractions varied considerably. For this reason, the measured enzymatic activities have to be taken with some caution.

Figure 4. Expression and purification of recombinant amidases. Hexahistidine tagged amidases were expressed in *E. coli* and purified as described in the experimental section. Aliquots of 12.5 μL of the corresponding elution fractions were used for SDS-polyacrylamide gel electrophoresis and subsequent coomassie-staining (**A**) and immunoblotting (**B**). For the immunoblot only the region between 75 and 37 kDa is shown.

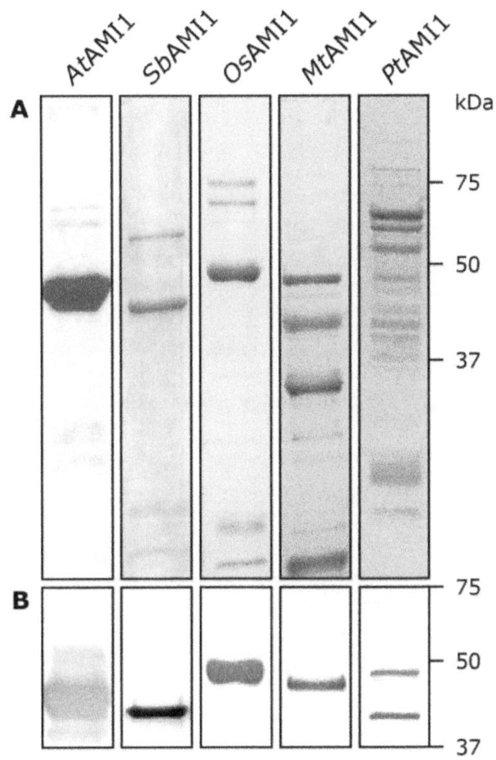

In order to compare the biochemical properties of the recombinant amidases to those of *At*AMI1 in terms of specific activity and substrate selectivity, equal amounts of total protein (5 μg each) from desalted elution fractions were utilized. In this series of experiments, substrate concentrations of 10 mM were used. The data for the substrate specificity are shown on a relative scale normalized to the activity towards IAM (100%) (Figure 5).

Generally, the substrate preferences of the amidases from *O. sativa*, *S. bicolor*, *M. truncatula*, and *P. trichocarpa* were similar, but not identical, to those of *At*AMI1. Under the used conditions, the best substrate found was phenyl-2-acetamide (PAM), a compound that is endogenous to *Phaseolus mungo* [42,43], while an occurrence in *A. thaliana* has not yet been described. *In planta*, PAM might arise from turnover or breakdown of glucotropaeolin, a L-phenylalanine derived benzylglucosinolate [44]. The product of the amidase catalyzed PAM conversion, phenyl-2-acetic acid (PAA), has been shown to occur in a number of different plant species, including *P. mungo* [45], *Solanum lycopersicum*, *N. tabacum*, and *Z. mays* [4], as well as *Tropaeolum majus* [5], but to our knowledge, occurrence of PAA in *A. thaliana* has not yet been reported.

Figure 5. Analysis of the relative substrate conversion of recombinant amidases.

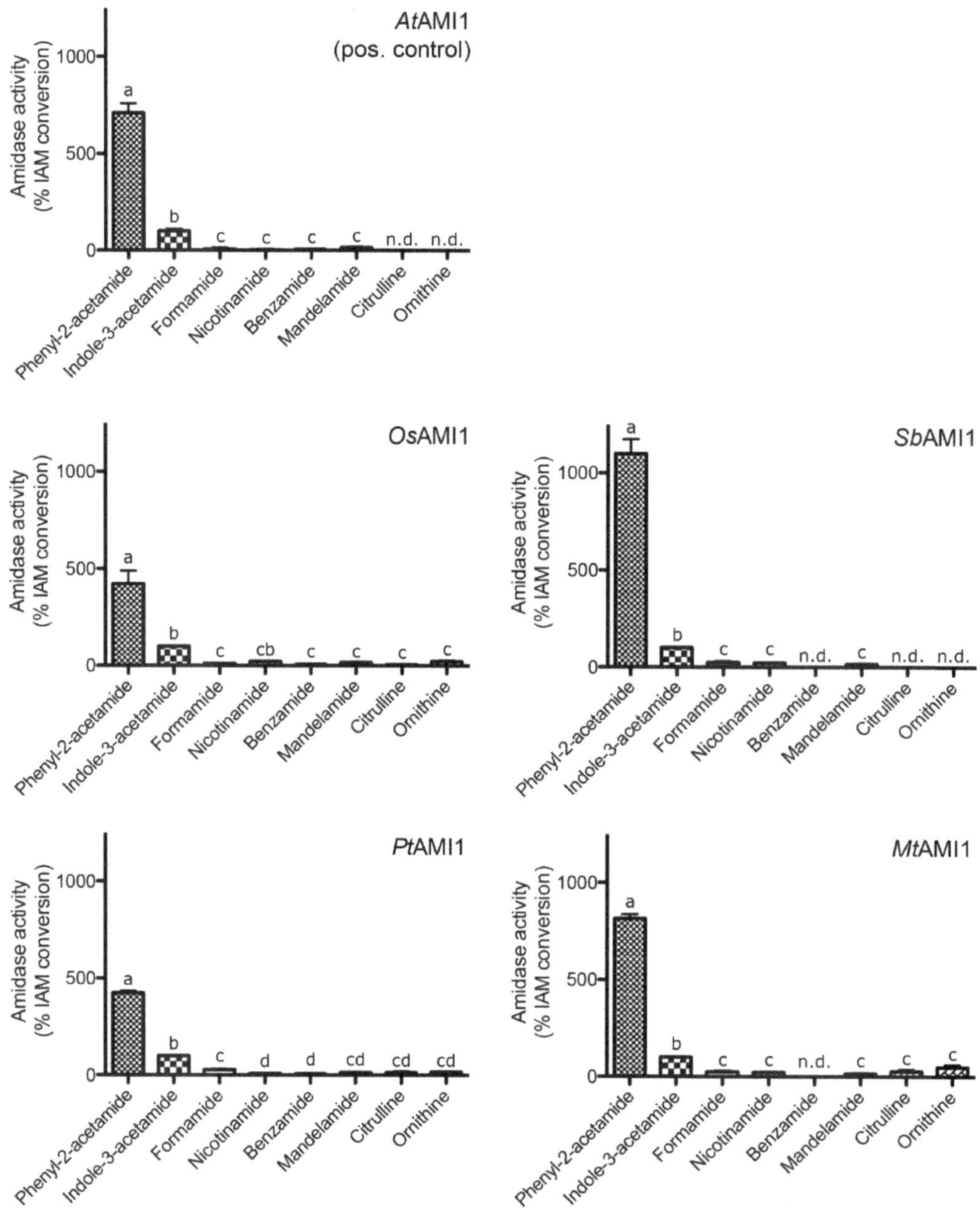

All assays were performed in a total reaction volume of 300 μL at pH 7.5 and 30 °C (substrate concentration 10 mM, reaction time 4 to 6 h). All activities were normalized to the conversion of IAM (100%). Catalytic activities are corrected for background activities obtained for the non-enzymatic breakdown of substrates in heat-inactivated and empty-vector control samples. Maximum specific activities for IAM were 3070 pkat (mg protein)$^{-1}$ (*At*AMI1), 2378 pkat (mg protein)$^{-1}$ (*Sb*AMI1), 375 pkat (mg protein)$^{-1}$ (*Os*AMI1), 37 pkat (mg protein)$^{-1}$ (*Mt*AMI1), and 329 pkat (mg protein)$^{-1}$ (*Pt*AMI1). The data shown are means ± SE derived from $n \geq 3$ experiments. n.d.: not detected. Mean values within a graph are significantly different ($p < 0.05$) where superscript letters differ.

To further explore the enzymatic parameters of the enzymes, pH, as well as temperature optima, were estimated. These data are given, together with the specific activities, in Table 1. All enzymes displayed highest conversion rates at a pH between 6 and 7.5, characteristic for cytosolic proteins, confirming the previously described intracellular localization of the proteins. Relative to the other compared polypeptides, *Os*AMI1 exhibited a slightly lower temperature optimum, at 27 °C, which might be explained by differences in the plants' habitat. Whereas *A. thaliana*, *S. bicolor*, *P. trichocarpa* and *M. truncatula* grow on dry land, *O. sativa*, as a helophyte, has a high tolerance towards flooding. For the latter, evaporation of water from the paddy fields, accounting for a slightly cooler microclimate [46], might be the reason for an evolutionary adjustment of the enzymatic parameters to higher activities at lower temperatures. However, when trying to determine the K_m values for IAM, it was found that none of the enzymes followed a classical Michaelis-Menten kinetics but rather followed a sigmoidal curve progression, which resemble the results obtained for *At*AMI1 [26], and might indicate an allosteric mode of action. Due to solubility constraints of the substrates and inhibitory effects of the organic solvent (either methanol or ethanol), substrate concentrations of more than 25 mM could not be assessed. At this concentration no saturation of the enzymatic reaction could be observed, thus, preventing the calculation of apparent K_m values (data not shown). In most instances, the analyzed amidases displayed quite similar characteristics to *At*AMI1, showing comparable pH and temperature optima, although some distinct features were found as well. Most strikingly, the amidases from both *O. sativa* (*Os*AMI1) and *P. trichocarpa* (*Pt*AMI1) showed an approximately ten-fold decrease in specific activity towards IAM, whereas the amidase of *M. truncatula* (*Mt*AMI1) displayed a 100-fold decrease in specific activity for the same substrate *in vitro*. Nevertheless, the examined polypeptides showed clear amidase activity, converting preferentially PAM and IAM relative to other tested amides. In assays with empty vector controls and heat-denatured protein fractions, respectively, no considerable substrate conversion was detected. The highest background activity for the non-enzymatic conversion of the amide substrates to the corresponding free acids was detected for formamide and benzamide, with 0.14% and 0.11% of the enzymatic IAM conversion, respectively. A summarized overview of the enzymatic parameters is given in Table 1.

Table 1. Comparison of the enzymatic parameters obtained for recombinant amidases from *S. bicolor*, *O. sativa*, *P. trichocarpa*, and *M. truncatula* with those of the *A. thaliana* AMI1 [a].

Parameter	*At*AMI1	*Sb*AMI1	*Os*AMI1	*Pt*AMI1	*Mt*AMI1
Amino acid identity to *At*AMI1 [%]	100	58	57	64	66
Intracellular localization	cytoplasm/ nucleoplasm	cytoplasm/ nucleoplasm	cytoplasm/ nucleoplasm	cytoplasm/ nucleoplasm	cytoplasm/ nucleoplasm
Specific activity (IAM) [pkat mg^{-1}]	3070 ± 520	2378 ± 324	375 ± 52	329 ± 22	37 ± 3
Temperature optimum [°C]	37	45	27	37	35
pH optimum	7.5	6	7.5	7.5	7.5
Calculated molecular weight [kDa]	46	45	50	53	57

[a] The data shown are means ± SE from at least three independent experiments.

In summary, besides a cytoplasmic/nucleoplasmic localization similar to that of *At*AMI1 (Figure 3), all of the examined enzymes are indeed functionally related with respect to their enzymatic activity. They all catalyze the conversion of IAM to IAA *in vitro*, albeit with different specific activities (Table 1). Nonetheless, our enzymatic studies provided clear evidence to support their recognition as plant PAM/IAM-amidohydrolases. Together with the observation that the *N. tabacum* homolog of *AtAMI1* also encodes for a functional IAM-hydrolase [27], the presented results highlight a widespread occurrence of IAM-hydrolases throughout the plant kingdom. This is supportive of a conserved and presumably basic role of AMI1-like IAM-hydrolases in plant development and a tentative contribution to auxin biosynthesis in particular.

2.4. Occurrence and Auxin Activity of Phenyl-2-acetic Acid in Arabidopsis

The determination of substrate specificities of the examined amidases revealed a particular preference for the conversion of PAM to PAA. To investigate this aspect in closer detail, we first tried to analyze whether PAA has an impact on the root development of sterilely grown *Arabidopsis* seedlings, as earlier described for *P. sativum* [47], *Avena sativa* [48], and *T. majus* [5]. As shown in Figure 6, PAA exerts an auxin-typical effect on primary root growth. The required dose to phenocopy the impact of IAA, however, was approximately 50-fold higher. This finding is in agreement with previous work of Koepfli and colleagues [47], assigning PAA 5% of IAA activity in pea.

Figure 6. Impact of PAA and IAA on primary root elongation.

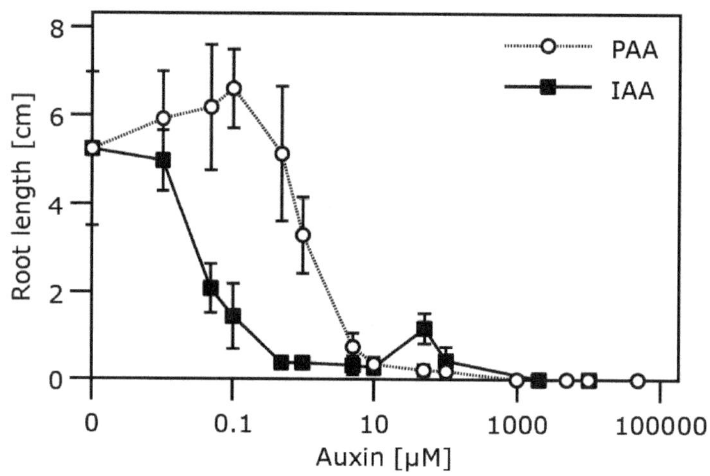

Root elongation was quantified in two-week old *A. thaliana* seedlings grown on medium containing indicated concentrations of either IAA or PAA. Seedlings were removed from the petri dishes, and root length was recorded. At least ten seedlings for each condition were measured. Shown is the absolute root length in cm for either IAA or PAA treated plants. The error bars indicate the standard error of the mean (SE).

Secondly, we determined the steady-state levels of PAA and IAA in leaf tissue of both sterilely grown and soil grown *Arabidopsis* plantlets. After purification and analysis of the methylated extracts, PAA was identified as an endogenous compound in *A. thaliana* by full-scan mass spectrometry (Figure 7).

Next, the content of PAA in two- and six-week old plants was investigated (Table 2). Previous reports give a free IAA content of approximately 30 to 60 pmol (g FW)$^{-1}$ for two-week old *Arabidopsis* seedlings [21,49]. In contrast, the PAA content in leaf tissue of sterile grown plants was slightly higher in two-week old plantlets. This is in contrast to an observed 10- to 100-fold lower concentration of PAA relative to IAA in *T. majus* [5]. The presented results may hint at the dynamic nature of auxin concentrations that have been demonstrated within the plant and throughout its life cycle.

Figure 7. Detection of endogenous PAA. Endogenous PAA in sterile-grown *A. thaliana* was analyzed by extracting the organic compounds from rosettes of two-week old plants with boiling methanol in the presence of 1 nmol [$^{13}C_2$]-PAA standard (Sigma). (**A**) The extract was pre-purified by solid-phase extraction and analyzed by GC-MS. Under these conditions, PAA elutes at 4:55 min. The upper panel shows the extracted ion chromatograms for endogenous PAA (*m*/*z* 91 and 151) and stable isotope labeled [$^{13}C_2$]-PAA (*m*/*z* 92 and 153); (**B**) The middle panel shows the characteristic full-scan mass spectrum for co-chromatographed endogenous PAA (*m*/*z* = 91, 151) and [$^{13}C_2$]-PAA (*m*/*z* = 92, 153); (**C**) Fragmentation and structure of the [$^{13}C_2$]-PAA methyl ester. As the two ^{13}C atoms in [$^{13}C_2$]-PAA are attached to the acetate side chain, which is cleaved during ionization, only the mass of the parent ion of the labelled standard shows a shift of +2 atomic mass units relative to the endogenous compound. In consequence of fragmentation, however, the mass of the fragment of the standard PAA is shifted by only +1 atomic mass unit.

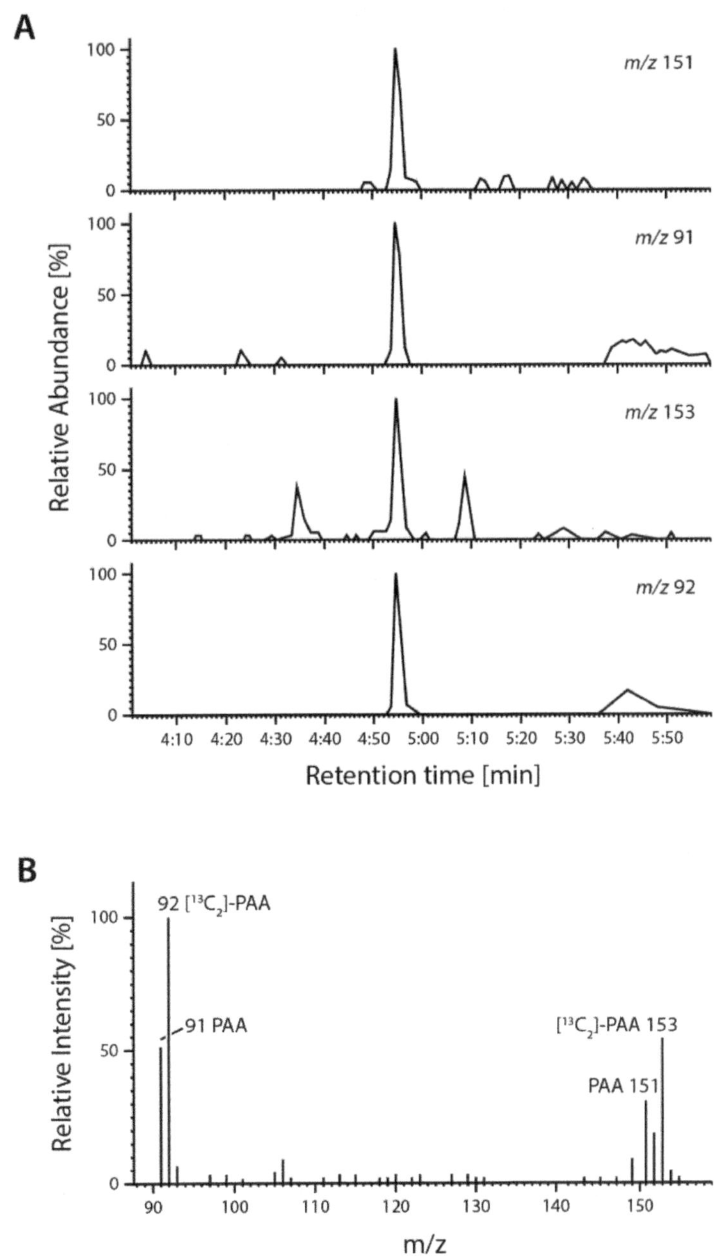

Figure 7. *Cont.*

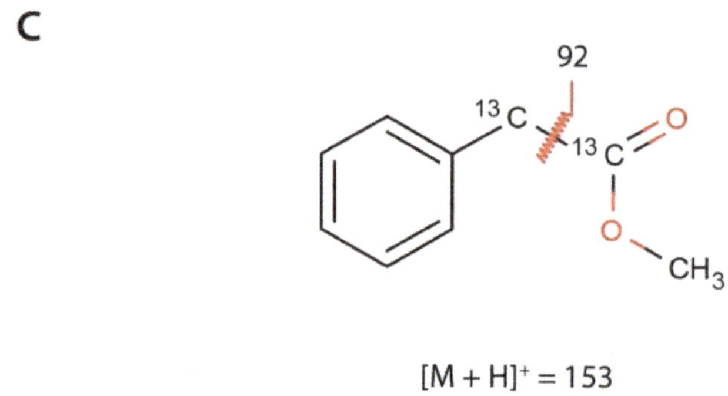

Table 2. PAA contents in two-week old and six-week old leaf tissue of *A. thaliana* [a].

Plant Age (Weeks)	Free PAA pmol (g FW)$^{-1}$	Free IAA pmol (g FW)$^{-1}$
Two weeks	98 ± 6.4	35 ± 2.7
Six weeks	23 ± 2	15 ± 1

[a] Shown are means ± SD from three independent experiments.

3. Experimental Section

3.1. Plant Material and Plant Growth Conditions

All experiments were carried out using either *A. thaliana* (L.) Heynh. ecotype Col-0 (originally from Nottingham *Arabidopsis* Stock Centre, NASC, stock N1092), *O. sativa* L. cv. Millin (obtained from Yanco Rice Research Institute, Yanco, NSW, Australia), *S. bicolor* (L.) Moench *var.* Redland, *M. truncatula* Gaert. cv. Jemalong A17, or *P. trichocarpa* Hook. (Botanical Garden, Ruhr-University Bochum, Germany). While leaf material from poplar was directly taken from mature trees grown in their natural environment, *S. bicolor* and *M. truncatula* seeds were sown onto a mixture of soil and sand (2:1) and cultivated in a greenhouse at 22 to 24 °C during daytime and 18 to 20 °C over night, with a 16 h light/8 h dark cycle. Unless stated otherwise, rice was grown in Murashige and Skoog medium [50] containing 3% (w/v) sucrose and 0.3% (w/v) gelrite (Duchefa, Netherlands). The photosynthetically active radiation was no less than 150 $\mu E \cdot m^{-2} \cdot s^{-1}$ (supplementary light, if required, from sodium-vapor lamps).

3.2. RNA Isolation and RT-PCR

Total RNA was extracted with TRIzol reagent (Invitrogen), and the first strand cDNA was synthesized from 2 µg of total RNA using Oligo(dT)$_{15}$ primer (Promega) and an AMV Reverse Transcriptase (Promega). The PCR parameters for the amplification of the amidase genes were 95 °C for 10 min followed by 35 cycles of 95 °C for 45 s, 58 °C for 45 s, and 72 °C for 90 s; for the amplification of the coding regions, a proofreading polymerase (Pfu, Fermetas, St. Leon-Rot, Germany) was used. The primers used for PCR are given in Table 3. Obtained fragments were either subcloned into the vector pGEM-T easy (Promega) or pBluescript SK(+) (Stratagene) and sequence integrity was subsequently verified by commercial sequencing (GATC, Konstanz, Germany; Sequencing Service, Ruhr-University Bochum, Dept. of Biochemistry I, Bochum, Germany).

Table 3. Primers used for cloning of the described amidases.

Primer Name	Sequence (5'–3')
*Os*AMI1-SacI-His2C-For	5'-TATGAGCTCTATGGCGATGGCGGGTGGAG-3'
*Os*AMI1-KpnI-His2C-Rev	5'-TATGGTACCGTGATTGGAGGACCAAGTTTTAG-3'
SbAMI1-SpeI-pUC-For	5'-TATACTAGTATGGCGATGGGCGGCGATTAC-3'
SbAMI1-SmaI-pUC-Rev	5'-TATCCCGGGGAGAGAGGAGTCTGGTGAGC-3'
*Mt*AMI1-XhoI-His2B-For	5'-TATCTCGAGATGGAAACAGCCTCAGACTATG-3'
*Mt*AMI1-XbaI-His2B-Rev	5'-TATTCTAGATATTTTTCAATGTTATCATAAATACTC-3'
*Pt*AMI1-XhoI-His2B-For	5'-TATCTCGAGATGGAACGAGACCCGGATTATG-3'
*Pt*AMI1-XbaI-His2B-Rev	5'-TATTCTAGATATTTTTCAGTGATCTCAACCTG-3'
*Mt*AMI1-BglII-pUC-For	5'-TATAGATCTATGGAAACAGCCTCAGACTATG-3'
*Mt*AMI1-SalI-pUC-Rev	5'-TATGTCGACCTATTTTTCAATGTTATCATAAATACTC-3'

Restriction sites are underlined.

3.3. Generation of Bacterial Expression Constructs

All DNA fragments were amplified of total cDNA of the different species using PCR with gene-specific primers (Table 3). In case of the rice amidase, the DNA fragment was inserted into the expression vector pTrcHis2C (Invitrogen) by using the *Sac*I/*Kpn*I sites. The DNA fragments of the amidases from *Medicago* and poplar were integrated into the *Xho*I/*Xba*I site of the vector pTrcHis2B (Invitrogen). The *S. bicolor* amidase cDNA fragment was inserted into the *XhoI*/*HindIII* site of pTrcHis2A (Invitrogen). In the four resulting constructs pTrcHis2C-*Os*AMI1, pTrcHis2B-*Mt*AMI1, pTrcHis2B-*Pt*AMI1, and pTrc2A-*Sb*AMI1 a translational fusion of the amidases with a myc/$(His)_6$ double-tag was generated.

3.4. Preparation of GFP Amidase Fusion Constructs

Constructs suitable for the analysis of the intracellular localization of the deduced amidase GFP fusion proteins were obtained by cloning the amidase fragments into pSP-EGFP [31]. The amidase fragment from pTrcHis2C-*Os*AMI1 was digested with *Sac*I/*Kpn*I and integrated into the same sites of pSP-EGFP. The *Sac*I/*Sal*I fragment from pTrcHis2B-*Pt*AMI1 was also cloned into pSP-EGFP, resulting in a *Pt*AMI1:GFP fusion. As all C-terminal GFP fusions of the *Medicago* amidase lacked *in vivo* fluorescence, a N-terminal GFP-tag was fused to this amidase by using the vector pUC-GFPn [51]. To obtain a suitable DNA fragment, the coding sequence was amplified by PCR, adding restriction sites for *Bgl*II and *Sal*I, respectively (Table 3). After subcloning of the resulting DNA fragment into pBluescript SK(+) and sequencing, the *MtAMI1 Spe*I/*Sal*I fragment was introduced into pUC-EGFPn. In case of the *S. bicolor* amidase gene, the coding sequence was amplified using primers that added *Spe*I/*Sma*I sites to the corresponding DNA fragment. After cloning into pBluescript SK(+) and sequence analysis the *Spe*I/*Sma*I fragment was introduced into the *Spe*I/*Sal*I site of pUC-EGFPn.

3.5. Heterologous Expression of Recombinant Amidases

To express hexahistidine tagged amidases 300 mL of 2xYT medium were freshly inoculated (1:10) with appropriate over night cultures and incubated at 37 °C under constant shaking. Protein expression was induced with 1 mM IPTG when an OD_{600} of either 0.6, in case of the *Mt*AMI1 construct, or 1.3, for the *Os*AMI1 and *Pt*AMI1 constructs, was reached. The cultures were transferred to either 4 °C (*Mt*AMI1) or 30 °C (*Os*AMI1, *Sb*AMI1, *Pt*AMI1) allowing protein expression to occur for at least 18 h. Thereafter, bacteria were harvested by centrifugation (4,000 *g*, 15 min, 4 °C) and pellets were resuspended in 1/10 of the culture volume of lysis buffer (50 mM sodium phosphate buffer pH 7.5, 300 mM NaCl, 10 mM imidazole). The samples were then snap frozen in liquid nitrogen and then kept at −80 °C until further use. Complete cell disruption was subsequently achieved by a combination of incubation with lysozyme (45 min, 4 °C) and ultrasound (120 s). The soluble protein fraction, obtained by centrifugation (35,000 *g*, 15 min, 4 °C), was sterile filtered (0.22 μm) and used for Ni^{2+}-affinity purification. The chromatography was carried out in accordance to the manufacturer's protocol (Qiagen).

3.6. Transient Expression in Plants and Confocal Laser Scanning Microscopy

Transient expression and subsequent microscopic analysis was carried out on the basis of Pollmann *et al.* [31]. In brief, leaves from two- to three-week old *Arabidopsis* plants, grown on ½ MS plates [50] under environmentally controlled conditions, were bombarded with 1 μm gold particles at 6.5 bar pressure. The particles were emitted from a particle inflow gun as described by Finer and coworkers [52]. After bombardment, the plants were kept for one day under constant conditions (8 h light at 24 °C, 16 h darkness at 20 °C, photosynthetically active radiation 105 $\mu E \cdot m^{-2} \cdot s^{-1}$ from standard white fluorescent tubes). The transformed leaves were then analyzed with a confocal laser scanning system (ZEISS LSM 510 Meta).

3.7. Assay for Amidase Activity

Amidase activity was estimated by measuring the ammonia released from the amide substrates during the reaction as previously described [30]. In short, the different substrates (10 mM) were incubated with 5 μg of purified protein in 50 mM potassium phosphate buffer at pH 7.5, at 30 °C in a total volume of 0.3 mL. For background control, heat-inactivated enzyme (20 min, 100 °C) and empty-vector control samples were used. After an incubation time of 4 to 6 h, aliquots of 100 μL were taken and the reaction was stopped by adding 100 μL each of 0.33 M sodium phenolate, 0.02 M sodium hypochlorite, and 0.01% (w/v) sodium pentacyanonitrosyl ferrate(III) (sodium nitroprusside). After incubating for 2 min at 100 °C, each sample was diluted with 600 μL of water, and the absorbance was read at 640 nm. Each experiment was calibrated using NH_4Cl solutions of known concentrations.

3.8. Root Growth Assay

The root growth bioassay was carried out according to the work of Zimmerman and Hitchcock [53]. Plants were grown aseptically on ½ MS medium [50] containing 1% sucrose, solidified with 0.8% gelrite (Duchefa, The Netherlands) or on the same medium supplemented with either 10 nM to 10 mM indole-3-acetic acid (IAA; from 100 mM stock in methanol) or 10 nM to 10 mM phenyl-2-acetic acid (PAA; from 100 mM stock in methanol). Plates were wrapped in gas-permeable surgical tape (BSN medical GmbH, Hamburg, Germany) and grown vertically under constant environmental conditions

(8 h light at 24 °C, 16 h darkness at 20 °C, photosynthetically active radiation 105 $\mu E \cdot m^{-2} \cdot s^{-1}$ from standard white fluorescent tubes) for two to three weeks. Subsequently, the root lengths of at least ten individual plants per treatment were determined.

3.9. Auxin Extraction and Purification

The extraction and purification of auxins and other organic acids was carried out according to a previously described method [54]. Approximately 0.1 g of two- and six-week old *A. thaliana* leaf material was added to 1 mL of pre-warmed (65 °C) methanol and the extraction conducted for another 60 min at room temperature under gentle shaking. Prior to the extraction process each sample was supplemented with 1 nmol of [$^{13}C_2$]-PAA and 50 pmol of [$^{2}H_2$]-IAA (internal standard). Cell-free supernatants were dried under vacuum and pre-purified for subsequent gas chromatography-mass spectrometry analysis. For this, the dried residues were dissolved in 30 μL methanol to which 200 μL diethyl ether was added, followed by ultrasonic treatment (Sonorex RK510S; Bandelin, Berlin, Germany). The particle-free sample was then applied to a custom-made microscale aminopropyl solid-phase extraction-cartridge [55]. The cartridge was washed with 250 μL of chloroform:2-propanol = 2:1 (v/v), and the PAA containing fraction subsequently eluted with 400 μL acidified diethyl ether (2% acetic acid (v/v)). The eluates were taken to dryness using a vacuum centrifuge, re-dissolved in 20 μL methanol, and afterwards treated with 100 μL ethereal diazomethane and transferred to autosampler vials (Chromacol 05-CTV(A) 116; Fisher Scientific, Waltham, MA, USA). Excess diazomethane and remaining solvent were removed under a gentle stream of nitrogen, and the methylated samples were then taken up in 15 μL of chloroform.

3.10. Quantification of Endogenous Phenyl-2-Acetic Acid and Indole-3-Acetic Acid from Arabidopsis thaliana

Analysis was performed by GC-MS on a Varian GC 3400 gas chromatograph coupled to a Finnigan MAT Magnum system. Compounds were separated on a VF5ms column (Varian), 30 m, 0.25 mm ID, 0.25 μm film. The temperature program was 50 °C for 1 min, followed by an increase of 30 $°C \cdot min^{-1}$ to 150 °C, and kept at this temperature for 4 min. The temperature was further increased to 250 °C with a velocity of 20 $°C \cdot min^{-1}$. Thereafter the temperature was kept at 250 °C for 5 min. An aliquot of 1 μL of each sample was injected into the GC-MS system. PAA eluted at 4:55 min under these conditions, while IAA eluted at 12:15 min. Full scan spectra were recorded for the peaks co-eluting with authentic methylated standards and the mass spectra compared to spectra obtained from authentic standards. For the determination of free PAA, the molecular and tropylium ions at *m/z* 151/153 and 91/92, respectively (ions deriving from endogenous and [$^{13}C_2$]-PAA) were monitored. For IAA, ions at *m/z* 189/191 and 130/132 (ions deriving from endogenous and [$^{2}H_2$]-IAA) were analyzed.

3.11. Gel Electrophoresis and Immunoblotting

Denaturing gel electrophoresis was performed according to Laemmli [56]. The discontinuous systems consisted of 4% stacking gels and 12.5% resolving gels. Protein blotting onto nitrocellulose was carried out electrophoretically overnight (4 °C, 60 mA) as described by Towbin and colleagues [57].

Immunodetection followed standard procedures [58] with goat anti-rabbit IgG-conjugated alkaline phosphatase as the secondary antibody and 4-nitrotetrazolium blue and 5-bromo-4-chloro-3-indolyl phosphate as substrates. As primary antibody an α-AMI1-antiserum, previously described [31], have been used.

3.12. Phylogenetic Analysis

The phylogenetic tree was inferred using the Neighbor-Joining method [59]. The percentage of replicate trees in which the associated taxa clustered together was calculated by a bootstrap test with 500 replicates [60]. The evolutionary distances were computed using the Poisson correction method [61] and are in the units of the number of amino acid substitutions per site. All positions containing gaps and missing data were eliminated from the dataset (Complete deletion option). Phylogenetic analyses were conducted in MEGA5 (Molecular Evolutionary Genetics Analysis, MEGA Software).

3.13. Statistic Analysis

The data were analyzed with one-way ANOVA followed by Tukey's B *post hoc* test to allow for comparisons among all means. Statistical analyses were conducted using PRISM version 5.03 (GraphPad Software).

4. Conclusions

On the basis of sequence similarity, it was possible to identify more than 40 AMI1-like candidate amidases in publicly available plant genome databases (Figure 2). Four *At*AMI1-orthologous, namely *Os*AMI1 (*O. sativa*), *Sb*AMI1 (*S. bicolor*), *Mt*AMI1 (*M. truncatula*), and *Pt*AMI1 (*P. trichocarpa*), were selected and closely analyzed.

Taking an *in vitro* approach, it was possible to provide evidence for their capability of converting auxin precursors like IAM, but also PAM, to naturally occurring auxins (Figure 5, Table 1). Our results infer that an amidase-dependent auxin biosynthesis pathway is likely to be widely established in plants to produce auxins. Intriguingly, our experiments revealed a specificity of the compared enzymes for PAM (Figure 5), which is a biosynthetic precursor of PAA, described to be endogenous to *P. mungo* [42,43]. PAA has been demonstrated to be endogenous to *A. thaliana* (Figure 7, Table 2), while we were not able to quantify PAM in the same plant with the equipment available in this study. Our findings confirm previous studies, which demonstrated that PAA is an endogenous constituent of many plant species including *P. mungo* [45], *S. lycopersicum*, *N. tabacum*, *Z. mays* [4], and *T. majus* [5].

PAA has long been known to trigger auxin-like effects (e.g., the initiation and stimulation of adventitious root growth [62,63]), although with considerably lower activity relative to IAA [64,65]. Nevertheless, PAA may act as an effective auxin under some developmental circumstances, emphasized by the observation that it was more efficient in lateral root formation in pea seedlings than IAA [66]. In addition, PAA is assumed to have both antifungal and antibacterial properties [67–69]; it was isolated from culture extracts of *Azospirillium brasilense*, pinpointing an involvement in defense mechanisms, protecting this bacterial strain from other soil inhabitants like for instance *Agrobacteria* and thus providing an advantage for *A. brasilense* to survive in its natural habitat. A comparable protection mechanism for the defense against microbial pathogens could also be effective in plants.

It is generally accepted that the major portion of physiologically active auxin is produced via the indole-3-pyruvic acid-pathway [9,10]. On the basis of our results, however, we suggest that IAM/PAM-amidohydrolases may also contribute to auxin biosynthesis, most likely in a pathway parallel to the main route. Genetic work from Zhao and coworkers [70] highlights a possible role for an IAM hydrolase-dependent IAA biosynthesis, as they demonstrated that *yuc* double and triple mutants were rescued by the tissue specific co-expression of a bacterial tryptophan 2-monooxygenase (iaaM), but not by exogenous application of auxin. The authors claim that iaaM converts tryptophan to IAA once it is expressed in plants. This statement is, however, not entirely correct because several studies clearly showed that overexpression of iaaM (e.g., in petunia) produced very high levels of IAM in these mutants [71,72]. While IAM contents in control lines were determined to be <1 pg/g fresh weight, the IAM levels reached 2.8 up to 25 μg/g fresh weight in leaf tissue of transformed plants. In addition, it has been shown that the petunia plants transformed with the *p19S*::*iaaM* construct also contained around 11- to 12-fold higher IAA levels, which implicates the action of downstream acting indole-3-acetamide hydrolases, functionally homologous to the bacterial iaaH enzyme, capable of converting the accumulated IAM to IAA. Thus, effective conversion of IAM to IAA in the *yuc* mutants co-expressing the *iaaM* gene, but no additional *iaaH* gene, is implicated. Alternatively, it may be concluded that IAM by itself, or any other compound that is produced from IAM is responsible for rescuing the *yuc* phenotype. Most intriguingly, it has recently been shown that YUC6 is capable of converting not only IPyA to IAA, but to also accept phenyl pyruvate as substrate to produce PAA [73]. The results presented here may point towards a co-evolution of enzymes involved in auxin production to establish dual activity for the production of IAA and PAA, respectively.

Currently, the elucidation of the *in vivo* function of *At*AMI1 is of utmost importance to our lab and preliminary data obtained from mutants overexpressing the protein support our conclusions. At this stage, however, we cannot yet postulate that the enzymatic activity of *At*AMI1 *in planta* is unambiguously linked to IAA production *in vivo*. Nevertheless, the results of the present study will help to further elucidate the enzymatic basis of auxin biosynthesis in plants.

Acknowledgments

The research was supported in part by a grant from the German Research Foundation (Grant No. SFB480-A10 to SP), a grant from the Spanish Ministry of Science and Innovation (Grant No. BFU2011-25925 to SP), and by the Marie-Curie Actions program from the European Community (Grant FP7-PEOPLE-CIG-2011-303744 to SP).

The authors would like to thank Jutta Ludwig-Müller, Technical University Dresden, Germany, for providing *M. truncatula* seeds.

Author Contributions

Beatriz Sánchez-Parra, Henning Frerigmann, Marta-Marina Pérez Alonso, Víctor Carrasco Loba, Ricarda Jost, and Mathias Hentrich all had substantial contributions to accomplish the presented experiments. Moreover, they critically reviewed the manuscript for its intellectual content. Stephan Pollmann designed the work and wrote the manuscript.

Conflicts of Interest

The authors declare no conflict of interest.

References

1. Thimann, K.V. *Hormone Action in the Whole Life of Plants*; Univ of Massachusetts Press: Amherst, MA, USA, 1977.
2. Davies, P.J. *Plant Hormones. Biosynthesis, Signal Transduction, Action!* Revised 3rd ed.; Kluwer Academic Publishers: London, UK, 2010.
3. Slovin, J.P.; Bandurski, R.S.; Cohen, J.D. Auxin. In *Biochemistry and Molecular Biology of Plant Hormones*; Hooykaas, P.J.J., Hall, M.A., Libbenga, K.R., Eds.; Elsevier: Amsterdam, The Netherlands, 1999; pp. 115–140.
4. Wightman, F.; Lighty, D.L. Identification of phenylacetic acid as a natural auxin in the shoots of higher plants. *Physiol. Plant.* **1982**, *55*, 17–24.
5. Ludwig-Müller, J.; Cohen, J.D. Identification and quantification of three active auxins in different tissues of *Tropaeolum majus*. *Physiol. Plant.* **2002**, *115*, 320–329.
6. Fawcett, C.H.; Wain, R.L.; Wightman, F. The metabolism of 3-indolylalkanecarboxylic acids, and their amides, nitriles and methyl esters in plant tissues. *Proc. R. Soc. Lond. B Biol. Sci.* **1960**, *152*, 231–254.
7. Epstein, E.; Chen, K.-H.; Cohen, J. Identification of indole-3-butyric acid as an endogenous constituent of maize kernels and leaves. *Plant Growth Regul.* **1989**, *8*, 215–223.
8. Ludwig-Müller, J.; Epstein, E. Occurrence and *in vivo* biosynthesis of indole-3-butyric acid in corn (*Zea mays* L.). *Plant Physiol.* **1991**, *97*, 765–770.
9. Won, C.; Shen, X.; Mashiguchi, K.; Zheng, Z.; Dai, X.; Cheng, Y.; Kasahara, H.; Kamiya, Y.; Chory, J.; Zhao, Y. Conversion of tryptophan to indole-3-acetic acid by TRYPTOPHAN AMINOTRANSFERASES OF *ARABIDOPSIS* and YUCCAs in *Arabidopsis*. *Proc. Natl. Acad. Sci. USA* **2011**, *108*, 18518–18523.
10. Stepanova, A.N.; Yun, J.; Robles, L.M.; Novak, O.; He, W.; Guo, H.; Ljung, K.; Alonso, J.M. The *Arabidopsis* YUCCA1 flavin monooxygenase functions in the indole-3-pyruvic acid branch of auxin biosynthesis. *Plant Cell* **2011**, *23*, 3961–3973.
11. Tivendale, N.D.; Ross, J.J.; Cohen, J.D. The shifting paradigms of auxin biosynthesis. *Trends Plant Sci.* **2014**, *19*, 44–51.
12. Pollmann, S.; Müller, A.; Weiler, E.W. Many roads lead to "auxin": Of nitrilases, synthases, and amidases. *Plant Biol.* **2006**, *8*, 326–333.
13. Lehmann, T.; Hoffmann, M.; Hentrich, M.; Pollmann, S. Indole-3-acetamide-dependent auxin biosynthesis: A widely distributed way of indole-3-acetic acid production? *Eur. J. Cell Biol.* **2010**, *89*, 895–905.

14. Brumos, J.; Alonso, J.M.; Stepanova, A.N. Genetic aspects of auxin biosynthesis and its regulation. *Physiol. Plant.* **2014**, *151*, 3–12.
15. Morris, R.O. Genes specifying auxin and cytokinin biosynthesis in phytopathogens. *Ann. Rev. Plant Physiol.* **1986**, *37*, 509–538.
16. Manulis, S.; Valinski, L.; Gafni, Y.; Hershenhorn, J. Indole-3-acetic acid biosynthetic pathways in Erwinia herbicola in relation to pathogenicity on *Gypsophila paniculata. Physiol. Mol. Plant Pathol.* **1991**, *39*, 161–171.
17. Spaepen, S.; Vanderleyden, J.; Remans, R. Indole-3-acetic acid in microbial and microorganism-plant signaling. *FEMS Microbiol. Rev.* **2007**, *31*, 425–448.
18. Takahashi, N.; Yamaguchi, I.; Kõno, T.; Igoshi, M.; Hirose, K.; Suzuki, K. Characterization of plant growth substances in *Citrus unshiu* and their change in fruit development. *Plant Cell Physiol.* **1975**, *16*, 1101–1111.
19. Saotome, M.; Shirahata, K.; Nishimura, R.; Yahaba, M.; Kawaguchi, M.; Syõno, K.; Kitsuwa, T.; Ishii, Y.; Nakamura, T. The identification of indole-3-acetic acid and indole-3-acetamide in the hypocotyls of japanese cherry. *Plant Cell Physiol.* **1993**, *34*, 157–159.
20. Rajagopal, R.; Tsurusaki, K.I.; Kannangara, G.; Kuraishi, S.; Sakurai, N. Natural occurrence of indoleacetamide and amidohydrolase activity in etiolated aseptically-grown squash seedlings. *Plant Cell Physiol.* **1994**, *35*, 329–339.
21. Pollmann, S.; Müller, A.; Piotrowski, M.; Weiler, E.W. Occurrence and formation of indole-3-acetamide in *Arabidopsis thaliana. Planta* **2002**, *216*, 155–161.
22. Sugawara, S.; Hishiyama, S.; Jikumaru, Y.; Hanada, A.; Nishimura, T.; Koshiba, T.; Zhao, Y.; Kamiya, Y.; Kasahara, H. Biochemical analyses of indole-3-acetaldoxime-dependent auxin biosynthesis in *Arabidopsis. Proc. Natl. Acad. Sci. USA* **2009**, *106*, 5430–5435.
23. Kawaguchi, M.; Kobayashi, M.; Sakurai, A.; Syõno, K. The presence of an enzyme that converts indole-3-acetamide into IAA in wild and cultivated rice. *Plant Cell Physiol.* **1991**, *32*, 143–149.
24. Arai, Y.; Kawaguchi, M.; Syono, K.; Ikuta, A. Partial purification of an enzyme hydrolyzing indole-3-acetamide from rice cells. *J. Plant Res.* **2004**, *117*, 191–198.
25. Kawaguchi, M.; Fujioka, S.; Sakurai, A.; Yamaki, Y.T.; Syõno, K. Presence of a pathway for the biosynthesis of auxin via indole-3-acetamide in *Trifoliata orange*. *Plant Cell Physiol.* **1993**, *34*, 121–128.
26. Pollmann, S.; Neu, D.; Weiler, E.W. Molecular cloning and characterization of an amidase from *Arabidopsis thaliana* capable of converting indole-3-acetamide into the plant growth hormone, indole-3-acetic acid. *Phytochemistry* **2003**, *62*, 293–300.
27. Nemoto, K.; Hara, M.; Suzuki, M.; Seki, H.; Muranaka, T.; Mano, Y. The *NtAMI1* gene functions in cell division of tobacco BY-2 cells in the presence of indole-3-acetamide. *FEBS Lett.* **2009**, *583*, 487–492.
28. Schröder, G.; Waffenschmidt, S.; Weiler, E.W.; Schröder, J. The T-region of Ti plasmids codes for an enzyme synthesizing indole-3-acetic acid. *Eur. J. Biochem.* **1984**, *138*, 387–391.
29. Yamada, T.; Palm, C.J.; Brooks, B.; Kosuge, T. Nucleotide sequences of the *Pseudomonas savastanoi* indoleacetic acid genes show homology with Agrobacterium tumefaciens T-DNA. *Proc. Natl. Acad. Sci. USA* **1985**, *82*, 6522–6526.

30. Neu, D.; Lehmann, T.; Elleuche, S.; Pollmann, S. *Arabidopsis* amidase 1, a member of the amidase signature family. *FEBS J.* **2007**, *274*, 3440–3451.
31. Pollmann, S.; Neu, D.; Lehmann, T.; Berkowitz, O.; Schäfer, T.; Weiler, E.W. Subcellular localization and tissue specific expression of amidase 1 from *Arabidopsis thaliana*. *Planta* **2006**, *224*, 1241–1253.
32. Aronsson, H.; Boij, P.; Patel, R.; Wardle, A.; Töpel, M.; Jarvis, P. Toc64/OEP64 is not essential for the efficient import of proteins into chloroplasts in *Arabidopsis thaliana*. *Plant J.* **2007**, *52*, 53–68.
33. Expósito-Rodríguez, M.; Borges, A.; Borges-Pérez, A.; Hernández, M.; Pérez, J. Cloning and biochemical characterization of *ToFZY*, a tomato gene encoding a flavin monooxygenase involved in a tryptophan-dependent auxin biosynthesis pathway. *J. Plant Growth Regul.* **2007**, *26*, 329–340.
34. LeClere, S.; Schmelz, E.A.; Chourey, P.S. Sugar levels regulate tryptophan-dependent auxin biosynthesis in developing maize kernels. *Plant Physiol.* **2010**, *153*, 306–318.
35. Tobeña-Santamaria, R.; Bliek, M.; Ljung, K.; Sandberg, G.; Mol, J.N.; Souer, E.; Koes, R. FLOOZY of petunia is a flavin mono-oxygenase-like protein required for the specification of leaf and flower architecture. *Genes Dev.* **2002**, *16*, 753–763.
36. Yamamoto, Y.; Kamiya, N.; Morinaka, Y.; Matsuoka, M.; Sazuka, T. Auxin biosynthesis by the *YUCCA* genes in rice. *Plant Physiol.* **2007**, *143*, 1362–1371.
37. Tivendale, N.D.; Davies, N.W.; Molesworth, P.P.; Davidson, S.E.; Smith, J.A.; Lowe, E.K.; Reid, J.B.; Ross, J.J. Reassessing the role of N-hydroxytryptamine in auxin biosynthesis. *Plant Physiol.* **2010**, *154*, 1957–1965.
38. Zhao, Y.; Hull, A.K.; Gupta, N.R.; Goss, K.A.; Alonso, J.; Ecker, J.R.; Normanly, J.; Chory, J.; Celenza, J.L. Trp-dependent auxin biosynthesis in *Arabidopsis*: Involvement of cytochrome P450s CYP79B2 and CYP79B3. *Genes Dev.* **2002**, *16*, 3100–3112.
39. Janowitz, T.; Trompetter, I.; Piotrowski, M. Evolution of nitrilases in glucosinolate-containing plants. *Phytochemistry* **2009**, *70*, 1680–1686.
40. Emanuelsson, O.; Brunak, S.; von Heijne, G.; Nielsen, H. Locating proteins in the cell using TargetP, SignalP and related tools. *Nat. Protoc.* **2007**, *2*, 953–971.
41. Cokol, M.; Nair, R.; Rost, B. Finding nuclear localization signals. *EMBO Rep.* **2000**, *1*, 411–415.
42. Isogai, Y.; Okamoto, T.; Koizumi, T. Isolation of 2-phenylacetamide, indole-3-acetamide, and indole-3-carboxaldehyde from etiolated seedling of phaseolus. *Chem. Pharm. Bull.* **1963**, *11*, 1217–1218.
43. Isogai, Y.; Okamoto, T.; Koizumi, T. Studies on plant growth regulators. I. Isolation of indole-3-acetamide, 2-phenylacetamide, and indole-3-carboxaldehyde from etiolated seedlings of phaseolus. *Chem. Pharm. Bull.* **1967**, *15*, 151–158.
44. Wittstock, U.; Halkier, B.A. Cytochrome P450 CYP79A2 from *Arabidopsis thaliana* L. Catalyzes the conversion of L-phenylalanine to phenylacetaldoxime in the biosynthesis of benzylglucosinolate. *J. Biol. Chem.* **2000**, *275*, 14659–14666.
45. Okamoto, T.; Koizumi, T.; Isogai, Y. Studies on plant growth regulators. II. Isolation of indole-3-acetic acid, phenylacetic acid, and several plant growth inhibitors from etiolated seedlings of phaseolus. *Chem. Pharm. Bull.* **1967**, *15*, 159–163.
46. Yokohari, M.; Brown, R.D.; Kato, Y.; Yamamoto, S. The cooling effect of paddy fields on summertime air temperature in residential Tokyo, Japan. *Landsc. Urban Plan.* **2001**, *53*, 17–27.

47. Koepfli, J.B.; Thimann, K.V.; Went, F.W. Phytohormones: Structure and physiological activity. I. *J. Biol. Chem.* **1938**, *122*, 763–780.
48. Muir, R.M.; Fujita, T.; Hansch, C. Structure-activity relationship in the auxin activity of mono-substituted phenylacetic acids. *Plant Physiol.* **1967**, *42*, 1519–1526.
49. Müller, A.; Weiler, E.W. IAA-synthase, an enzyme complex from *Arabidopsis thaliana* catalyzing the formation of indole-3-acetic acid from (*S*)-tryptophan. *Biol. Chem.* **2000**, *381*, 679–686.
50. Murashige, T.; Skoog, F. A revised medium for rapid growth and bio assays with tobacco tissue cultures. *Physiol. Plant.* **1962**, *15*, 473–497.
51. Lehmann, T.; Pollmann, S. Gene expression and characterization of a stress-induced tyrosine decarboxylase from *Arabidopsis thaliana*. *FEBS Lett.* **2009**, *583*, 1895–1900.
52. Finer, J.J.; Vain, P.; Jones, M.W.; McMullen, M.D. Development of the particle inflow gun for DNA delivery to plant cells. *Plant Cell Rep.* **1992**, *11*, 323–328.
53. Zimmerman, P.W.; Hitchcock, A.E. Substituted phenoxy and benzoic acid growth substances and the relation of structure to physiological activity. *Contrib. Boyce Thompson Inst.* **1942**, *12*, 321–343.
54. Müller, A.; Düchting, P.; Weiler, E.W. A multiplex GC-MS/MS technique for the sensitive and quantitative single-run analysis of acidic phytohormones and related compounds, and its application to *Arabidopsis thaliana*. *Planta* **2002**, *216*, 44–56.
55. Müller, A.; Düchting, P.; Weiler, E.W. Hormone profiling in *Arabidopsis*. *Methods Mol. Biol.* **2006**, *323*, 449–457.
56. Laemmli, U.K. Cleavage of structural proteins during the assembly of the head of bacteriophage T4. *Nature* **1970**, *227*, 680–685.
57. Towbin, H.; Staehelin, T.; Gordon, J. Electrophoretic transfer of proteins from polyacrylamide gels to nitrocellulose sheets: Procedure and some applications. *Proc. Natl. Acad. Sci. USA* **1979**, *76*, 4350–4354.
58. Parets-Soler, A.; Pardo, J.M.; Serrano, R. Immunocytolocalization of plasma membrane H-ATPase. *Plant Physiol.* **1990**, *93*, 1654–1658.
59. Saitou, N.; Nei, M. The neighbor-joining method: A new method for reconstructing phylogenetic trees. *Mol. Biol. Evol.* **1987**, *4*, 406–425.
60. Felsenstein, J. Confidence limits on phylogenies: An approach using the bootstrap. *Evolution* **1985**, *39*, 783–791.
61. Zuckerkandl, E.; Pauling, L. Evolutionary divergence and convergence in proteins. In *Evolving Genes and Proteins*; Bryson, V., Vogel, H.J., Eds.; Academic Press: New York, NY, USA, 1965; pp. 97–166.
62. Haagen-Smit, A.J.; Went, F.W. A physiological analysis of the growth substance. *Proc. K. Ned. Akad. Wet.* **1935**, *38*, 852–857.
63. Zimmerman, P.W.; Wilcoxon, F. Several chemical growth substances which cause initiation of roots and other responses in plants. *Contrib. Boyce Thompson Inst.* **1935**, *7*, 209–229.
64. Thimann, K.V.; Schneider, C.L. The relative activities of different auxins. *Am. J. Bot.* **1939**, *26*, 328–333.
65. Leuba, V.; LeTourneau, D. Auxin activity of phenylacetic acid in tissue culture. *J. Plant Growth Regul.* **1990**, *9*, 71–76.

66. Wightman, F.; Schneider, E.A.; Thimann, K.V. Hormonal factors controlling the initiation and development of lateral roots. *Physiol. Plant.* **1980**, *49*, 304–314.
67. Burkhead, K.D.; Slininger, P.J.; Schisler, D.A. Biological control bacterium *Enterobacter cloacae* S11:T:07 (NRRL B-21050) produces the antifungal compound phenylacetic acid in Sabouraud maltose broth culture. *Soil Biol. Biochem.* **1998**, *30*, 665–667.
68. Hwang, B.K.; Lim, S.W.; Kim, B.S.; Lee, J.Y.; Moon, S.S. Isolation and *in vivo* and *in vitro* antifungal activity of phenylacetic acid and sodium phenylacetate from *Streptomyces humidus*. *Appl. Environ. Microbiol.* **2001**, *67*, 3739–3745.
69. Kim, Y.; Cho, J.Y.; Kuk, J.H.; Moon, J.H.; Cho, J.I.; Kim, Y.C.; Park, K.H. Identification and antimicrobial activity of phenylacetic acid produced by *Bacillus licheniformis* isolated from fermented soybean, Chungkook-Jang. *Curr. Microbiol.* **2004**, *48*, 312–317.
70. Cheng, Y.; Dai, X.; Zhao, Y. Auxin biosynthesis by the YUCCA flavin monooxygenases controls the formation of floral organs and vascular tissues in *Arabidopsis*. *Genes Dev.* **2006**, *20*, 1790–1799.
71. Klee, H.J.; Horsch, R.B.; Hinchee, M.A.; Hein, M.B.; Hoffmann, N.L. The effects of overproduction of two *Agrobacterium tumefaciens* T-DNA auxin biosynthetic gene products in transgenic petunia plants. *Genes Dev.* **1987**, *1*, 86–96.
72. Inzé, D.; Follin, A.; Velten, J.; Velten, L.; Prinsen, E.; Rüdelsheim, P.; van Onckelen, H.; Schell, J.; van Montagu, M. The *Pseudomonas savastanoi* tryptophan-2-mono-oxygenase is biologically active in *Nicotiana tabacum*. *Planta* **1987**, *172*, 555–562.
73. Dai, X.; Mashiguchi, K.; Chen, Q.; Kasahara, H.; Kamiya, Y.; Ojha, S.; Dubois, J.; Ballou, D.; Zhao, Y. The biochemical mechanism of auxin biosynthesis by an arabidopsis YUCCA flavin-containing monooxygenase. *J. Biol. Chem.* **2013**, *288*, 1448–1457.

Appendix

Figure A1. Sequence alignment of five members of the amidase-signature family from plants. The analyzed sequences are *At*AMI1 (At1g08980), *Mt*AMI1 (Mt1g099370), *Os*AMI1 (Os04g02780), *Pt*AMI1 (Pt13g02200), and *Sb*AMI1 (Sb02g039510). Amino acid positions are indicated. Sequence conservation among the polypeptides is color-coded in the alignment. Asterisks mark the catalytic triad; the glycine- and serine-rich amidase-signature motif is underlined.

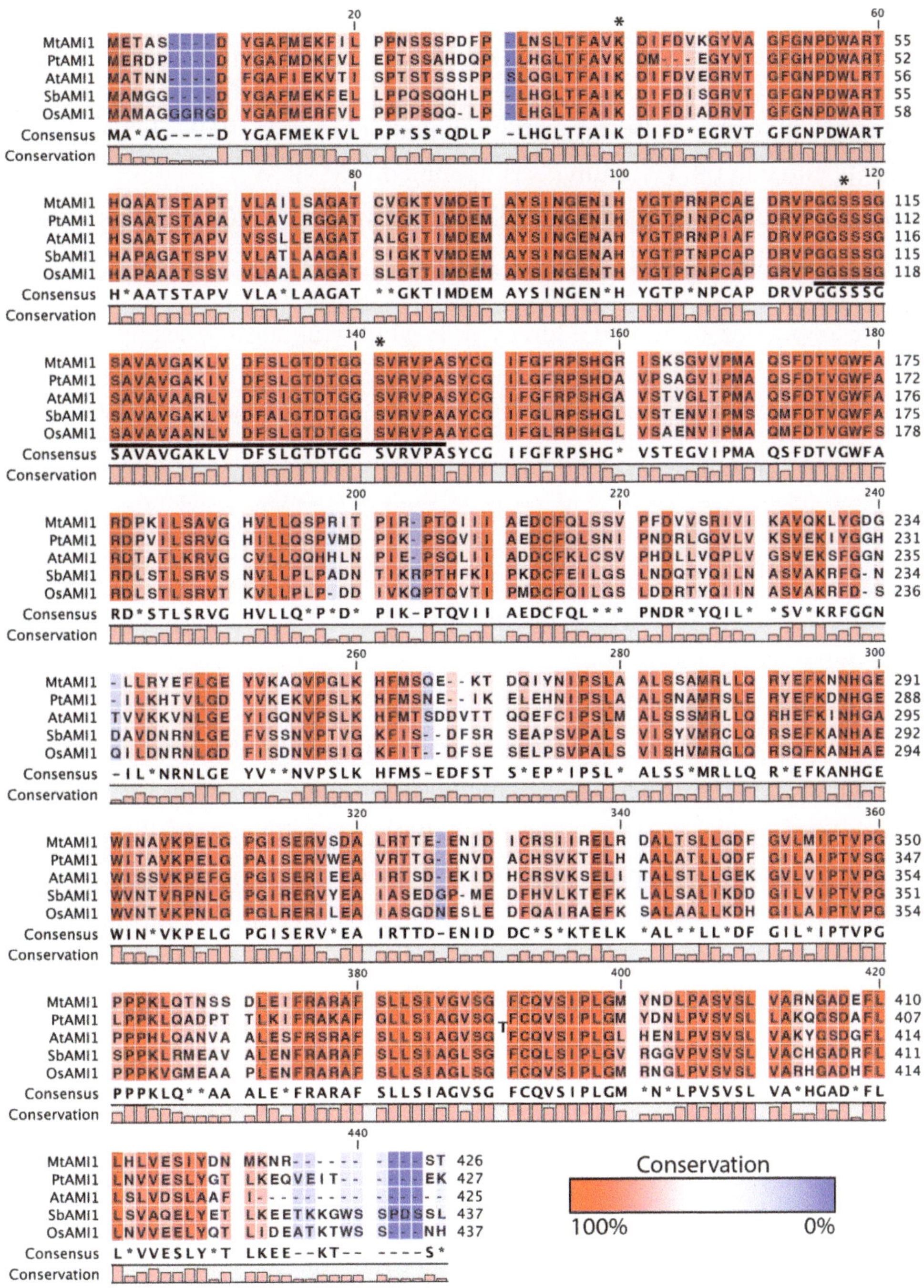

7

LEAFY and Polar Auxin Transport Coordinately Regulate *Arabidopsis* Flower Development

Nobutoshi Yamaguchi [1], Miin-Feng Wu [1], Cara M. Winter [1,2] and Doris Wagner [1,*]

[1] Department of Biology, University of Pennsylvania, Philadelphia, PA 19104, USA; E-Mails: nobuy@sas.upenn.edu (N.Y.); miin@sas.upenn.edu (M.-F.W.)

[2] Department of Biology, Duke University, Box 90338, Durham, NC 27708, USA; E-Mail: cara.winter@duke.edu

* Author to whom correspondence should be addressed; E-Mail: wagnerdo@sas.upenn.edu

---

Abstract: The plant specific transcription factor LEAFY (LFY) plays a pivotal role in the developmental switch to floral meristem identity in *Arabidopsis*. Our recent study revealed that LFY additionally acts downstream of AUXIN RESPONSE FACTOR5/MONOPTEROS to promote flower primordium initiation. LFY also promotes initiation of the floral organ and floral organ identity. To further investigate the interplay between LFY and auxin during flower development, we examined the phenotypic consequence of disrupting polar auxin transport in *lfy* mutants by genetic means. Plants with compromised LFY activity exhibit increased sensitivity to disruption of polar auxin transport. Compromised polar auxin transport activity in the *lfy* mutant background resulted in formation of fewer floral organs, abnormal gynoecium development, and fused sepals. In agreement with these observations, expression of the auxin response reporter *DR5rev::GFP* as well as of the direct LFY target *CUP-SHAPED COTYLEDON2* were altered in *lfy* mutant flowers. We also uncovered reduced expression of *ETTIN*, a regulator of gynoecium development and a direct LFY target. Our results suggest that *LFY* and polar auxin transport coordinately modulate flower development by regulating genes required for elaboration of the floral organs.

Keywords: *Arabidopsis thaliana*; auxin transport; CUP-SHAPED COTYLEDON2; ETTIN; flower development; LEAFY; PIN-FORMED1; PINOID

---

1. Introduction

The phytohormone auxin is a central regulator of lateral organ initiation [1–4]. Auxin accumulates in a graded and dynamic manner with the sites of auxin maxima correlating with the sites of primordium initiation [3,5]. The formation of auxin gradients is established by local auxin biosynthesis and polar auxin transport [1–8]. Direct transport is controlled by the PINFORMED (PIN) proteins, which encode auxin efflux carriers and exhibit polarized plasma membrane localization [1–5]. The Ser/Thr protein kinase PINOID (PID) catalyzes PIN phosphorylation and contributes to the regulation of apical-basal PIN polarity [9,10]. *NAKED PINS IN YUC MUTANT* (*NPY*) family genes, which encode NONPHOTOTROPIC HYPOCOTYL 3-like proteins, regulate PIN endocytosis and control auxin accumulation in incipient organ primordia [11–13]. On the other hand, YUCCA (YUC) flavin monooxygenase catalyzes a rate-limiting step in a tryptophan-dependent auxin biosynthesis pathway [14–16]. Mutations in *PIN1*, *PINOID*, *NPY*, and *YUC* result in inflorescences that form only a few flowers and grow as a pin-like structure [1,12,16,17].

Subsequent to determination of the primordium initiation site by an auxin maximum, the incipient flower primordium undergoes extensive growth. The AUXIN RESPONSE FACTOR5/MONOPTEROS (ARF5/MP) has a central role translating local auxin concentration into specific gene expression outputs and flower initiation [18,19]. In the absence of auxin, MP activity is inhibited by the physical interaction between MP and Aux/IAA proteins such as BODENLOS/IAA12 [20,21], this represses transcription of downstream target genes involved in flower formation. Auxin sensing promotes the degradation of BDL/IAA12, resulting in MP-dependent transcriptional activation of target genes [20–22]. MP directly binds to the regulatory region of *LEAFY* (*LFY*), which specifies floral fate [23,24], and to two transcription factors, *AINTEGUMENTA* and *AINTEGUMENTA-LIKE6/PLETHORA3* (*AIL6/PLT3*), key regulators of floral meristem outgrowth [25–28]. The *lfy ant ail6* triple mutant is defective in flower primordium initiation, while reintroduction of LFY and ANT activity into *mp* mutants partially rescues the organogenesis defect [22]. These results suggest that upregulation of *LFY*, *ANT*, and *AIL6/PLT3* by MP contributes to flower primordium initiation.

In addition to regulating floral primordium initiation, auxin also regulates floral organ development [29]. Plant harboring mutations in *PIN1*, *PID*, *YUC*, and *NPY* produce a few abnormal flowers [1,12,16,17]. These flowers typically have fewer sepals and stamens, more petals, fused floral organs, and valveless gynoecia. The observed alterations in gynoecium patterning are similar to those resulting from mutations in *ETTIN* (*ETT*), which encodes AUXIN RESPONSE FACTOR3, or from treatment with an auxin transport inhibitor [30]. Floral organ fusion phenotypes are caused by loss-of-function mutations in the *CUP-SHAPED COTYLEDON* (*CUC*) genes [31,32]. *lfy* single mutants do not display any defects in floral primordium initiation or in floral organ development, similarly to *pin* or *pid* mutants [1,17,23]. However, LFY interacts genetically with PID in lateral organ primordium initiation [33].

Here, we further probe the interactions between LFY and polar auxin transport. We have examined the consequences of loss-of-LFY function in plants compromised in polar auxin transport by genetic means. These experiments demonstrate that *LFY* promotes not only flower primordium initiation, but also subsequent floral organ initiation and development, in concert with polar auxin transport. LFY executes this role likely via regulating downstream direct targets, such as *CUC2* and *ETT*, whose activity is also controlled by polar auxin transport. Our study uncovers a complex set of interactions between LFY and the auxin pathway in flower development.

2. Results and Discussion

2.1. Loss-of-LFY Function Enhances Floral Primordium Initiation Defects of pin1 and pid Mutants

Because *pin1* and *pid* single mutants make several flowers, they provide a sensitized background in which to study the interplay between *LFY* and auxin transport during flower primordium initiation. We first introduced the *lfy-1* (Col) null mutant into *pin1* mutants. The strong *pin1-8* allele formed naked inflorescences with a few flowers, as reported previously (Figure 1a,g) [1,34]. In *pin1-8 lfy-1* double mutants, no flowers were produced (Figure 1b,g; $p < 10^{-3}$). By contrast, the weak *pin1-5* allele formed approximately 8 flowers per inflorescence, as described previously (Figure 1c,h) [17]. While flowers were produced in double mutants between *pin1-5* and the weak *lfy-5* or *lfy-6* (L*er*) null mutants, they were dramatically reduced in number compared to the *pin1-5* single mutant (Figure 1d,h). These genetic interactions suggest that *LFY* acts synergistically with polar auxin transport to promote flower primordium initiation.

Figure 1. *lfy* enhances the floral primordium initiation defects of auxin transport mutants. (**a**–**f**) Close-up view of the inflorescences formed in a strong *pin1* mutant allele, *pin1-8* (**a**); and *pin1-8 lfy-1* (**b**); a weak *pin1* mutant allele, *pin1-5* (**c**); and *pin1-5 lfy-6* (**d**); *pid-16* (**e**) and *pid-16 lfy-6* (**f**); (**g**–**i**) Quantification of the flower initiation defects in *pin1-8* and *pin1-8 lfy-1* (**g**); *pin1-5*, *pin1-5 lfy-5*, and *pin1-5 lfy-6* (**h**); and *pid-16* and *pid-16 lfy-6* (**i**). Note that *lfy-1* and *lfy-6* carry the exact same mutation (Q32stop) but are the Columbia (Col) and Landsberg *erecta* (L*er*) cultivar, respectively. We employ both alleles as to eliminate mixed genetic backgrounds when generating double mutants. Scale bar, 200 μm (**a**,**b**,**d**,**f**), 5 mm (**c**,**e**).

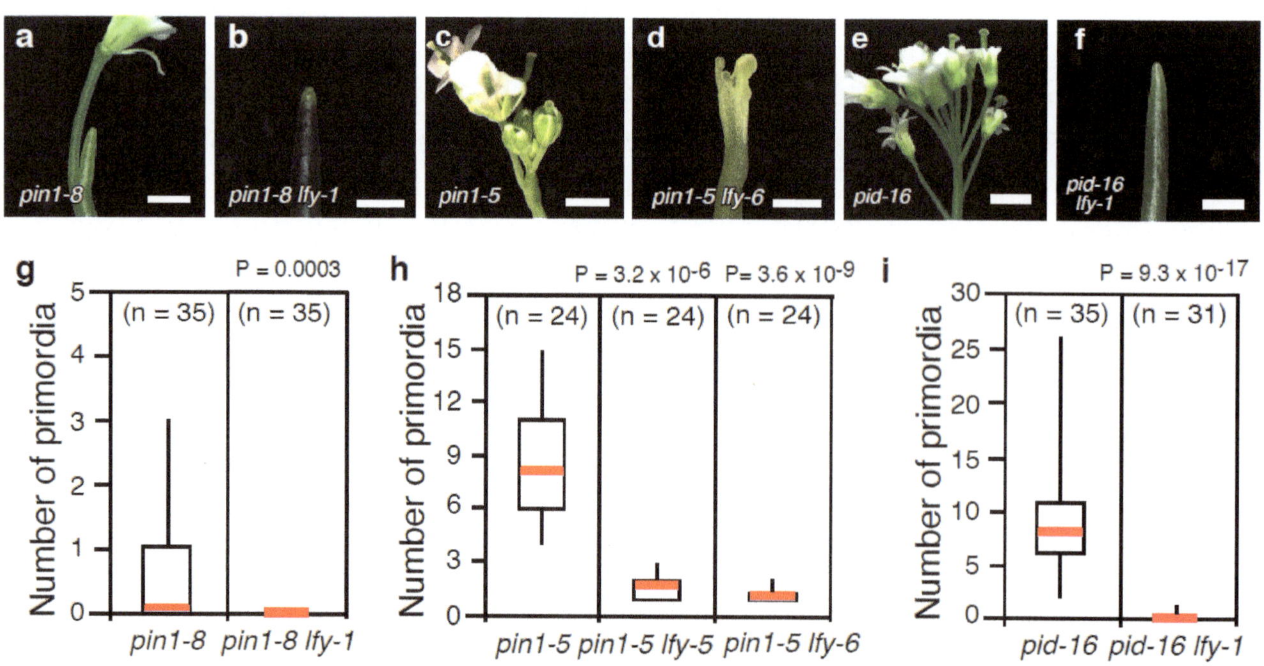

We next investigated the effect of introducing *lfy-1* into *pid* mutant background using the strong *pid-16* allele (Figure 1e,i). *pid-16 lfy-1* produced only a few flowers compared to the *pid-16* single mutant (Figure 1i; ($p < 10^{-16}$)). Taken together, the inability of *lfy* mutants to initiate flower primordia in genetic backgrounds defective in polar auxin transport confirms the proposed role for LFY in auxin-mediated flower primordium initiation [22]. Consistent with this finding, a recent genetic screen identified a *pid* mutation as a second site enhancer of weak *lfy-5* mutants [33].

Previously, it has been reported that pin-like apices of *pin1* single and *pin1 lfy* double mutants retain auxin responsiveness with respect to formation of lateral organs such as flowers and cauline leaves in the inflorescence [2,4]. By contrast, strong *mp* mutants are no longer responsive to auxin [2]. These findings are consistent with the observation that additional direct MP targets act in parallel with LFY in flower primordium initiation [22]. Our data are consistent with the strikingly enhanced defects in lateral organ initiation observed in *mp pid* or *mp pin* double mutants [35], which suggest that MP and MP targets like LFY act, in part, in parallel with these regulators of polar auxin transport.

2.2. Mutations in LFY Enhance Floral Organ Initiation and Growth Defects of pin1 *Mutants*

Because only *pin1-5 lfy-6* inflorescences make a reasonable number of flowers prior to a naked pin, we first examined the effect of loss of *LFY* function on floral organ development in the *pin1-5* mutant. Floral organs in wild-type plants are arranged in a series of whorls: four sepals in the outer whorl, followed by four petals, six stamens, and two carpels (Figure 2a,r). As reported previously, *lfy* mutant flowers display floral homeotic defects. *lfy* null mutants rarely form petals and stamens and largely consist of sepal- and carpel-like organs [23] (Figure 2b,r). This is consistent with molecular and genetic data, which show that LFY plays a major role in the up-regulation of the class B and class C floral homeotic genes [36–38]. On the other hand, the weak *pin1-5* flowers formed all four types of floral organs, like wild-type flowers (Figure 2c,r). However, *pin1-5* flowers showed an increase in the number of petals, and a decrease in the number of stamens formed (Figure 2r) [17]. In addition, *pin1-5* resulted in a reduction of the valve region and expansion of the stylar and stigmatic regions of the gynoecium (Figure 2f–h). Finally, *pin1-5* sepals are often fused (Figure 2l–n). These phenotypes are common in flowers of auxin defective mutants [15,16,30].

pin1-5 lfy-6 exhibited a reduction in the number of all four floral organs formed, compared with the wild-type or parental lines and generally lacked petals and stamens (Figure 2d,r). The sepal fusion and abnormal gynoecium phenotype of *pin1-5* was enhanced in the double *pin1-5 lfy-6* mutants (Figure 2h,i,n,o). The length of the fused sepal margin increased when compared with *pin1-5* single mutant flowers. The incidence of sepal fusion also increased in *pin1-5 lfy-6* flowers (32%) compared to *pin1-5* flowers (8.9%) (Figure 2s). In individual plants, sepal fusions became more severe and frequent acropetally (Figure 2n,o). Additionally, the size of the stigmatic and ovary regions in the gynoecium increased and decreased, respectively when compared with *pin1-5* single mutant flowers (Figure 2h,i). Although *pin1-5* formed a few seeds, *pin1-5 lfy-6* flowers were infertile, like *lfy-6* mutants (data not shown).

Figure 2. *lfy* enhances the flower developmental defects of auxin transport mutants. (**a–e**) Side view of the flowers formed in wild type (**a**), *lfy-6* (**b**), *pin1-5* (**c**), *pin1-5 lfy-6* (**d**), *pid-16* (**e**); (**f–k**) Side view of the gynoecium formed in wild type (**f**), *lfy-6* (**g**), *pin1-5* (**h**), *pin1-5 lfy-6* (**i**), *pid-16* (**j**), *pid-16 lfy-1* (**k**); (**l–q**) Side view of the sepal formed in wild type (**l**), *lfy-6* (**m**), *pin1-5* (**n**), *pin1-5 lfy-6* (**o**), *pid-16* (**p**), *pid-16 lfy-1* (**q**); (**r**) Floral organ number in wild-type, *lfy-6*, *pin1-5*, and *lfy-6 pin1-5* flowers; (**s**) Incidence of sepal fusion in wild-type, *lfy-6*, *pin1-5*, and *lfy-6 pin1-5* flowers. Scale bar, 2 mm.

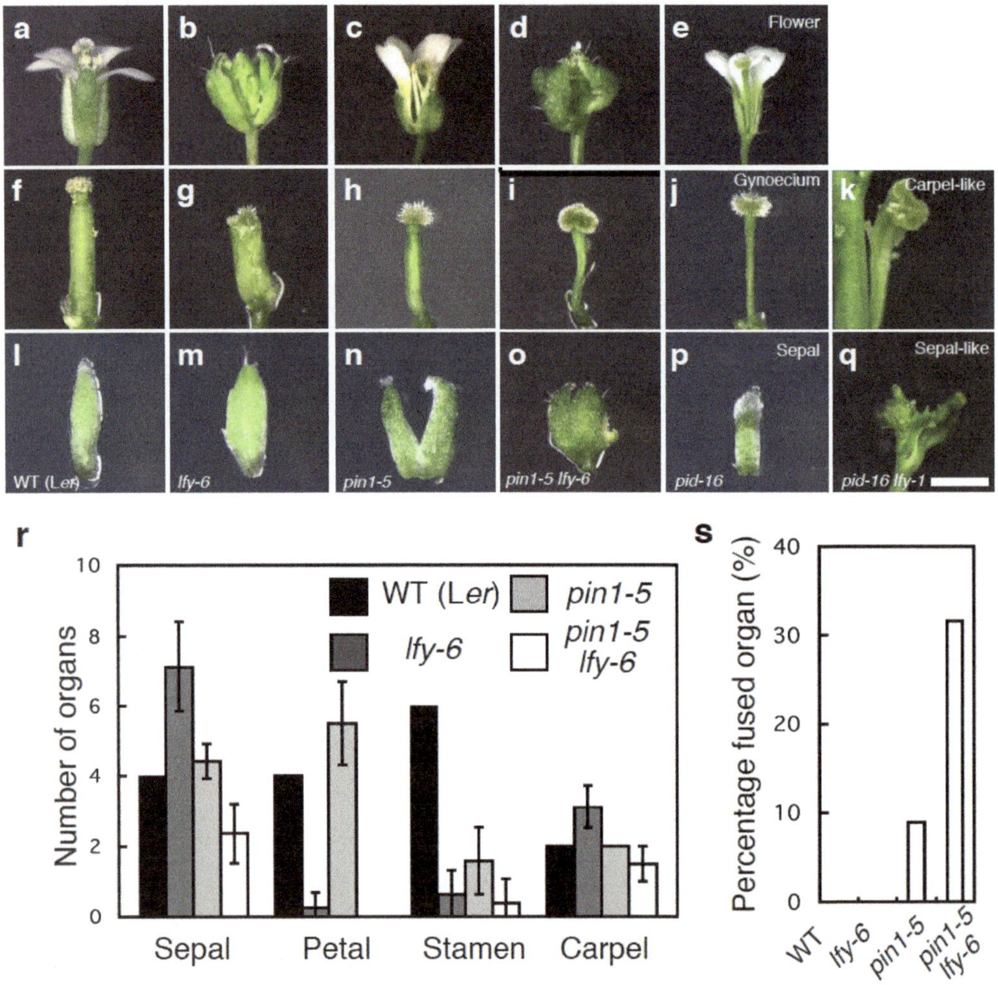

To further confirm the role of *LFY* in auxin transport-mediated flower development, we examined genetic interaction between *LFY* and *PID*. *pid-16* single mutants produced several flowers with fewer floral organs (Figure 2e). The flowers produced by *pid-16 lfy-1* plants exhibited more severe phenotypic defects than those observed in *pid-16* alone (Figure 2e,j,k,p,q). *pid-16 lfy-1* flowers never formed petals and stamens (Figure 2k). Flowers of *pid-16 lfy-1* double mutants were comprised of either several fused carpel-like or sepal-like organs (Figure 2k,q). In addition these abnormal flowers often displayed fusion with the pedicel (Figure 2k,q). Thus, *LFY* may act together with polar auxin transport in sepal and gynoecium development.

2.3. The Relationship between LFY and Auxin-Mediated Sepal Initiation

The sepal phenotype observed in *pin1 lfy* flowers may be due to an alteration in the positioning of sepal primordia in the floral meristem and/or the inability to establish boundaries between adjacent organs. To better understand how loss of *LFY* function affects sepal formation, we examined sepal primordium positioning by monitoring expression of the auxin efflux carrier protein, PIN1, and the auxin response promoter *DR5*. In stage 2 wild-type flowers, PIN1-GFP was expressed in the incipient sepal primordia at symmetrical positions within the flower meristem (Figure 3a). The PIN1-GFP positive sites of sepal primordium initiation in stage 2 flowers of *lfy* mutants were not properly spaced, as in the wild type. Frequently two incipient sepal primordia formed in close proximity (Figure 3a,b). In wild-type plants, *DR5rev::GFP* was expressed in the sepal primordia, as reported previously [3,39]. On the basis of *DR5rev::GFP*, wild-type sepals developed four discrete sepal primordia at symmetrical positions (Figure 3c). Like *lfy* floral primordia [22,33], *lfy* sepal primordia showed slightly reduced *DR5* expression, especially at the abaxial and the lateral side of the sepals (Figure 3c,d insets). Although we saw an increase in the number of sepals in *lfy* mutants at later stages, stage 3 *lfy-1* mutant flowers only had four sepal primordia based on *DR5* expression (Figure 3c,d). These data suggest that additional sepal primordia form after stage 3 in *lfy* flowers. As seen for the PIN1-GFP expression, the sites of sepal initiation marked by *DR5* in stage 3 flowers of *lfy* mutants were not properly spaced in the floral primordium; frequently two sepal primordia formed in close proximity (Figure 3b,d). At stage 4 of flower development, *DR5* expression in *lfy* sepals was often not perfectly cruciform, but somewhat twisted (Figure 3e,f). Hence, LFY is required for proper positioning of sepal primordia.

Figure 3. Floral organ primordium initiation in *lfy* loss-of-function mutant flowers. (**a**,**b**) pPIN1::PIN1-GFP expression in wild-type (**a**) and *lfy* (**b**) flowers at stage 2. (**c**–**f**) *DR5rev::GFP* expression in wild-type (**c**,**e**) and *lfy* (**d**,**f**) flowers at stage 3 (**c**,**d**) and stage 4 (**e**,**f**). Strongly increased GFP signal was observed in sepal primordia (arrowheads), as reported previously [3,39]. Insets show color-coded GFP intensity, with background levels being purple and blue, green, yellow, and red to signifying increasing *DR5rev::GFP* signal. ab, abaxial; ad, adaxial. Scale bar, 25 μm.

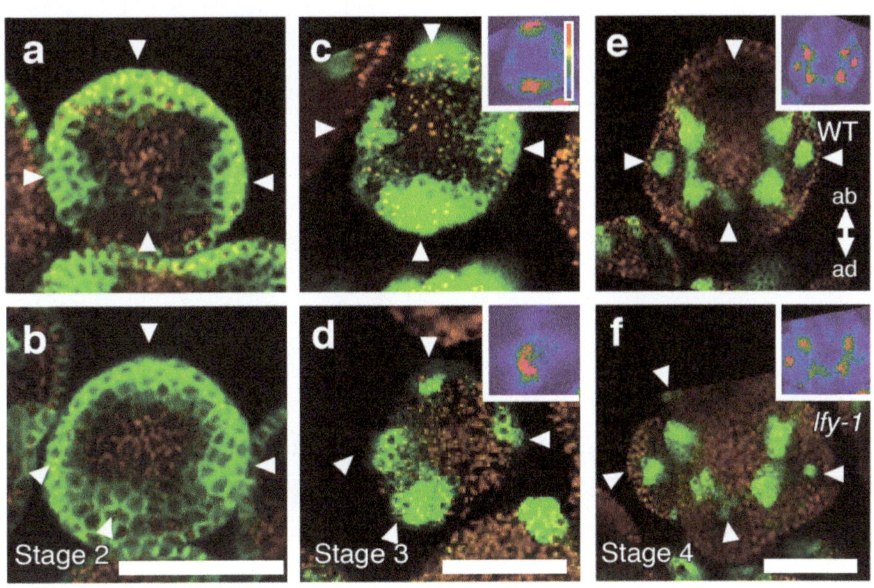

2.4. A Link between LFY and Polar Auxin Transport in Sepal Boundary Formation

A further connection between LFY and sepal development was suggested by genome-wide identification of LFY binding sites [37,40]. The most highly enriched Gene Ontology terms among direct LFY targets in the inflorescence were "organ development" and "flower development" [37,40]. Amongst the candidate direct LFY targets identified by these studies, one gene with a known role in organ boundary formation stood out, *CUC2*. Although LFY is already expressed prior to flower formation [41], LFY only bound to the 5'-regulatory region of the *CUC2* gene in inflorescences (Figure 4a), suggesting that LFY regulates *CUC2* expression specifically at this stage. We further found that *CUC2* mRNA levels rapidly increased upon 35S::LFY-GR activation in inflorescences (Figure 4b). In addition, *CUC2* showed a subtle, but reproducible reduction of expression in *lfy-6* null mutants relative to the wild type (Figure 4c). *CUC2* mRNA levels were also reduced in *pin1-5* mutants. Furthermore, a more dramatic reduction in *CUC2* expression was observed in *pin1-5 lfy-6* compared to the wild type or the parental lines, suggesting that both *LFY* and auxin transport are important for proper *CUC2* induction in the flower.

Figure 4. LFY regulates *CUC2* gene expression. (**a**) LFY binds to the regulatory regions of *CUC2* gene based on a published ChIP-chip data set [37]. Screenshots of the binding peaks (blue vertical lines). Significant LFY-binding peaks are indicated by a red horizontal line below the peaks. Green boxes below the graphs denote exons, blue arrows indicate the direction of transcription; (**b**) Expression of the *CUC2* gene based on qRT-PCR three hours after mock or dexamethasone (dex) treatment of *lfy-6* 35S::LFY-GR inflorescences; (**c**) Expression of the *CUC2* gene based on qRT-PCR in *lfy-6* null mutant, the weak *pin1-5* mutant, and *pin1-5 lfy-6* double mutant inflorescence apices compared to wild-type inflorescence apices. RNA from floral bud clusters (containing stages 1–6 flowers) was used in this study.

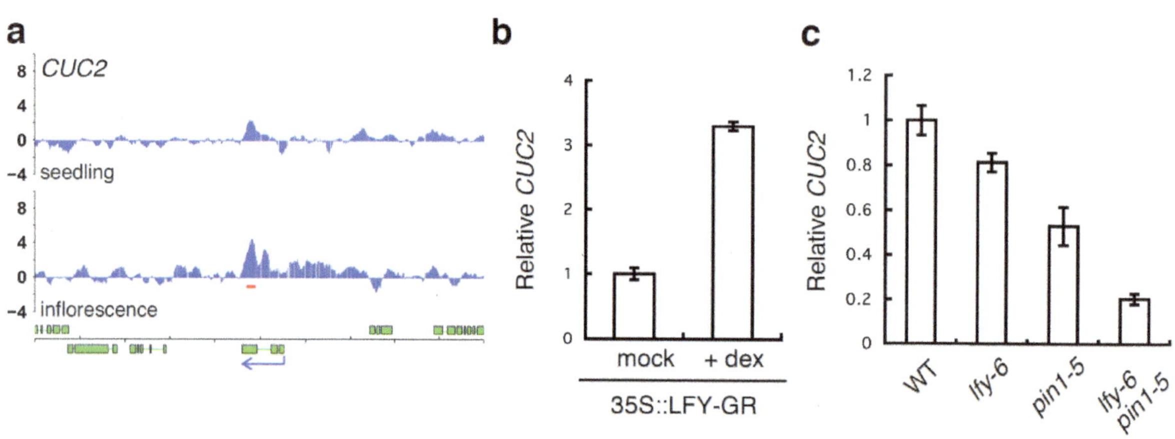

To further test the relationship between LFY and *CUC2*, we performed *in situ* hybridization. In wild-type plants, *CUC2* was expressed in the sepal boundary of stage 3 flowers, as reported for *CUC1* or *CUC3* (Figure 5a) [42,43]. *CUC2* expression pattern and levels were essentially the same in stage 3 wild type and *lfy-1* flowers (Figure 5b). In stage 4 wild-type flowers, strong *CUC2* expression was observed in the boundary region between the sepal primordia and the floral meristem dome, where growth is retarded (Figure 5c,e). However, *CUC2* expression in the boundary between the sepal primordia and the floral meristem dome was frequently weaker in stage 4 *lfy-1* flowers and *CUC2* expression was overall more diffuse (Figure 5d). Occasionally, *CUC2* expression was entirely missing from a boundary region (Appendix Figure A1).

Figure 5. Altered *CUC2* gene expression in *LFY* loss-of-function flowers. (**a**–**d**) *CUC2* expression in wild-type (**a**,**c**) and *lfy* (**b**,**d**) flowers at stage 3 (**a**,**b**) and stage 4 (**c**,**d**). Arrowheads denote the boundary between the floral meristem and the sepals. Bars indicate diffused *CUC2* expression. Scale bar, 25 μm.

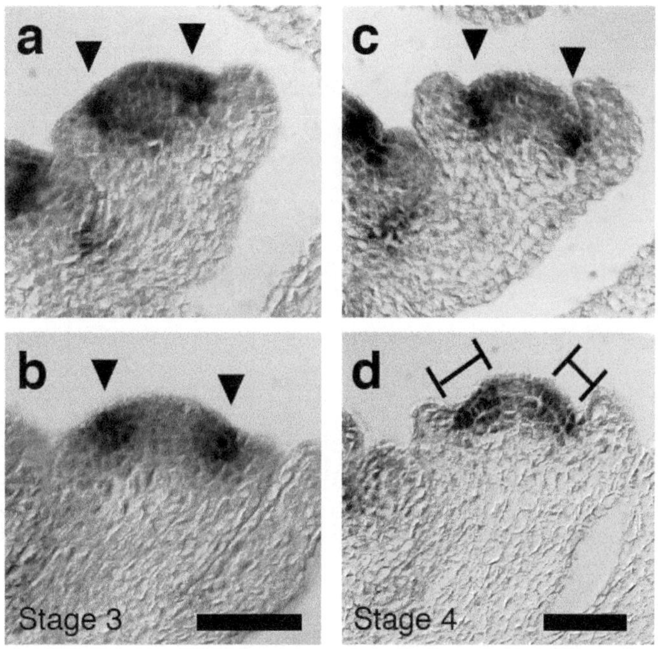

It has been reported that *CUC1* and *CUC2* genes act redundantly in regulating formation of the sepal boundary [31], and that *CUC2* expression is affected by auxin transport [44]. Since we could observe both a sepal fusion phenotype and a dramatic reduction of *CUC2* mRNA levels in *pin1 lfy* double mutants, the combined data are consistent with the idea that LFY and polar auxin transport may modulate *CUC2* expression in the sepal boundary.

2.5. A Link between LFY and Auxin Transport in Gynoecium Development

Among the LFY-bound genes identified by ChIP-Seq or ChIP-chip are many genes in the auxin pathway [22,37,40], several of which have been implicated in gynoecium development [12,15,30]. Thus, several potential direct LFY target genes may underlie the gynoecium formation defects observed in *pin lfy* mutants. Because the single *ett* null mutant displays gynoecium defects very similar to those we observed here [30], we focused on the relationship between LFY and *ETT*. Two strong LFY binding peaks were present at the *ETT* locus in inflorescences, one in the promoter and one in the ninth intron (Figure 6a). When we tested *ETT* expression by qRT-PCR, we found that *ETT* mRNA was reduced in *lfy* mutants and rapidly increased upon 35S::LFY-GR activation in inflorescences (Figure 6b,c). These results suggest that LFY may activate *ETT* to regulate gynoecium development. However, we did not see a strong reduction of *ETT* expression in *pin1-5* mutants compared to wild type (Figure 6c). Moreover, no further change in *ETT* expression was observed in *pin1-5 lfy-6* mutant relative to *lfy-6* mutant inflorescences (Figure 6c). Thus, additional regulators of gynoecium development besides *ETT* may contribute to the developmental defects observed in *pin1 lfy* double mutants.

Figure 6. LFY regulates the *ETT* gene. (**a**) LFY binds to the regulatory regions of the *ETT* gene based on a published ChIP-chip data set [37]. Screenshots of the LFY binding peaks (blue vertical lines). Significant LFY-binding peaks are indicated by a red horizontal line below the peaks. Green boxes below the graphs denote exons, blue arrows indicate the direction of transcription; (**b**) Expression of the *ETT* gene based on qRT-PCR three hours after mock or dexamethasone (dex) treatment of *lfy-6* 35S::LFY-GR inflorescences (**c**) Expression of the *ETT* gene based on qRT-PCR in *lfy-6* null mutant, weak *pin1-5* mutant, and *pin1-5 lfy-6* double mutant inflorescence apices compared to wild-type inflorescence apices. RNA from floral bud clusters (containing stages 1-6 flowers) was used in this study.

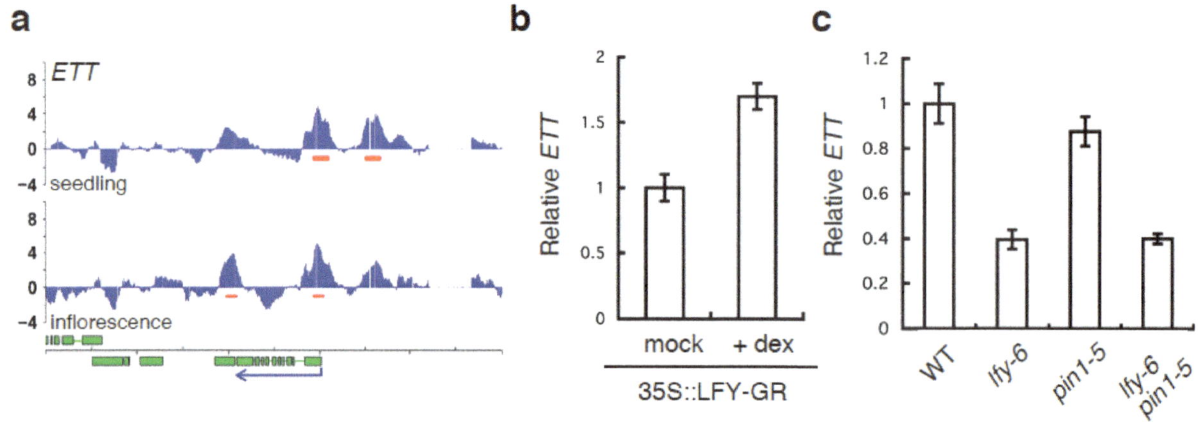

Alternatively, reduced *ETT* accumulation in *lfy* mutants combined with reduced ETT activity in *pin1-5* may cause the observed gynoecium defect. Although the ETT protein lacks the conserved domains III and IV for interaction with AUX/IAA proteins [45], it has been reported that the transcriptional repressor ETT suppresses a synthetic AuxRE-based reporter in protoplasts in an auxin-dependent manner [46], and *pin1* mutants have reduced auxin accumulation [1]. In agreement with this study, treatment with the auxin transport inhibitor NPA enhanced the gynoecium defects of weak *ett* mutants to phenocopy *ett* null mutants, while *ett* null mutant gynoecium defects were not enhanced by this treatment [30]. Thus, we favor a model in which the reduction of *ETT* mRNA level (due to the *lfy* mutation) and attenuated ETT protein activity (due to the *pin1* mutation/low auxin level) cause gynoecium developmental defects in *pin1 lfy* double mutants. Further studies are required to resolve how auxin modulates ETT activity in the gynoecium.

3. Experimental Section

3.1. Plant Materials and Growth Conditions

Plants were grown at 23 °C in a 16 h light/8 h dark cycle. *Arabidopsis thaliana* accessions Columbia (Col) or Landsberg *erecta* (L*er*) were used in this study. Mutants employed were: *lfy-1* [23], *lfy-6* [23], *pin1-5* [17], *pin1-8* (SALK_97114) [34], *pid-16* (SALK_082564), 35S::LFY-GR [47], and

DR5rev::GFP [3]. For dexamethasone treatments, plants were grown in soil. Dexamethasone treatments were performed by spraying 30-day-old plants once with 5 μM dexamethasone.

3.2. Statistical Tests

The Student *t*-test (two-tailed) was used for all experiments that displayed a normal distribution based on the Kolmogorov–Smirnov test [48].

3.3. Genetics and PCR Genotyping

lfy/+ plants were crossed to *pin1-5*, *pin1-8/+*, or *pid-16/+* plants. Double mutants were identified in the F2 or later generations as plants with new phenotypes and confirmed by PCR genotyping. *lfy* mutations were PCR genotyped as described previously. Genotyping primers are shown in Table 1.

Table 1. Primer sets used in this study.

Primer name	Sequence
Genotyping	
lfy-1/-6-FW	AAGCAGCCGTCTGCGGTGTCAGCAGCTGTT
lfy-1/-6-RV	CTGTCAATTTCCCAGCAAGACAC
pid-16-LP	TCCGTCATAGACAACCTCACC
pid-16-RP	GAGTAAGCGTACGAATGAGCG
pin1-8-LP	AACTGGCTTCACAGCAGAAAG
pin1-8-RP	TCAACAAAAAGGGCATTGTTC
LBb1.3	ATTTTGCCGATTTCGGAAC
Primer name	**Sequence**
qRT-PCR	
CUC2-FW	GGAAGAGCTCCGAAAGGAGA
CUC2-RV	TCCGGTGCTAGCTAAAGTGG
ETT-FW	CAGGGACATTTGGAACAAGC
ETT-RV	CAGGAAGAAGAGAGACTTGAGCA
EIF4-FW	AAACTCAATGAAGTACTTGAGGGACA
EIF4-RV	TCTCAAAACCATAAGCATAAATACCC
***In situ* hybridization**	
CUC2-FW	CGGAATTCATGGACATTCCGTATTACCA
CUC2-RV	CCACTAGTTCAGTAGTTCCAAATACAGT

3.4. Confocal Microscopy

Imaging of GFP signal was performed as described by [22]. For imaging of GFP signal, inflorescence apices were dissected to remove older flowers and imaged for green and red fluorescence using a Leica confocal microscope (Leica, LCS SL) equipped with an argon-krypton ion laser with the appropriate filter sets for visualizing GFP and propidium iodide. For comparisons of GFP fluorescence, the same offset and gain settings were used in the analysis of plants that were transformed with the same transgene. At least ten inflorescences were prepared from different plants, and representative images are shown.

3.5. Quantitative RT-PCR

RNA from floral bud clusters (containing stages 1–6 flowers) was used in this study. Total RNA was extracted from inflorescences using the RNeasy mini kit (Qiagen) and was treated with DNase (Qiagen) prior to RT. cDNA was synthesized from 0.5 ng of total RNA with the Super Script III (Invitrogen). qRT-PCR was performed as previously described by Yamaguchi *et al.* [22]. Real-time PCR was performed with the 7100 Real-Time PCR system and Power SYBR Green PCR Master Mix (Applied Biosystems), as described by the manufacturer. The means and standard errors were determined using two to three biological replicates with three technical replicates each. Plant materials were grown and harvested at different times. One representative experiment is shown. Gene-specific signals were normalized over that of the *EUKARYOTIC TRANSLATION INITIATION FACTOR 4A-1* (*EIF4; At1g54270*). Primers are shown in Table 1.

3.6. Screenshot

Significant LFY binding to regulatory regions of *CUC2* and *ETT* in seedlings and inflorescences were generated by Winter *et al.* [37].

3.7. In situ Hybridization

In situ hybridization was performed as described previously [49,50]. The *CUC2* probe consisted of base pairs 29 to 1137 (TSS = 1). Cloning primers are shown in Table 1. Probes were cloned into pGEM-T Easy (Promega). Antisense *CUC2* probe was digested with *Eco*RI and transcribed with the T7 polymerase. The Riboprobe Combination System (Promega) and DIG RNA labeling mix (Roche) were used for probe synthesis.

4. Conclusions

The plant specific transcription factor LFY is necessary and sufficient for the developmental transition to flower formation [23,24] Previous genomic, genetic, and molecular analyses have suggested a role for LFY downstream of AUXIN RESPONSE FACTOR5/MONOPTEROS in flower primordium initiation [22,37,40]. In addition, LFY feedback regulates the auxin pathway [22,33]. Finally, like MP, LFY also acts in a pathway parallel to the polar auxin transport regulators PID and PIN1 in lateral organ initiation (this study) [33,35]. We have uncovered enhanced sensitivity of *lfy* mutants to alterations in polar auxin transport in other aspects of flower morphogenesis such as floral organ initiation and sepal and gynoecium development. It is difficult to know whether the interactions between auxin transport and LFY are direct, via changes in auxin concentrations, or indirect via combinatorial effects on gene expression. Disruption of polar auxin transport has a dramatic effect on expression of a spectrum of genes [51]. We provide evidence that transcriptional regulation of the organ boundary regulator *CUC2* and the gynoecium development regulator *ETT* by LFY (this study) and auxin [1–17] may contribute to defects in sepal and gynoecium development observed in double mutants between polar auxin transport regulators and *lfy*. Our study combined with prior investigations reveals a complex set of interactions between LFY and polar auxin transport in developing flowers.

Acknowledgments

We thank Jiri Friml for pPIN1::PIN1-GFP and *DR5rev::GFP* seeds, and the *Arabidopsis* Biological Resource Center for insertion seeds. This work was supported by NSF grant IOS-1257111 to D.W. and JSPS postdoctoral fellowships for research abroad to N.Y.

Conflicts of Interest

The authors declare no conflict of interest.

References

1. Okada, K.; Ueda, J.; Komaki, M.K.; Bell, C.J.; Shimura, Y. Requirement of the Auxin Polar Transport System in Early Stages of Arabidopsis Floral Bud Formation. *Plant Cell* **1991**, *3*, 677–684.
2. Reinhardt, D.; Mandel, T.; Kuhlemeier, C. Auxin regulates the initiation and radial position of plant lateral organs. *Plant Cell* **2000**, *12*, 507–518.
3. Benkova, E.; Michniewicz, M.; Sauer, M.; Teichmann, T.; Seifertova, D.; Jurgens, G.; Friml, J. Local, efflux-dependent auxin gradients as a common module for plant organ formation. *Cell* **2003**, *115*, 591–602.
4. Reinhardt, D.; Pesce, E.R.; Stieger, P.; Mandel, T.; Baltensperger, K.; Bennett, M.; Traas, J.; Friml, J.; Kuhlemeier, C. Regulation of phyllotaxis by polar auxin transport. *Nature* **2003**, *426*, 255–260.
5. Heisler, M.G.; Ohno, C.; Das, P.; Sieber, P.; Reddy, G.V.; Long, J.A; Meyerowitz, E.M. Patterns of auxin transport and gene expression during primordium development revealed by live imaging of the Arabidopsis inflorescence meristem. *Curr. Biol.* **2005**, *15*, 1899–1911.
6. Stepanova, A.N.; Rovertson-Hoyt, J.; Yun, J.; Benavente, L.M.; Xie, D.Y.; Dolezal, K.; Schlereth, A.; Jurgens, G.; Alonso, J.M. TAA-mediated auxin biosynthesis is essential for hormone crosstalk and plant development. *Cell* **2008**, *133*, 177–191.
7. Zhao, Y. Auxin biosynthesis and its role in plant development. *Annu. Rev. Plant Biol.* **2010**, *61*, 49–64.
8. Robert, H.S.; Grones, P.; Stepanova, A.N.; Robles, L.M.; Lokerse, A.S.; Alonso, J.M.; Weijers, D.; Friml, J. Local auxin sources orient the apical-basal axis in Arabidopsis embryos. *Curr. Biol.* **2013**, *23*, 2506–2512.
9. Friml, J.; Yang, X.; Michniewicz, M.; Weijers, D.; Quint, A.; Tietz, O.; Benjamins, R.; Ouwerkerk, P.B.; Ljung, K.; Sandberg, G.; *et al.* A PINOID-dependent binary switch in apical-basal PIN polar targeting directs auxin efflux. *Science* **2004**, *306*, 862–865.
10. Michniewicz, M.; Zago, M.K.; Abas, L.; Weijers, D.; Schweighofer, A.; Meskiene, I.; Marcus, G.H.; Ohno, C.; Zhang, J.; Huang, F.; *et al.* Antagonistic regulation of PIN phosphorylation by PP2A and PINOID directs auxin flux. *Cell* **2007**, *130*, 1044–1056.
11. Cheng, Y.; Qin, G.; Dai, X.; Zhao, Y. NPY1, a BTB-NPH3-like protein, plays a critical role in auxin-regulated organogenesis in *Arabidopsis*. *Proc. Natl. Acad. Sci. USA* **2007**, *104*, 18825–18829.

12. Cheng, Y.; Qin, G.; Dai, X.; Zhao, Y. NPY genes and AGC kinases define two key steps in auxin-mediated organogenesis in *Arabidopsis*. *Proc. Natl. Acad. Sci. USA* **2008**, *105*, 21017–21022.
13. Furutani, M.; Nakano, Y.; Tasaka, M. MAB4-induced auxin sink generates local auxin gradients in *Arabidopsis* organ formation. *Proc. Natl. Acad. Sci. USA* **2014**, *11*, 1198–1203.
14. Zhao, Y.; Christensen, S.K.; Fankhauser, C.; Cashman, J.R.; Cohen, J.D.; Weigel, D.; Chory, J. A role for flavin monooxygenase-like enzymes in auxin biosynthesis. *Science* **2001**, *291*, 306–309.
15. Cheng, Y.; Dai, X.; Zhao, Y. Auxin biosynthesis by the YUCCA flavin monooxygenase controls the formation of floral organs and vascular tissues in *Arabidopsis*. *Genes Dev.* **2006**, *20*, 1790–1799.
16. Cheng, Y.; Dai, X.; Zhao, Y. Auxin synthesized by the YUCCA flavin monooxygenases is essential for embryogenesis and leaf formation in Arabidopsis. *Plant Cell* **2007**, *19*, 2430–2439.
17. Bennett, S.R.M.; Alvarez, J.; Bossinger, G.; Smyth, D.R. Morphogenesis in pinoid mutants of *Arabidopsis thaliana*. *Plant J.* **1995**, *8*, 505–520.
18. Przemeck, G.K.; Mattsson, J.; Hardtke, C.S.; Sung, Z.R.; Berleth, T. Studies on the role of the *Arabidopsis* gene MONOPTEROS in vascular development and plant cell axialization. *Planta* **1996**, *200*, 229–237.
19. Hardtke, C.S.; Berleth, T. The *Arabidopsis* gene MONOPTEROS encodes a transcription factor mediating embryo axis formation and vascular development. *EMBO J.* **1998**, *17*, 1405–1411.
20. Hamann, T.; Benkova, E.; Baurle, I.; Kientz, M.; Jurgens, G. The *Arabidopsis* BODENLOS gene encodes an auxin response protein inhibiting MONOPTEROS-mediated embryo patterning. *Genes Dev.* **2002**, *16*, 1610–1615.
21. Weijers, D.; Schlereth, A.; Ehrismann, J.S.; Schwank, G.; Kientz, M.; Jurgens, G. Auxin triggers transient local signaling for cell specification in *Arabidopsis* embryogenesis. *Dev. Cell* **2006**, *10*, 265–270.
22. Yamaguchi, N.; Wu, M-F.; Winter, C.; Berns, M.; Nole-Wilson, S.; Yamaguchi, A.; Coupland, G.; Krizek, B.; Wagner, D. A molecular framework for auxin-mediated initiation of floral promordia. *Dev. Cell* **2013**, *24*, 271–282.
23. Weigel, D.; Alvarez, J.; Smyth, D.R.; Yanofsky, M.F.; Meyerowitz, E.M. LEAFY controls floral meristem identity in *Arabidopsis*. *Cell* **1992**, *69*, 843–859.
24. Weigel, D.; Nilsson, O. A developmental switch sufficient for flower initiation in diverse plants. *Nature* **1995**, *377*, 495–500.
25. Elliott, R.C.; Betzner, A.S.; Huttner, E.; Oakes, M.P.; Tucker, W.Q.J.; Gerentes, D.; Perez, P.; Smyth, D.R. *AINTEGUMENTA*, an *APETALA2*-like gene of Arabidopsis with pleiotropic roles in ovule development and floral organ growth. *Plant Cell* **1996**, *8*, 155–168.
26. Krizek, B.A. Ectopic expression of AINTEGUMENTA in *Arabidopsis* plants results in increased growth of floral organs. *Dev. Genet.* **1999**, *25*, 224–236.
27. Mizukami, Y.; Fischer, R.L. Plant organ size control: AINTEGUMENTA regulates growth and cell numbers during organogenesis. *Proc. Natl. Acad. Sci. USA* **2000**, *97*, 942–947.
28. Krizek, B.A. AINTEGUMENTA and AINTEGUMENTA-LIKE6 act redundantly to regulate Arabidopsis floral growth and patterning. *Plant Physiol.* **2009**, *150*, 1916–1929.
29. Krizek, B.A. Auxin regulation of *Arabidopsis* flower development involves members of the AINTEGUMENTA-LIKE/PHETHORA (AIL/PLT) family. *J. Exp. Bot.* **2011**, *62*, 3311–3319.

30. Nemhauser, J.L.; Feldman, L.J.; Zambryski, P.C. Auxin and ETTIN in *Arabidopsis* gynoecium morphogenesis. *Development* **2000**, *127*, 3877–3888.
31. Aida, M.; Ishida, T.; Fukaki, H.; Fujisawa, H.; Tasaka, M. Genes involved in organ separation in Arabidopsis: An analysis of the cup-shaped cotyledon mutant. *Plant Cell* **1997**, *9*, 841–857.
32. Sieber, P.; Wellmer, F.; Gheyselinck, J.; Riechmann, J.L.; Meyerowitz, E.M. Redundancy and specillization among plant microRNAs: Role of the *MIR164* familiy in developmental robstness. *Development* **2007**, *134*, 1051–1060.
33. Li, W.; Zhou, Y.; Liu, X.; Yu, P.; Cohen, J.D.; Meyeriwitz, E.M. LEAFY controls auxin response pathways in floral primordium formation. *Sci. Signal.* **2013**, *6*, ra23.
34. Kawamura, E.; Horiguchi, H.; Tsukaya, H. Mechanisms of leaf tooth formation in *Arabidopsis*. *Plant J.* **2010**, *62*, 429–441.
35. Schuetz, M.; Berleth, T.; Mattsson, J. Multiple MONOPTEROS-dependent pathways are involved in leaf initiation. *Plant Physiol.* **2008**, *148*, 870–880.
36. Weigel, D.; Meyerowitz, E.M. Activation of floral homeotic genes in *Arabidopsis*. *Science* **1993**, *261*, 1723–1726.
37. Winter, C.M.; Austin, R.S.; Blanvillain-Baufume, S.; Reback, M.A.; Monniaux, M.; Wu, M.F.; Sang, Y.; Yamaguchi, A.; Yamaguchi, N.; Parker, J.E.; *et al.* LEAFY Target Genes Reveal Floral Regulatory Logic, *cis* motifs, and a link to biotic stimulus response. *Dev. Cell* **2011**, *20*, 430–443.
38. Wu, M.F.; Sang, Y.; Bezhani, S.; Yamaguchi, N.; Han, S.K.; Li, Z.; Su, Y.; Slewinski, T.L.; Wagner, D. SWI2/SNF2 chromatin remodeling ATPases overcome polycomb repression and control floral organ identity with the LEAFY and SEPALLATA3 transcription factors. *Proc. Natl. Acad. Sci. USA* **2012**, *109*, 3576–3581
39. Lampugnani, E.R.; Kilinc, A.; Smyth, D.R. *PETAL LOSS* is a boundary gene that inhibits growth between developing sepals in *Arabidopsis thaliana*. *Plant J.* **2012**, *71*, 724–735.
40. Moyroud, E.; Minguet, E.G.; Ott, F.; Yant, L.; Pose, D.; Monniaux, M.; Blanchet, S.; Bastien, O.; Thevenon, E.; Weigel, D.; *et al.* Prediction of regulatory interactions from genome sequences using a biophysical model for the *Arabidopsis* LEAFY transcription factor. *Plant Cell* **2011**, *23*, 1293–1306.
41. Blazquez, M.A.; Soowal, L.N.; Lee, I.; Weigel, D. LFY expression and flower initiation in *Arabidopsis*. *Development* **1998**, *121*, 3835–3844.
42. Takada, S.; Hibara, K.; Ishida, T; Tasaka, M. The *CUP-SHAPED COTYLEDON1* gene of *Arabidopsis* regulates shoot apical meristem formation. *Development* **2001**, *128*, 1127–1135.
43. Hibara, K.; Karim, M.R.; Takada, S.; Taoka, K.; Furutani, M.; Aida, M.; Tasaka, M. *Arabidopsis* CUP-SHAPED COTYLEDON3 regulates postembryonic shoot meristem and organ boundary formation. *Plant Cell* **2006**, *18*, 2946–2957.
44. Bilsborough, G.D.; Runions, A.; Barkoulas, M.; Jenkins, H.W.; Hasson, A.; Galinha, C.; Laufs, P.; Hay, A.; Prusinkiewicz, P.; Tsiantis, M. Model for the regulation of *Arabidopsis thaliana* leaf margin development. *Proc. Natl. Acad. Sci. USA* **2011**, *108*, 3424–3429.
45. Hagen, G.; Guilfoyle, T. Auxin-responsive gene expression: Genes, promoters and regulatory factors. *Plant Mol. Biol.* **2002**, *49*, 373–385.
46. Tiwari, S.B.; Hagen, G.; Guilfoyle, T. The roles of auxin response factor domains in auxin responsive transcription. *Plant Cell* **2003**, *15*, 533–543.

47. Wagner, D.; Sablowski, R.W.; Meyerowitz, E.M. Transcriptional activation of APETALA1 by LEAFY. *Science* **1999**, *285*, 582–584.
48. Massay, F.J. The Kolmogorov-Smirnov test for goodness of fit. *J. Am. Stat. Assoc.* **1951**, *46*, 68–78.
49. Long, J.A.; Barton, M.K. The development of apical embryonic pattern in *Arabidopsis*. *Development* **1998**, *125*, 3027–3035.
50. Wu, M.F.; Wagner, D. RNA *in situ* hybridization in *Arabidopsis*. *Methods Mol. Biol.* **2012**, *883*, 75–86.
51. Nemhauser, J.L.; Hong, F.; Chory, J. Different plant hormones regulate similar processes through largely nonoverlapping transcriptional responses. *Cell* **2006**, *126*, 467–475.

Appendix

Figure A1. Altered *CUC2* gene expression in *LFY* loss-of-function flowers. (**a**,**b**) *CUC2* expression in wild-type (**a**) and *lfy* (**b**) flowers at stage 4. Arrowheads denote the boundary between the floral meristem and the sepals. Scale bar, 25 μm.

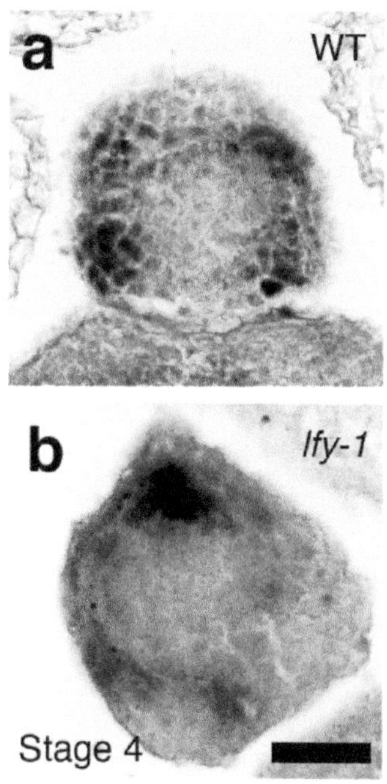

8

Ca^{2+}-Transport through Plasma Membrane as a Test of Auxin Sensitivity

Anastasia A. Kirpichnikova [1,†], Elena L. Rudashevskaya [1,†,‡], Vladislav V. Yemelyanov [1,2] and Maria F. Shishova [1,*]

1 Department of Plant Physiology and Biochemistry, St. Petersburg State University, Universitetskaya emb. 7/9, St. Petersburg 199034, Russia; E-Mails: nastin1972@mail.ru (A.A.K.); bootika@mail.ru (V.V.Y.)

2 Department of Genetics and Biotechnology, St. Petersburg State University, Universitetskaya emb. 7/9, St. Petersburg 199034, Russia

† These authors contributed equally to this work.

‡ Current address: Institute of Medical Chemistry, Medical University of Vienna; Waehringer strasse 10, Vienna 1090, Austria; E-Mail: elena.rudashevskaya@meduniwien.ac.at.

* Author to whom correspondence should be addressed; E-Mail: mshishova@mail.ru

---

Abstract: Auxin is one of the crucial regulators of plant growth and development. The discovered auxin cytosolic receptor (TIR1) is not involved in the perception of the hormone signal at the plasma membrane. Instead, another receptor, related to the ABP1, auxin binding protein1, is supposed to be responsible for the perception at the plasma membrane. One of the fast and sensitive auxin-induced reactions is an increase of Ca^{2+} cytosolic concentration, which is suggested to be dependent on the activation of Ca^{2+} influx through the plasma membrane. This investigation was carried out with a plasmalemma enriched vesicle fraction, obtained from etiolated maize coleoptiles. The magnitude of Ca^{2+} efflux through the membrane vesicles was estimated according to the shift of potential dependent fluorescent dye diS-C_3-(5). The obtained results showed that during coleoptiles ageing (3rd, 4th and 5th days of seedling etiolated growth) the magnitude of Ca^{2+} efflux from inside-out vesicles was decreased. Addition of ABP1 led to a recovery of Ca^{2+} efflux to the level of the youngest and most sensitive cells. Moreover, the efflux was more sensitive, responding from 10^{-8} to 10^{-6} M 1-NAA, in vesicles containing ABP1, whereas native vesicles showed the highest efflux at 10^{-6} M 1-NAA. We suggest that auxin increases plasma membrane permeability to Ca^{2+} and that ABP1 is involved in modulation of this reaction.

Keywords: ABP1; auxin perception; Ca^{2+} signaling; elongation growth

---

Abbreviations

ABP1, auxin binding protein 1; AFB, auxin F-box protein; AU, arbitrary units; $[Ca^{2+}]_{cyt}$, concentration of Ca^{2+} in cytoplasm; cyt, cytosol; diS-C_3-(5), 3,3'-dipropylthiodicarbocyanine iodide; ER, endoplasmic reticulum; IAA, indole-3-acetic acid; MP, membrane potential; 1-NAA, naphthalene-1-acetic acid; PLA2, phospholipase A2; PM, plasma membrane; SEM, standard error of the mean; TIR1, transport inhibitor resistant 1.

1. Introduction

One of the most frequent questions, which are faced by investigators of auxin effects is: How can this simple molecule trigger such a huge diversity of physiological reactions? During the last decades, an enormous amount of data on auxin signaling has accumulated. A big step in understanding the auxin mechanism of action was made after elucidation of the soluble hormone-receptor TIR1 and the closely related AFB proteins [1,2]. Nevertheless, not all auxin-induced reactions are mediated by activation of TIR1-signaling. A number of primary reactions, e.g., changes in membrane potential, cytosol acidification, elevation in cytosolic Ca^{2+} concentration, activation of PLA2, protoplast swelling, protein cycling, *etc.*, require another type of receptor, localized in the plasma membrane [3–7]. Nevertheless, the process of auxin sensing at cell surface is still under investigation. The auxin binding protein 1 (ABP1) is considered to be a part of plasma membrane (PM) receptor or closely linked to it [7–11]. The ABP1 appears to be a dimer of 22-kDa subunit, and can easily be solubilized by detergent or acetone from membranes [12–14]. A number of *ABP1* genes are known to encode the protein in different plants [15–19]. The ABP1 protein has a single N-glycosylation site, which binds a mannose type glycan [14,20,21]. Two conservative domains (Box A, responsible for auxin binding, and Box B) and an ER targeting marker *C*-terminal KDEL tetrapeptide were determined in the ABP1 structure [21–23]. The main ABP1 pool is localized in ER where it is supposed to be inactive. Only about 2% of the protein is secreted to extracellular space to fulfill its physiological function [24].

It was supposed that ABP1 might be coupled to a transmembrane docking protein [25,26]. Unfortunately, up to now little is known about the nature and function of this protein. Based on the knowledge of different animal hormone-receptor complexes at the PM, several models were developed for the auxin receptor. The first one suggested that a kinase could fulfill the role of a transmembrane domain [27]. Another model implicates G-protein-coupled receptor as the auxin one [28]. The third model assumes that ABP1 binds/interacts with a Ca^{2+}-permeable ion channel, which fulfills the role of a docking protein [29]. It is possible that ABP1 has more than one binding partner, differing in plant tissues and stage of development.

The absence of ABP1 or its reduction leads to significant changes: arrest of embryo development and elongation intensity [30,31]. Even heterozygous *ABP1/abp1* insertion mutants show a number of developmental disturbances confirmed by reduction of sensitivity to auxin and shift in the intensity of early auxin-regulated *Aux/IAAs* genes expression [32,33]. Decrease in ABP1 via antisense transformation leads to significant decrease in elongation intensity [31] and cell enlargement/protoplast swelling [34–36].

It was shown earlier that addition of exogenous ABP1 to a model system like protoplasts increased the amplitude of auxin-induced PM hyperpolarization [37]. Recently, a fast ABP1-related auxin-induced shift in the membrane potential (MP) was shown in a similar model system, by use of a sensitive fluorescent dye [38]. The advantage of the latter investigation was the ascertainment that the effect was triggered even by the *C*-terminal peptide of ABP1 and was blocked by antibodies against it. Overexpression of *ABP1* enhances the K^+-transport by activation of K^+-channels and quantity of their expression [39,40]. Thus, it could be concluded that ABP1 is an important modulator of cell sensitivity to the hormone at plasma membrane, but the mechanism of this regulation is still debated.

One of the fast and sensitive reactions triggered by auxin is an elevation of Ca^{2+} concentration in the cytosol. This reaction was estimated for different plant cells, including maize coleoptile parenchyma cells [9,41] Most probably it reflects the activation of plasma membrane channels, permeable for Ca^{2+} [9]. The coleoptile is a juvenile organ, the main function of which is to protect the first leaf at the initial stages of grass seedling development. Coleoptiles are very sensitive to auxin [42]. In maize coleoptiles, the native growth slows down tremendously from the 3rd to 5th day of seedling development [43]. The most intensive growth decrement appears at transition from the 3rd to 4th day of seedling development [44]. This phenomenon coincides with a loss of auxin-induced growth of coleoptile segments [43] and a significant decrease of auxin induced $[Ca^{2+}]_{cyt}$ elevation [44]. Thus, a possible reduction in cell sensitivity to the hormone is due to probable changes in auxin signal perception and early transduction. The current investigation focuses on the involvement of a plasma membrane Ca^{2+}-transport system in auxin signal perception under the control of ABP1.

2. Results and Discussion

The intensity of Ca^{2+} transport through vesicle membranes, obtained from maize coleoptiles of different ages was estimated as ΔMP, determined by a shift in fluorescence of diS-C_3-(5) dye, commonly used to test transmembrane potential not only in purified vesicles, but also at whole cell level, like protoplast or bacterial cell [45,46].

Our model system contained two types of vesicles: right-side-out, which copy the native cell orientation, and inside-out ones. Only Ca^{2+} ions had a gradient across the vesicle membrane (Figure 1a). Addition of IAA into the incubation medium led to a fast shift of dye fluorescence (Figure 1b), similar to our earlier results [47]. The detected shift in MP was due to Ca^{2+} efflux from the vesicles. We assume that right-side-out vesicles do not participate in ΔMP generation because transport of Ca^{2+} out of the cell is carried out by active systems like Ca^{2+}-ATPase and by the Ca^{2+}/proton antiporter systems (for review see [48]). Conditions for activation of these transporters were absent; therefore, the estimated ΔMP was due to flux of Ca^{2+} ions across membranes of inverted vesicles, which correspond to the flow directed into cell *in vivo*. The revealed IAA-induced shift in MP was similar to the effect obtained after addition of 1-NAA, an active synthetic auxin, but not after 2-NAA addition, a non-active synthetic analogue (Figure 1b).

Figure 1. Auxin-induced generation of the membrane potential in a model system represented by plasma membrane vesicles from maize coleoptile cells isolated at the 3rd day of seedling development. (**a**) Scheme of vesicles loading (inside-out orientation of vesicle), cyt—cytosolic side of vesicle, arrow—direction of Ca^{2+} flux; (**b**) Single traces of diS-C_3-(5) fluorescence shift after auxin addition, arrow—addition of auxin.

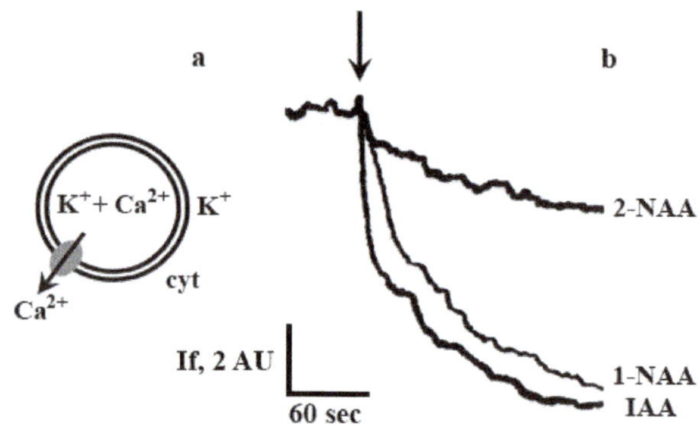

The fluorescence intensity of diS-C_3-(5) is calibrated mainly in a model system with gradients of K^+ and Na^+ ions. The maximal amplitude of dye fluorescence in our model system decreases to −110 mV of the K^+-diffusion potential as calculated by the Nernst equation when we added valinomycin into the Na^+ incubation medium containing K^+-loaded vesicles [47]. According to earlier results the IAA-induced change in dye fluorescence did not exceed −30 to 33 mV in case of membrane vesicles purified from coleoptiles at 4th day of seedling development. In the absence of Na^+, detected changes of the amplitude of the auxin-induced fluorescence signal are presented in Figure 2 in arbitrary units.

Figure 2. Auxin-induced generation of the membrane potential in a model system. Plasma membrane vesicles from maize coleoptile cells were isolated at the 3rd, 4th and 5th day of seedling development. Histograms represent mean values of the shift in diS-C_3-(5) intensity of fluorescence (If) in arbitrary units (AU) ± SEM.

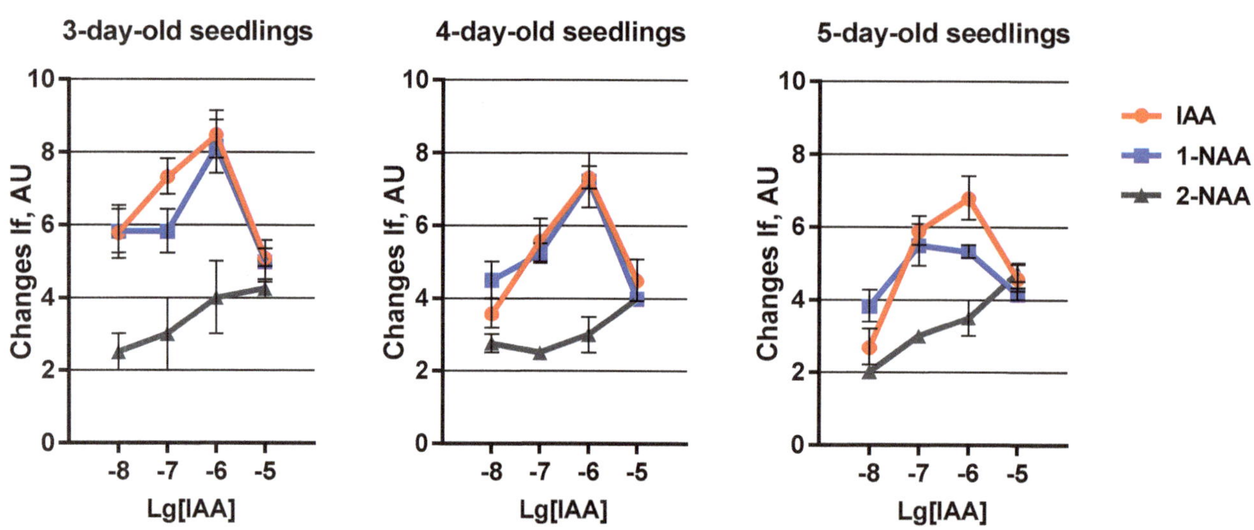

The distinct auxin concentration dependence of the fluorescence difference was determined. In our experiments, IAA and 1-NAA caused a maximum effect by addition of hormone at 10^{-6} M (Figure 2). Similar results were obtained with vesicles isolated from seedling of all tested ages. However, the amplitude of the fluorescence signal decreased in vesicles from older seedlings. Only in the youngest seedlings the effect of IAA was slightly higher than that of 1-NAA. The 2-NAA-induced shift of diS-C_3-(5) fluorescence increased slightly but almost linearly with concentration. The reaction did not exceed 4 arbitrary units (Figure 2). Moreover, no difference was obtained for 2-NAA during seedling ageing. Thus we conclude that this reaction is auxin-specific.

The obtained results show that physiologically active native and synthetic auxins trigger transport of Ca^{2+} through plasma membrane. The magnitude of this transport decreases within seedlings ageing. That coincides well with earlier results on protoplasts from maize-coleoptile cells, which showed a decrement of $[Ca^{2+}]_{cyt}$ elevation after auxin addition correlated with age of the coleoptile [44]. All this points out a possible decrease of auxin sensitivity at the plasma membrane and early transduction steps, like cytosolic Ca^{2+} elevation, over ageing [44]. If the idea is correct, that ABP1 is an important component of a hormone receptor at the plasma membrane then addition of this protein to the model system might restore the sensitivity for auxin lost within ageing.

Therefore, we modified the model system by loading of ABP1 (10^{-9} M) into the vesicles (Figure 3a). Addition of IAA 10^{-6} M to this system did not change the dynamics of fluorescent response and almost did not affect the amplitude in case of the youngest seedlings (Figure 3b).

Figure 3. Effect of auxin binding protein 1 (ABP1) on auxin-induced generation of the membrane potential in a model system represented by plasma membrane vesicles from maize coleoptile cells at the 3rd day of seedling development. (**a**) Scheme of vesicles loading (inside-out orientation of vesicle), cyt—cytosolic side of vesicle, arrow—direction of Ca^{2+} flux; (**b**) Single traces of diS-C_3-(5) fluorescence shift after auxin addition.

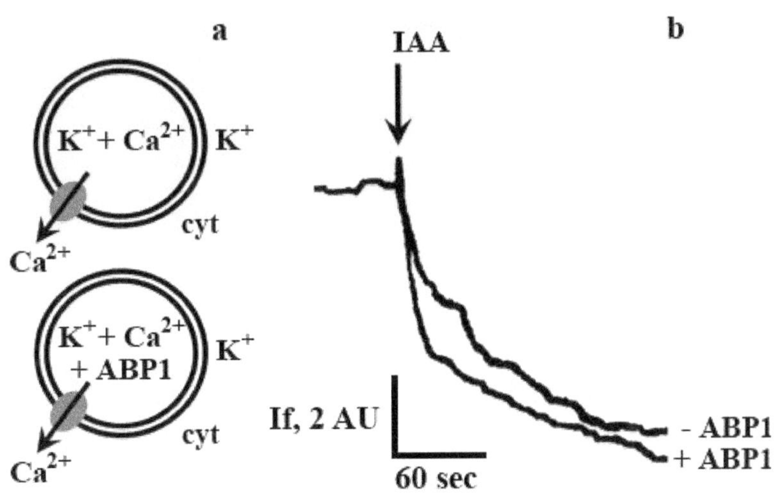

The presence of ABP1 inside the vesicles led to significant changes in concentration dependence of the response. IAA and 1-NAA-triggered Ca^{2+} efflux at the 3rd day of development did not increase the maximum amplitude but it reached a maximum sensitivity at 10^{-8} M (Figure 4). Thus, we found a significant increase in PM sensitivity to auxins.

Figure 4. The role of ABP1 in determination of auxin-induced generation of the membrane potential in a model system, represented by plasma membrane vesicles from maize coleoptile cells at the 3rd, 4th and 5th day of seedling development. Histograms represent mean values of the shift in diS-C_3-(5) intensity of fluorescence (If) in arbitrary units (AU) ± SEM.

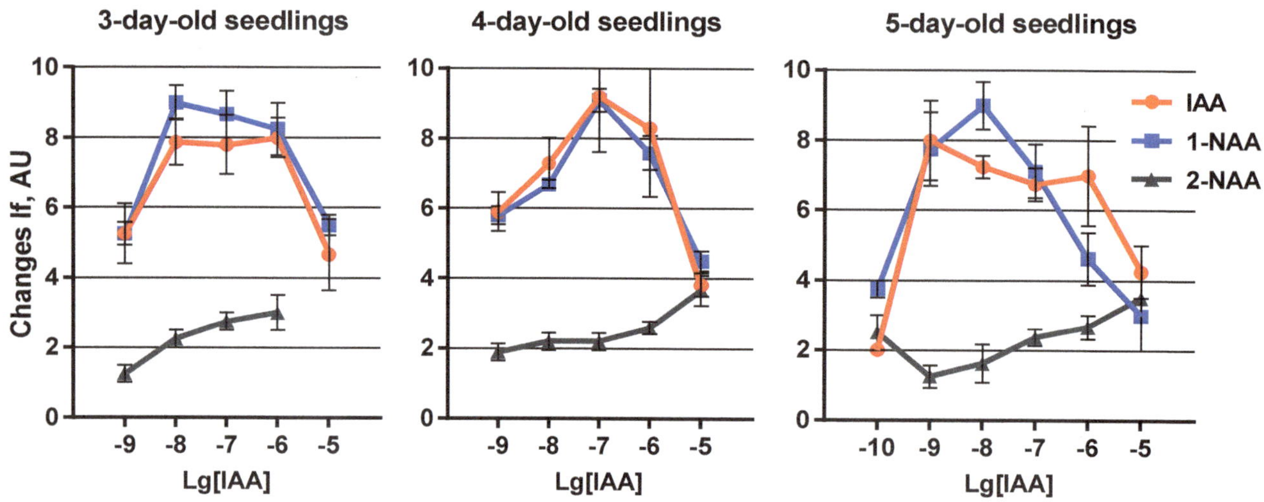

A change in auxin sensitivity was shown in older seedlings at 4th and especially at 5th day of seedling development (Figure 4). However, in none of the cases did ABP1 addition increase the maximum value of ΔMP generated in vesicles obtained at the 3rd day of seedling development, when the sensitivity to the hormone had its highest value. Thus, it might be concluded that ABP1 is a limiting factor, which determines the sensitivity of the plant cell at the cell surface. A developmental decrease in ABP1 concentration may occur with seedling ageing toward the end of coleoptile physiological function, and may be coincided with a slowdown of the Ca^{2+} transport through the plasma membrane. Supplementary ABP1 restored the amplitude of Ca^{2+} transport and, as we assume, increased sensitivity to the hormone.

Special attention was paid to ABP1 because recently more evidences indicate an important role of this protein in the perception of the auxin signal at the plasma membrane. In a number of electrophysiological investigations done on tobacco leaf protoplasts maize ABP1 and antibodies against it strongly affected sensitivity to auxin [26,27,49]. Addition of ABP1 to protoplasts from plants transformed with *rol* genes of *Agrobacterium rhizogenes* raised sensitivity to auxin 100- to 1000-fold. It was supposed that ABP1 might increase the number of active perception units at plasma membrane, which shift protoplast sensitivity [27]. Both stimulatory effects of ABP1 and inhibitory effects of antibodies to this protein were also found in another electrophysiological model system—the whole cell patch clamp [50]. ABP1 mediated an auxin-induced shift in cytosolic pH and a flux of K^+ [51,52]. Recent investigations showed that overexpression of ABP1 enhanced sensitivity of guard cells to auxin [40] and affected enlargement and division of plant cells [31,53,54].

Besides, the whole ABP1 protein, also its binding domain, as represented by the surface of an antibody D16 having auxin activity and the *C*-terminal peptide of ABP1 have physiological activity. D16 was able to trigger MP hyperpolarisation [55] and stimulation of the anion channel [37,56] in the absence of auxin. On the other hand, a synthetic peptide containing 12 residues of the *C*-terminus of ABP1, could mimic the action of high auxin concentrations in regulation of the K^+ current, MP value and cytosol alkalinization in guard cells [38,51,52]. Thus, the suggestion was that this peptide played an important role in auxin-ABP1 coupling to intracellular signal cascades [57].

All listed results indicate the importance of ABP1 at the plasma membrane level but does not reveal the mechanism of its action. Recent publications frequently assume that ABP1 participates in regulation of endocytosis [58,59]. Endomembrane trafficking is a process of great importance, which maintains the intracellular re-localization of macromolecules and membranes by secretion and endocytosis. In plant cells traffic of newly synthesized proteins, translocation to the endoplasmic reticulum (ER), and subsequent protein processing and targeting occur via vesicle trafficking through the secretory pathway. Vesicle secretion is a very important process in plant cell growth and signaling [60]. Several publications showed that during auxin-dependent formation of lateral roots, the endomembrane system plays an important role, establishing cellular localization and polarity of the auxin transporters [61–63]. Polarity of auxin transporters leads to the formation of auxin gradients determining both initiation of lateral root and the establishment of the primordium cell patterning [64,65]. According to the recently accepted opinion, ABP1 effects are due to ROP GTPase signaling [59]. Nevertheless, the mechanism of transition of auxin signal from extracellular ABP1 to ROP-signaling in the cytosol is still under investigation.

In this manuscript we present data on a fast auxin activation of Ca^{2+} transport through the plasma membrane. This effect is paralleled by the increase of cytosolic Ca^{2+} concentration found before [44]. The latter event might trigger another process, namely vesicle secretion. A fast initiation of exocytosis by elevation of $[Ca^{2+}]_{cyt}$ is well known for neurons and now is shown for plant cells [66–69]. Even if taking into consideration that exocytosis in animal cells is much more sensitive to cytosolic Ca^{2+} in comparison to cereal coleoptile protoplasts, the phenomenon might indicate common mechanisms of signaling. Analysis of electronic micrographs of oat coleoptile cells reveals a ~12% difference in number of vesicles which are correlated with alterations in growth capacity of the tested cells, preferentially in cells which start elongation [70]. We assume the secretory mechanism is the same in cells from the same tissue, whereas the pool size of vesicles may vary from cell to cell in identical cell types. Elevation of $[Ca^{2+}]$ induced by auxin is estimated to be in the range of approximate 100 nM after auxin treatment [71,72]. This is enough to activate the secretory system within 10–20 s after the increase in Ca^{2+} concentration [66].

The importance of exocytosis in the mechanism of auxin action nicely correlates with an earlier suggestion about hormone-induced increase in a number of H^+-ATPases in plasma membrane via vesicle secretion [73], which will lead along with post-transcriptional phosphorylation [74] to membrane hyperpolarization, shift in ion transport through plasmalemma, increase in acidification of cell wall and cell extension further on. The amplitude of Ca^{2+} rise might be considered as a threshold. Our data show that it depends on the amount of ABP1 and the external concentration of Ca^{2+} ([72], data presented here). Exocytosis is usually accompanied by recycling of vesicles. Electron micrograph data shows that about 60% of delivered vesicles to plasma membrane are recycled [75]. The intensity of recycling will depend on ROP GTPase signaling [76] and concentration of the cytosolic Ca^{2+} concentration [77]. Recently, ROP GTPase signaling was suggested to be linked to ABP1 [78] in support to our model (Figure 5). If cells already pass a step of elongation and have a loss or significant decrease in ABP1 perception facility then this could be another reason for the suggested scheme modulation.

Figure 5. Hypothetical scheme of auxin perception at the plasma membrane and primary signal transduction events in coleoptile cells at early stages of development. In physiologically young coleoptile cell auxin interacts with extracellular ABP1 and triggers Ca^{2+} transport inside the cell ([72], data presented here). Cytosolic Ca^{2+} elevation might initiate exocytosis [66]. This will result in the increase in H^+-ATPases protein in the plasma membrane [73]. Further Rho-like GTPase from plants (ROPs) signaling cascade, which is suggested to be linked to ABP1 [78], leads to endocytosis and redistribution of plasma membrane (PM) proteins, including auxin transporters PINs.

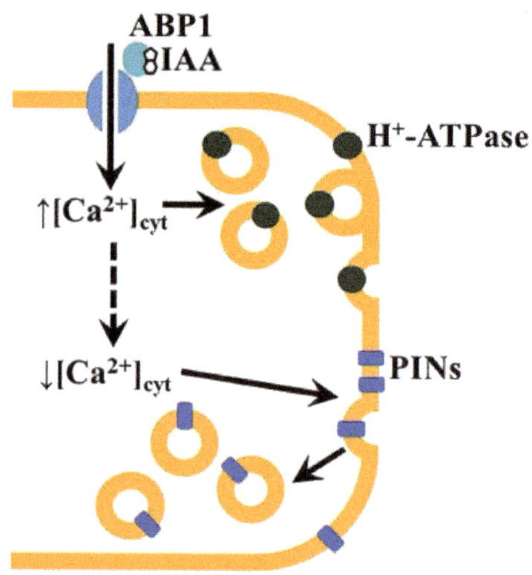

3. Experimental

3.1. Plant Material

Maize seedlings (*Zea mays* L., cv. Moldavsky-215) were grown in darkness at 26 ± 1 °C and 95% relative humidity. Seedlings were illuminated for 4 min every 24 h with a weak light in order to get straight coleoptiles and decrease mesocotyl growth. Seedlings were grown 48 h on wet filter paper and were thereafter transferred on glass tube raft on 1/10 Chesnokov nutrient solution. The ratios between different organs of the seedlings (leaf: coleoptile: mesocotyl: root) in cm were approximately 1.4:1.9:1.6:4.4; 2.3:4.2:2.9:9.1; 7.5:4.9:5.6:10.4, for 3-, 4- and 5- day-old seedlings, respectively.

3.2. Measurement of Calcium Transport Rate across Plasma Membrane Vesicles from Maize Coleoptile Cells

The fraction enriched with plasma membranes was isolated from the decapitated coleoptiles of 3, 4 and 5-day-old etiolated maize seedlings. The membrane fraction was prepared by differential centrifugation with subsequent purification in a sucrose density gradient [47]. The purified membranes were collected at the interface between sucrose layers of 1.13 and 1.17 g/cm^3 density. Inhibitor analysis showed no appreciable contamination with vacuolar membranes. The membrane fraction contained vesicles of right-side-out and inside-out orientation in the proportion 1:1, according to the enzymatic test with alamethicin.

Membrane vesicles were loaded by osmotic shock with a medium containing 150 mM K_2SO_4, 1 mM Tris-MES, 150 mM sucrose, pH 6.8 (K^+-medium) or with 1 mM $CaCl_2$ in addition (Ca^{2+} + K^+-medium).

ABP1 10^{-9} M, was added to the loading medium depending on the experiment scheme. After loading, the vesicles were concentrated by centrifugation (99,500× *g* for 1 h). The pellet was re-suspended in the K^+-medium.

The change in membrane potential (MP), generated by ion transport through the membrane vesicles due to Ca^{2+} ion gradients, was recorded by the change in fluorescence intensity of 3,3'-dipropylthiodicarbocyanine iodide (diS-C_3-(5)) fluorescent probe (Molecular Probes, Eugene, OR, USA) [79]. This dye is highly sensitive and can be used at a low concentration (0.8 μM). The fluorescence intensity is proportional to MP in a wide range. Fluorescence was measured with a spectrofluorometer constructed similar to MPF44-a Hitachi-Perkin-Elmer model (for details see [47]). Experiments were performed with a standard 1-cm quartz cuvette (volume of 0.3 mL). Excitation and emission wavelengths were 570 and 668 nm, respectively. Fluorescence was measured at an angle of 90 to the direction of excitation light. Aliquots of the dye solution and vesicle preparations were added to the cuvette with K^+-incubation medium. Auxins were added after mixing the membrane fraction with the incubation medium and reaching a steady level in MP.

Experiments were performed in 6–7 replicates in 3–4 independent assays. Figures are made with GraphPad Prism 6 and represent mean values and standard error of the mean (SEM) for $n = 4–8$.

4. Conclusions

In summary, we would like to suggest the following hypothetical scheme for auxin perception at the plasma membrane and further primary transduction (Figure 5). The sensitivity of coleoptile cells to auxin is determined by the amount of ABP1, which is postulated to vary during elongation growth. At high external Ca^{2+} concentration auxin induces a fast (10–20 s) elevation of the cytosolic concentration of this ion. In young cells, containing a vesicle pool, this may lead to activation of exocytosis. A further intensification of H^+-ATPase activity, a shift in membrane potential and magnitude of ion transport, cell wall acidification and initiation of elongation growth is supposed to occur. The cytosolic concentration of Ca^{2+} decreases rapidly and, in turn, auxin may induce a shift in endocytosis activity, which is followed by redistribution of PIN proteins and further formation of a new local auxin gradient.

Acknowledgments

Authors particularly want to thank R. Napier for a kind supply with ABP1 samples. Financial support from Russian Foundation of Basic Research (projects 13-04-00945-a and 12-04-01029-a) and Ministry of Education and Science of Russian Federation (agreement No. 8093 from 23.07.2012, project 2012-1.2.1-12-000-1013-003) is greatly appreciated.

Author Contributions

Anastasia A. Kirpichnikova performed the experiments, summarize a state of the art in the field and prepared a manuscript; Elena L. Rudashevskaya conducted the experiments, analyzed the data and prepared the manuscript; Vladislav V. Yemelyanov analyzed and discussed the data, prepared the manuscript, Maria F. Shishova designed the experiments, analyzed the data, prepared the manuscript and suggested the final scheme.

Conflicts of Interest

The authors declare no conflict of interest.

References

1. Dharmasiri, N.; Dharmasiri, S.; Estelle, M. The F-box protein TIR1 is an auxin receptor. *Nature* **2005**, *435*, 441–445.
2. Kepinski, S.; Leyser, O. The Arabidopsis F-box protein TIR1 is an auxin receptor. *Nature* **2005**, *435*, 446–451.
3. Badescu, G.O.; Napier, R.M. Receptors for auxin: Will it all end in TIRs? *Trends Plant Sci.* **2006**, *11*, 217–223.
4. Scherer, G.F.; Zahn, M.; Callis, J.; Jones, A.M. A role for phospholipase A in auxin-regulated gene expression. *FEBS Lett.* **2007**, *581*, 4205–4211.
5. Mockaitis, K.; Estelle, M. Auxin receptors and plant development: A new signaling paradigm. *Annu. Rev. Cell Dev. Biol.* **2008**, *24*, 55–80.
6. Vanneste, S.; Friml, J. Auxin: A trigger for change in plant development. *Cell* **2009**, *136*, 1005–1016.
7. Scherer, G.F. Auxin-binding-protein1, the second auxin receptor: What is the significance of a two-receptor concept in plant signal transduction? *J. Exp. Bot.* **2011**, *62*, 3339–3357.
8. Napier, R.M.V.; Venis, M.A. Tansley review No-79—Auxin action and auxin-binding proteins. *New Phytol.* **1995**, *129*, 167–201.
9. Shishova, M.; Lindberg, S. A new perspective on auxin perception. *J. Plant Physiol.* **2010**, *167*, 417–422.
10. Perrot-Rechenmann, C. Cellular responses to auxin: Division *versus* expansion. *Cold Spring Harb. Perspect. Biol.* **2010**, *2*, doi:10.1101/cshperspect.a001446.
11. Sauer, M.; Kleine-Vehn, J. Auxin binding protein1: The outsider. *Plant Cell* **2011**, *23*, 2033–2043.
12. Venis, M.A. Solubilisation and partial purification of auxin-binding sites of corn membranes. *Nature* **1977**, *266*, 268–269.
13. Shimomura, S.; Sotobayashi, T.; Futai, M.; Fukui, T. Purification and properties of an auxin-binding protein from maize shoot membranes. *J. Biochem.* **1986**, *99*, 1513–1524.
14. Napier, R.M.; Venis, M.A.; Bolton, M.A.; Richardson, L.I.; Butcher, D.W. Preparation and characterisation of monoclonal and polyclonal antibodies to maize membrane auxin-binding protein. *Planta* **1988**, *176*, 519–526.
15. Palme, K. Molecular analysis of plant signaling elements: Relevance of eukaryotic signal transduction models. *Int. Rev. Cytol.* **1992**, *132*, 223–283.
16. Shimomura, S.; Liu, W.; Inohara, N.; Watanabe, S.; Futai, M. Structure of the gene for an auxin-binding protein and a gene for 7SL RNA from *Arabidopsis thaliana*. *Plant Cell Physiol.* **1993**, *34*, 633–637.
17. Shimomura, S.; Watanabe, S.; Ichikawa, H. Characterization of auxin-binding protein 1 from tobacco: Content, localization and auxin-binding activity. *Planta* **1999**, *209*, 118–125.
18. Lazarus, C.M.; Macdonald, H. Characterization of a strawberry gene for auxin-binding protein, and its expression in insect cells. *Plant Mol. Biol.* **1996**, *31*, 267–277.

19. Watanabe, S.; Shimomura, S. Cloning and expression of two genes encoding auxin-binding proteins from tobacco. *Plant Mol. Biol.* **1998**, *36*, 63–74.
20. Löbler, M.S.K.; Hesse, T.; Klämbt, D. Auxin receptors in target tissues. In *Molecular Biology of Plant Growth Control*; Fox, J.E.J.M., Alan, R., Eds.; Liss: New York, NY, USA, 1987; pp. 279–288.
21. Hesse, T.; Feldwisch, J.; Balshusemann, D.; Bauw, G.; Puype, M.; Vandekerckhove, J.; Lobler, M.; Klambt, D.; Schell, J.; Palme, K. Molecular cloning and structural analysis of a gene from *Zea mays* (L.) coding for a putative receptor for the plant hormone auxin. *EMBO J.* **1989**, *8*, 2453–2461.
22. Pelham, H.R. Control of protein exit from the endoplasmic reticulum. *Annu. Rev. Cell Biol.* **1989**, *5*, 1–23.
23. Jones, A.M. Auxin-binding proteins. *Annu. Rev. Plant Physiol. Plant Mol. Biol.* **1994**, *45*, 393–420.
24. Diekmann, W.; Venis, M.A.; Robinson, D.G. Auxins induce clustering of the auxin-binding protein at the surface of maize coleoptile protoplasts. *Proc. Natl. Acad. Sci. USA* **1995**, *92*, 3425–3429.
25. Klambt, D. A view about the function of auxin-binding proteins at plasma membranes. *Plant Mol. Biol.* **1990**, *14*, 1045–1050.
26. Barbier-Brygoo, H.; Ephritikhine, G.; Klambt, D.; Maurel, C.; Palme, K.; Schell, J.; Guern, J. Perception of the auxin signal at the plasma membrane of tobacco mesophyll protoplasts. *Plant J.* **1991**, *1*, 83–93.
27. Barbier-Brygoo, H. Tracking auxin receptors using functional approaches. *Crit. Rev. Plant Sci.* **1995**, *14*, 1–25.
28. MacDonald, H. Auxin perception and signal transduction. *Physiol. Plant* **1997**, *100*, 423–430.
29. Shishova, M. Membrane Mechanism of Auxin Action on Plant Cell. Ph.D. Thesis, St. Petersburg University, Moscow, Russia, 1999.
30. Chen, J.G.; Shimomura, S.; Sitbon, F.; Sandberg, G.; Jones, A.M. The role of auxin-binding protein 1 in the expansion of tobacco leaf cells. *Plant J.* **2001**, *28*, 607–617.
31. Chen, J.-G.; Ullah, H.; Young, J.C.; Sussman, M.R.; Jones, A.M. ABP1 is required for organized cell elongation and division in *Arabidopsis* embryogenesis. *Genes Dev.* **2001**, *15*, 902–911.
32. Effendi, Y.; Rietz, S.; Fischer, U.; Scherer, G.F. The heterozygous *abp1/ABP1* insertional mutant has defects in functions requiring polar auxin transport and in regulation of early auxin-regulated genes. *Plant J.* **2011**, *65*, 282–294.
33. Effendi, Y.; Scherer, G.F. Auxin binding-protein1 (ABP1), a receptor to regulate auxin transport and early auxin genes in an interlocking system with PIN proteins and the receptor TIR1. *Plant Signal. Behav.* **2011**, *6*, 1101–1103.
34. Steffens, B.; Feckler, C.; Palme, K.; Christian, M.; Bottger, M.; Luthen, H. The auxin signal for protoplast swelling is perceived by extracellular ABP1. *Plant J.* **2001**, *27*, 591–599.
35. Christian, M.S.B.; Schenck, D.; Burmester, S.; Böttger, M.; Lüthen, H. How does auxin enhance cell elongation? Roles of auxin-binding proteins and potassium channels in growth control. *Plant Biol.* **2006**, *8*, 346–352.
36. Yamagami, M.; Haga, K.; Napier, R.M.; Iino, M. Two distinct signaling pathways participate in auxin-induced swelling of pea epidermal protoplasts. *Plant Physiol.* **2004**, *134*, 735–747.

37. Barbier-Brygoo, H.; Zimmermann, S.; Thomine, S.; White, I.R.; Millner, P.; Guern, J. Elementary response chains at the plasma membrane involve external ABP1 and multiple electrogenic ion transport proteins. *Plant Growth Regul.* **1996**, *18*, 23–28.
38. Dahlke, R.I.; Luethen, H.; Steffens, B. ABP1: An auxin receptor for fast responses at the plasma membrane. *Plant Signal. Behav.* **2010**, *5*, 1–3.
39. Philippar, K.; Fuchs, I.; Luthen, H.; Hoth, S.; Bauer, C.S.; Haga, K.; Thiel, G.; Ljung, K.; Sandberg, G.; Bottger, M.; *et al.* Auxin-induced K^+ channel expression represents an essential step in coleoptile growth and gravitropism. *Proc. Natl. Acad. Sci. USA* **1999**, *96*, 12186–12191.
40. Bauly, J.M.; Sealy, I.M.; Macdonald, H.; Brearley, J.; Droge, S.; Hillmer, S.; Robinson, D.G.; Venis, M.A.; Blatt, M.R.; Lazarus, C.M.; *et al.* Overexpression of auxin-binding protein enhances the sensitivity of guard cells to auxin. *Plant Physiol.* **2000**, *124*, 1229–1238.
41. Monshausen, G.D.; Miller, N.D.; Murphy, A.S.; Gilroy, S. Dynamics of auxin-dependent Ca^{2+} and pH signaling in root growth revealed by integrating high-resolution imaging with automated computer vision-based analysis. *Plant J.* **2011**, *65*, 309–318.
42. Cleland, R.E. The outer epidermis of *Avena* and maize coleoptiles is not a unique target for auxin in elongation growth. *Planta* **1991**, *186*, 75–80.
43. Rudashevskaya, E.L.; Yemelyanov, V.V.; Kirpichnikova, A.A.; Burova, E.A.; Bobinova, O.A.; Shishova, M.F. The dependence of age changes in maize coleoptile and mesocotyl growth activity on indole-3-acetic acid content. *Bull. St. Petersburg Univ.* **2002**, *3*, 99–106.
44. Shishova, M.; Yemelyanov, V.; Rudashevskaya, E.; Lindberg, S. A shift in sensitivity to auxin within development of maize seedlings. *J. Plant Physiol.* **2007**, *164*, 1323–1330.
45. Oka, K.; Naitou, S.; Yoshida, M.; Ishikawa, H.; Ohta, E.; Sakata, M. Membrane potential measurement of protoplasdts isolated from *Vigna mungo* hypocotil using a fluorescent probe, diS-C_3-(5). *Plant Cell Physiol.* **1987**, *28*, 843–849.
46. Suzuki, H.; Wang, Z.Y.; Yamakoshi, M.; Kobayashi, M.; Nozawa, T. Probing the transmembrane potential of bacterial cells by voltage-sensitive dyes. *Anal. Sci.* **2003**, *19*, 1239–1242.
47. Shishova, M.F.; Inge-Vechtomova, N.I.; Vykhvalov, K.A.; Rudashevskaya, E.L.; Polevoi, V.V. Auxin-dependent transport of K^+ and Ca^{2+} across the membrane of plasmalemma vesicles from coleoptile cells. *Russ. J. Plant Physiol.* **1998**, *45*, 67–73.
48. Kudla, J.; Batistic, O.; Hashimoto, K. Calcium signals: The lead currency of plant information processing. *Plant Cell* **2010**, *22*, 541–563.
49. Barbier-Brygoo, H.; Ephritikhine, G.; Klambt, D.; Ghislain, M.; Guern, J. Functional evidence for an auxin receptor at the plasmalemma of tobacco mesophyll protoplasts. *Proc. Natl. Acad. Sci. USA* **1989**, *86*, 891–895.
50. Rück, A.P.K.; Venis, M.A.; Napier, R.M.; Felle, H. Patch-clamp analysis establishes a role for an auxin-binding protein in the auxin stimulation of plasma membrane current in *Zea mays* protoplasts. *Plant J.* **1993**, *4*, 41–46.
51. Thiel, G.; Blatt, M.R.; Fricker, M.D.; White, I.R.; Millner, P. Modulation of K^+ channels in *Vicia* stomatal guard cells by peptide homologs to the auxin-binding protein *C*-terminus. *Proc. Natl. Acad. Sci. USA* **1993**, *90*, 11493–11497.
52. Gehring, C.A.; McConchie, R.M.; Venis, M.A.; Parish, R.W. Auxin-binding-protein antibodies and peptides influence stomatal opening and alter cytoplasmic pH. *Planta* **1998**, *205*, 581–586.

53. Fellner, M.E.G.; Barbier-Brygoo, H.; Guern, J. An antibody raised to a maize auxin-binding protein has inhibitory effects on cell division of tobacco mesophyll protoplasts. *Plant Physiol. Biochem.* **1996**, *34*, 133–138.
54. Jones, A.M.; Im, K.H.; Savka, M.A.; Wu, M.J.; DeWitt, N.G.; Shillito, R.; Binns, A.N. Auxin-dependent cell expansion mediated by overexpressed auxin-binding protein 1. *Science* **1998**, *282*, 1114–1117.
55. Venis, M.A.; Napier, R.M.; Barbier-Brygoo, H.; Maurel, C.; Perrot-Rechenmann, C.; Guern, J. Antibodies to a peptide from the maize auxin-binding protein have auxin agonist activity. *Proc. Natl. Acad. Sci. USA* **1992**, *89*, 7208–7212.
56. Zimmermann, S.; Thomine, S.; Guern, J.; Barbier-Brygoo, H. An anion current at the plasma membrane of tobacco protoplasts shows ATP-dependent voltage regulation and is modulated by auxin. *Plant J.* **1994**, *6*, 707–716.
57. Leblanc, N.; Perrot-Rechenmann, C.; Barbier-Brygoo, H. The auxin-binding protein Nt-ERabp1 alone activates an auxin-like transduction pathway. *FEBS Lett.* **1999**, *449*, 57–60.
58. Robert, S.; Kleine-Vehn, J.; Barbez, E.; Sauer, M.; Paciorek, T.; Baster, P.; Vanneste, S.; Zhang, J.; Simon, S.; Covanova, M.; *et al.* ABP1 mediates auxin inhibition of clathrin-dependent endocytosis in *Arabidopsis*. *Cell* **2010**, *143*, 111–121.
59. Murphy, A.S.; Peer, W.A. Vesicle trafficking: ROP-RIC roundabout. *Curr. Boil.* **2012**, *22*, R576–R578.
60. Bassham, D.C.; Blatt, M.R. SNAREs: Cogs and coordinators in signaling and development. *Plant Physiol.* **2008**, *147*, 1504–1515.
61. Geldner, N.; Friml, J.; Stierhof, Y.D.; Jurgens, G.; Palme, K. Auxin transport inhibitors block PIN1 cycling and vesicle trafficking. *Nature* **2001**, *413*, 425–428.
62. Dhonukshe, P.; Aniento, F.; Hwang, I.; Robinson, D.G.; Mravec, J.; Stierhof, Y.D.; Friml, J. Clathrin-mediated constitutive endocytosis of PIN auxin efflux carriers in *Arabidopsis*. *Curr. Boil.* **2007**, *17*, 520–527.
63. Friml, J. Subcellular trafficking of PIN auxin efflux carriers in auxin transport. *Eur. J. Cell Boil.* **2010**, *89*, 231–235.
64. Benkova, E.; Michniewicz, M.; Sauer, M.; Teichmann, T.; Seifertova, D.; Jurgens, G.; Friml, J. Local, efflux-dependent auxin gradients as a common module for plant organ formation. *Cell* **2003**, *115*, 591–602.
65. Perez-Henriquez, P.; Raikhel, N.V.; Norambuena, L. Endocytic trafficking towards the vacuole plays a key role in the auxin receptor SCF(TIR)-independent mechanism of lateral root formation in *A. thaliana*. *Mol. Plant* **2012**, *5*, 1195–1209.
66. Sutter, J.U.; Homann, U.; Thiel, G. Ca^{2+}-stimulated exocytosis in maize coleoptile cells. *Plant Cell* **2000**, *12*, 1127–1136.
67. Sutter, J.U.; Denecke, J.; Thiel, G. Synthesis of vesicle cargo determines amplitude of Ca^{2+}-sensitive exocytosis. *Cell Calcium* **2012**, *52*, 283–288.
68. Thiel, G.; Sutter, J.U.; Homann, U. Ca^{2+}-sensitive and Ca^{2+}-insensitive exocytosis in maize coleoptile protoplasts. *Pflug. Archiv. Eur. J. Physiol.* **2000**, *439*, R152–R153.
69. Campanoni, P.; Blatt, M.R. Membrane trafficking and polar growth in root hairs and pollen tubes. *J. Exp. Bot.* **2007**, *58*, 65–74.

70. Quaite, E.; Parker, R.E.; Steer, M.W. Plant cell extension: Structural implications for the origin of the plasma membrane. *Plant Cell Environ.* **1983**, *6*, 429–432.
71. Felle, H. Auxin causes oscillations of cytosolic free calcium and pH in *Zea mays* coleoptiles. *Planta* **1988**, *174*, 495–499.
72. Shishova, M.; Lindberg, S. Auxin induces an increase of Ca^{2+} concentration in the cytosol of wheat leaf protoplasts. *J. Plant Physiol.* **2004**, *161*, 937–945.
73. Hager, A.; Debus, G.; Edel, H.G.; Stransky, H.; Serrano, R. Auxin induces exocytosis and the rapid synthesis of a high-turnover pool of plasma-membrane H^+-ATPase. *Planta* **1991**, *185*, 527–537.
74. Takahashi, K.; Hayashi, K.; Kinoshita, T. Auxin activates the plasma membrane H^+-ATPase by phosphorylation during hypocotyl elongation in *Arabidopsis*. *Plant Physiol.* **2012**, *159*, 632–641.
75. Phillips, G.D.; Preshaw, C.; Steer, M.W. Dictyosome vesicle production and plasma membrane turnover in auxin-stimulated outer epidermal cells of coleoptile segments from *Avena sativa* (L.). *Protoplasma* **1988**, *145*, 59–65.
76. Yalovsky, S.; Bloch, D.; Sorek, N.; Kost, B. Regulation of membrane trafficking, cytoskeleton dynamics, and cell polarity by ROP/RAC GTPases. *Plant Physiol.* **2008**, *147*, 1527–1543.
77. Hwang, J.U.; Gu, Y.; Lee, Y.J.; Yang, Z. Oscillatory ROP GTPase activation leads the oscillatory polarized growth of pollen tubes. *Mol. Biol. Cell* **2005**, *16*, 5385–5399.
78. Xu, T.; When, M.; Nagawa, S.; Fu, Y.; Chen, J.-G.; Wu, M.-J.; Perrot-Rechenmann, K.; Friml, J.; Jones, A.; Yang, Z. Cell surface- and Rho GTPase-based auxin signaling controls cellular interdigitation in *Arabidopsis*. *Cell* **2011**, *143*, 99–110.
79. Ivkova, M.N.; Pechatnikov, V.A.; Ivkov, V.G.; Pletnev, V.V. Mechanism of the fluorescent response of carbocyanine probe diS-C_3-(5) to membrane potential change. *Biofizika* **1983**, *28*, 160–170.

9

Glycoside Hydrolase Activities in Cell Walls of Sclerenchyma Cells in the Inflorescence Stems of *Arabidopsis thaliana* Visualized *in Situ*

Alicja Banasiak [1], Farid M. Ibatullin [2,3], Harry Brumer [2,4,5] and Ewa J. Mellerowicz [6,*]

1 Institute of Experimental Biology, University of Wroclaw, 50-328 Wroclaw, Poland; E-Mail: alicja.banasiak@uni.wroc.pl

2 Division of Glycoscience, School of Biotechnology, Royal Institute of Technology (KTH), AlbaNova University Centre, 106 91 Stockholm, Sweden; E-Mails: ibatullin@omrb.pnpi.spb.ru (F.M.I.); brumer@msl.ubc.ca (H.B.)

3 Biophysics Division, Petersburg Nuclear Physics Institute, National Research Center Kurchatov Institute, Gatchina 188300, Russia

4 Michael Smith Laboratories, University of British Columbia, 2185 East Mall, Vancouver, BC V6T 1Z1, Canada

5 Department of Chemistry, University of British Columbia, 2185 East Mall, Vancouver, BC V6T 1Z1, Canada

6 Department of Forest Genetics and Plant Physiology, Swedish University of Agricultural Sciences, Umea Plant Science Centre, 90183 Umea, Sweden

* Author to whom correspondence should be addressed; E-Mail: ewa.mellerowicz@slu.se

External Editor: Philip Harris

Abstract: Techniques for *in situ* localization of gene products provide indispensable information for understanding biological function. In the case of enzymes, biological function is directly related to activity, and therefore, knowledge of activity patterns is central to understanding the molecular controls of plant development. We have previously developed a novel type of fluorogenic substrate for revealing glycoside hydrolase activity *in planta*, based on resorufin β-glycosides Here, we explore a wider range of such substrates to visualize glycoside hydrolase activities in *Arabidopsis* inflorescence stems in real time, especially highlighting distinct distribution patterns of these activities in the secondary cell walls of sclerenchyma cells. The results demonstrate that β-1,4-glucosidase, β-1,4-glucanase and β-1,4-galactosidase activities accompany secondary wall deposition. In contrast, xyloglucanase activity follows a different pattern, with the highest signal observed in mature cells, concentrated in the middle lamella. These data further the understanding of the process of cell wall deposition and function in sclerenchymatic tissues of plants.

Keywords: cell wall; xylem; wood; sclerenchyma; glycoside hydrolase activity; *in situ* activity

1. Introduction

Knowledge of the precise localization of specific enzymatic activities in distinct cell types and cell compartments of plants is a prerequisite for understanding the biological functions of these enzymes and their encoding genes [1]. Plant genomes are renowned for encoding an abundance of glycoside hydrolases and transglycosidases, which probably reflects the need for cell wall remodeling [2–7]. Most work on cell wall remodeling by these enzymes has focused on changes during primary cell wall expansion [8–12] and fruit ripening [13]. However, there is an increasing awareness of the importance of secondary cell wall remodeling for plant development by enzymes, such as xyloglucan *endo*-transglycosylase (XET) [14–16] and xylan *endo*-transglycosylase [17]. For example, xyloglucan *endo*-transglycosylase activity in gelatinous fibers has been found to play an essential role in the tension wood function and was proposed to be involved in the generation of contractile forces by gelatinous fibers [15,18–20].

The abundance of different classes of glycoside hydrolases in secondary cell walls has become evident from proteomic analyses [21,22]. Here, we present *in situ* activity tests using a range of fluorogenic substrates to determine the tissue and cell distribution of major classes of cell wall-active enzymes in mature inflorescence stems of *Arabidopsis*, including β-1,4-glucosidases, β-1,4-glucanases, β-1,4-galactosidases and xyloglucanases. Most of these activities were associated with cell wall deposition in sclerenchyma cells, but were notably also found in the walls of mature cells. These results indicate that sclerenchyma cell wall remodeling is tightly associated with cell wall biosynthesis and that it may continue *post mortem*.

2. Results

2.1. Biosynthesis of Different Fluorogenic Substrates

A set of resorufin β-glycosides (Figure 1), serving as substrates for different glycoside hydrolases, was synthesized according to previously published protocols [1]. The release of the resorufin aglycone (pK_a 5.8) by glycoside hydrolase activity results in a significant increase in fluorescence due to the resorufin anion (extinction: 62,000 $m^{-1}\cdot cm^{-1}$; excitation maximum: 571 nm; emission maximum: 585 nm; quantum yield: 0.74). Five different resorufin glycosides were synthesized: (1) glucose-β-resorufin (Glc-β-Res, substrate for β-1,4-glucosidases); (2) cellobiose-β-resorufin (Glc-Glc-β-Res, substrate for β-1,4-glucanases [23], or β-1,4-glucosidases, via sequential cleavage); (3) thio-cellobiosyl-β-resorufin (Glc-*S*-Glc-β-Res, substrate for β-1,4-glucanases with a non-hydrolyzable internal glycosidic bond); (4) galactose-β-resorufin (Gal-β-Res, substrate for β-1,4-galactosidases); and (5) XXXG-β-resorufin

(XXXG-β-Res, heptasaccharide-containing substrate for xyloglucan *endo*-hydrolases [1]; see Tuomivaara *et al.* [24] for xyloglucan oligosaccharide nomenclature).

Figure 1. Structures of the fluorogenic substrates used in this study (Glc-β-Res; Glc-Glc-β-Res; Glc-*S*-Glc-β-Res; Gal-β-Res; XXXG-β-Res).

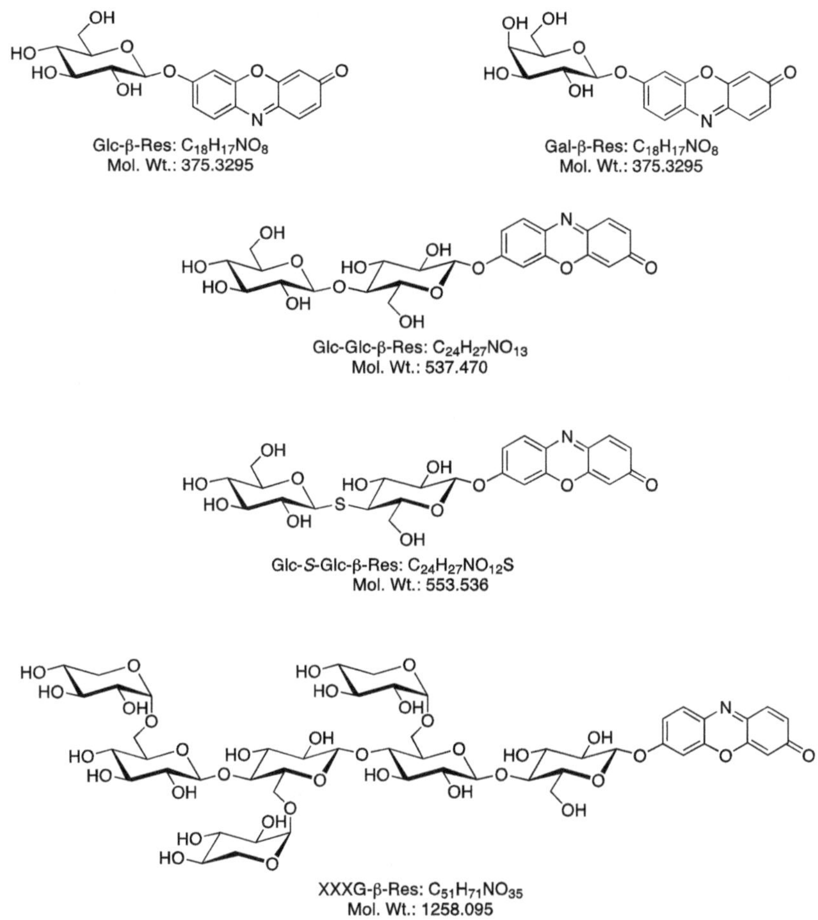

2.2. Optimization of in Situ Reaction Conditions for the Different Fluorogenic Substrates

Measurements of enzyme activity in real time were carried out on cross-sections through the basal part of *Arabidopsis thaliana* inflorescence stems. Previous *in vitro* studies using resorufin glycosides have established conditions when the fluorescence signal due to resorufin release depended linearly on time and enzyme concentration [1]. For the semi-quantitative analysis using confocal microscopy performed here, it was also necessary to determine the parameters under *in situ* conditions to ensure that signal saturation was avoided.

The optimal incubation time of tissue with the substrate needed for the enzymatic release of resorufin was first established based on the observation of time series. For this purpose, the substrates were dissolved in 25 mM MES buffer at pH 6.5 to a final concentration of 4.7×10^{-4} M and incubated with fresh hand sections of inflorescence stems for 1.5 h with periodic scanning by the confocal microscope. The signal from each substrate first increased with time, then reached a plateau or slightly decreased, possibly due to diffusion of the resorufin anion, or self-quenching or bleaching of the fluorophore. The time after which the fluorescence intensity was the strongest was different for each substrate, but the substrates could be split into fast and slow groups. The fast group included Glc-β-Res, Glc-Glc-β-Res and

Gal-β-Res, which gave clear signals already after 2 min, with signal saturation occurring after about 20 min. The slow group included Glc-*S*-Glc-β-Res and XXXG-β-Res, which required approximately 10 min to observe the first apparent signals, with the maximum signal typically arising after 60–90 min. These differences in the reaction times reflect differences in the abundance of the different enzymatic activities in inflorescence stem tissues. Thus, reaction times of either 20 or 60 min were chosen, one factor in the optimized reaction conditions for the fast- and slow-consumed substrates, respectively (Table 1).

Table 1. Optimal reaction conditions (reaction times, substrate concentrations) for the signal strength and dependence of the signal strength on pH for the different glycoside resorufin substrates in *Arabidopsis* stem sections.

Reaction Times	Substrates	Concentration (M)	pH				
			5.0	5.5	6.0	6.5	7.0
20 min	Glc-β-Res	9.3×10^{-5}	++	++	+++	+++	++
	Glc-Glc-β-Res	9.3×10^{-5}	++	++	+++	+++	++
	Gal-β-Res	4.7×10^{-5}	++	+++	+++	++++	++
60 min	Glc-*S*-Glc-β-Res	9.3×10^{-4}	-	-	+	+	++
	XXXG-β-Res	9.3×10^{-4}	-	-	+	++	+

Free resorufin has a pK_a value of 5.8, and the observable fluorescence is due to the resorufin anion (presenting a 50% abundance at pH 5.8), while many glycoside hydrolases exhibit acidic pH-rate optima. Therefore, five different pH values, between pH 5.0 and pH 7.0, were tested for each substrate (Table 1). The highest signal was typically observed when the pH was slightly acidic (6.0–6.5), likely reflecting a compromise between resorufin ionization and enzyme activity, with the exception of Glc-*S*-Glc-β-Res, which had the strongest signal at pH 7.0.

A third factor that is important for the precise location of a particular enzymatic activity is the substrate concentration. Four different concentrations were tested for each of the substrates at optimal pH with the time series: 4.7×10^{-5} M, 9.3×10^{-5} M, 4.7×10^{-4} M and 9.3×10^{-4} M, and the "signal-to noise" was visually evaluated. The lowest concentrations of acceptable signals are listed in Table 1. Substrates with long reaction times required the highest concentrations, likely to ensure enzyme saturation. In the case of Gal-β-Res, the fluorescence was very intense, and the diffusion of released resorufin was observed; therefore, the lowest of the concentrations tested, 4.7×10^{-5} M, was indicated for this substrate.

2.3. In Situ Enzymatic Activity Distribution Observed in Real Time at the Tissue Level in Inflorescence Stems of Arabidopsis thaliana

The fluorescence signal reflecting glycoside hydrolase activity using all five substrates was detected primarily in the cell walls of interfascicular fibers and xylem cells, *i.e.*, in a cylinder of sclerenchyma tissues (Figures 2 and 3). Some intracellular signals, associated with plastids, were also observed in the cortical tissues. However, this signal was observed without any substrate (Supplemental Material 1), and all of these signals were absent in heat-treated sections (Figures 2 and 3), which suggests that they were dependent on heat-labile compounds in plant tissues. Unlike the cell wall-localized signals, the intracellular signals in the cortex did not display the time-dependent increase expected for enzymatically-released resorufin. We therefore interpret these intracellular signals as resulting from autofluorescence of heat-labile pigments and discounted them from further analysis.

Figure 2. Distribution of signals in the inflorescence stem sections of *Arabidopsis* from Glc-β-Res, Glc-Glc-β-Res and Gal-β-Res substrates after incubations for the specified durations. The control sections were boiled for 30 min before incubation with the specified substrate for 20 min. Upper rows represent the transmitted light channel and the superimposed yellow fluorescence channel corresponding to resorufin signals, and the lower rows represent the yellow channel only. Blue arrows show differentiating sclerenchyma cells. if, interfascicular fibers; x, xylem. Bar = 50 μm.

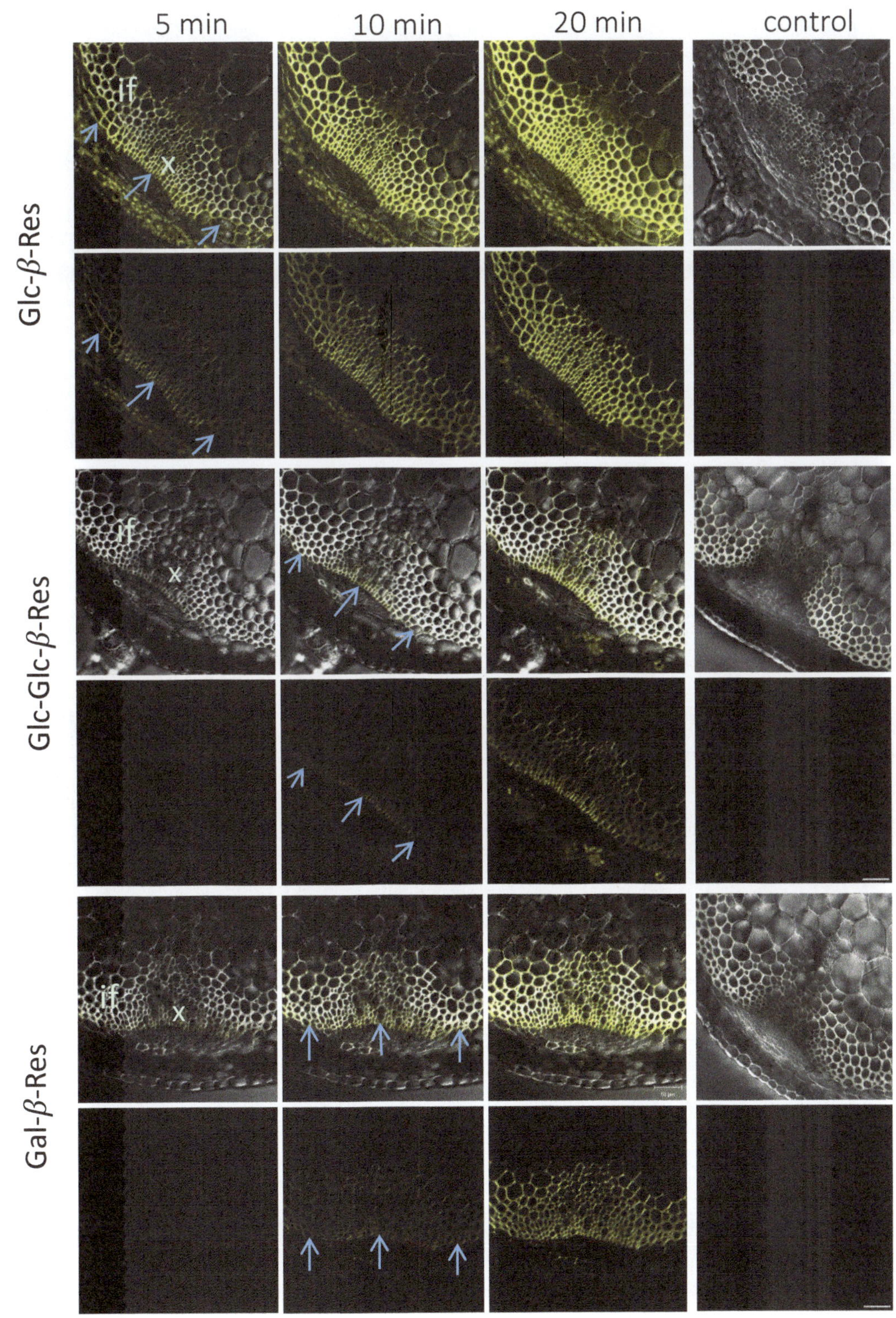

Figure 3. Distribution of signals in the inflorescence stem sections of *Arabidopsis* from Glc-*S*-Glc-β-Res and XXXG-β-Res substrates after incubations for the specified durations. The controls were boiled for 30 min before incubation for 60 min. Upper rows represent the transmitted light channel and the superimposed yellow fluorescence channel corresponding to resorufin signals, and the lower rows represent the yellow channel only. Blue arrows show differentiating sclerenchyma cells. if, interfascicular fibers; x, xylem. Bar = 50 μm.

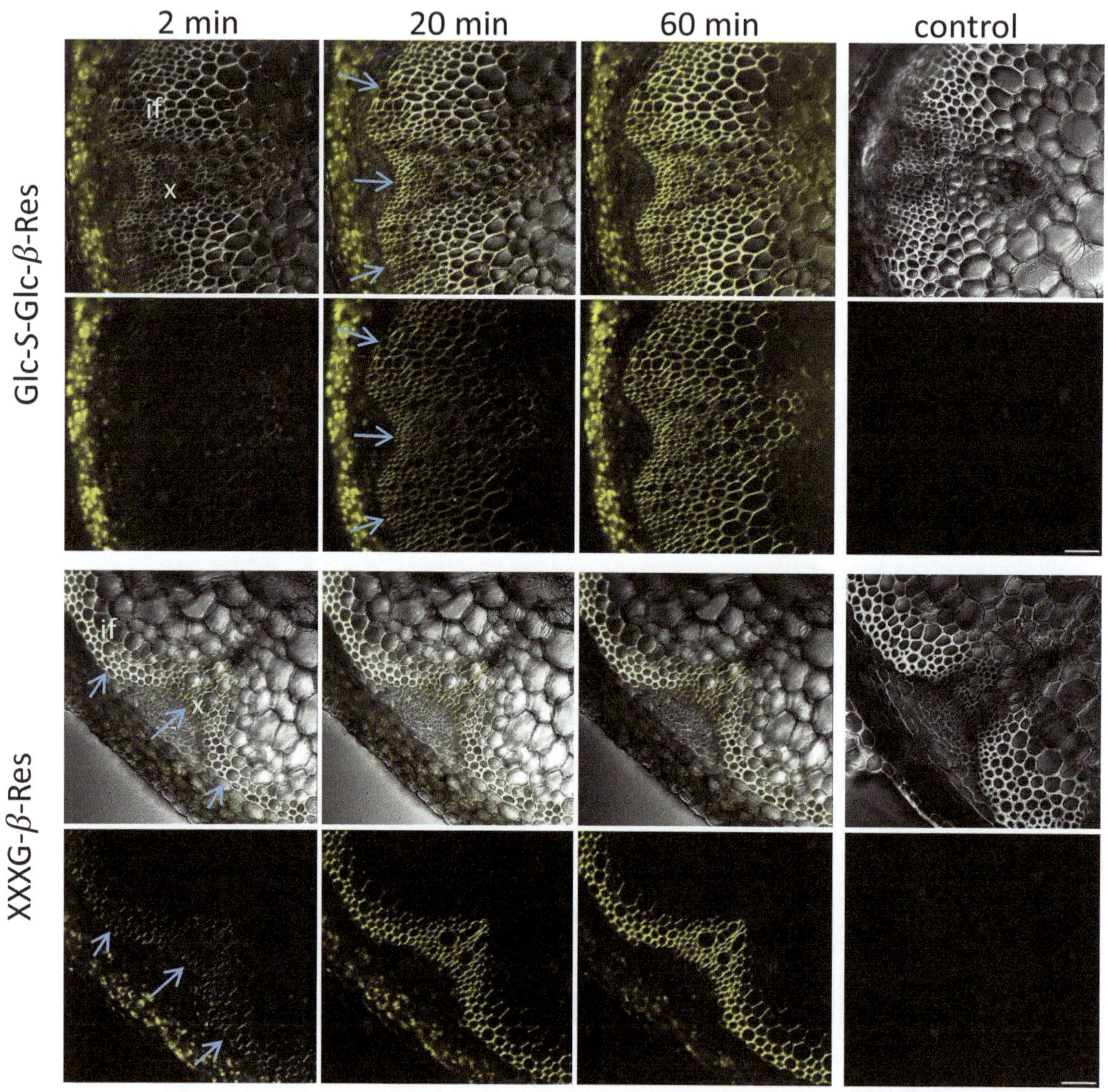

The signal evolution with Glc-β-Res, Glc-Glc-β-Res, Glc-*S*-Glc-β-Res and Gal-β-Res in stem sections was similar. Signals were first detected in the outer cylinder of sclerenchyma tissues, corresponding to the youngest cells developing a secondary cell wall (blue arrows in Figures 2 and 3). With ongoing incubation, the signal became more intense and gradually appeared throughout all of the cells of the sclerenchyma. This labeling pattern was distinct from that observed with XXXG-β-Res, which produced a signal only in the mature sclerenchyma cells and had an unchanged distribution pattern throughout the whole incubation time (Figure 3). The difference in signal resulting from XXXG-β-Res and the other substrates can be also distinctly seen in time-lapse movies (Supplemental Materials 2–6).

2.4. In Situ Enzymatic Activity Observed in Real Time on the Cell Level in Interfascicular Fibers of Inflorescence Stems in Arabidopsis thaliana

The distribution of signals reflecting the different glycoside hydrolytic activities was additionally analyzed at the cellular level in the interfascicular fibers of the *Arabidopsis* inflorescence stems. For all of the substrates, activity was observed in several cell wall layers (Figure 4), with prominent signals observed in the compound middle lamella (white arrows) and in the innermost, freshly deposited cell wall layer (red arrows). The relative intensity between these two positions varied among the substrates. Glc-Glc-β-Res and Glc-*S*-Glc-β-Res gave the strongest signals in the innermost layer of secondary wall, whereas the signal from XXXG-β-Res was most prominent in the compound middle lamella. The signals after Gal-β-Res and Glc-β-Res application were intermediate between these two divergent patterns. These results highlight that resorufin-β-glycosides, differing in their glycon component, can be used to identify individual glycoside hydrolase activities in real time, not only at the tissue level, but also at the cellular level.

Figure 4. Distribution of signals from different glycoside hydrolase activities in the cell walls of interfascicular fibers in *Arabidopsis* stems. (**A**) Glc-β-Res; (**B**) Glc-Glc-β-Res; (**C**) Glc-*S*-Glc-β-Res; (**D**) Gal-β-Res; (**E**) XXXG-β-Res. All reactions were carried out in optimal conditions (Table 1). The yellow fluorescence channel with resorufin signals is shown. Signals from the compound middle lamella are shown with the white arrows, and signals from the recently formed secondary wall layers are shown with red arrows. Intensity profiles through the cell wall area, shown by the white arrows, are included to visualize the labeling patterns in the cell wall better. Bar = 50 μm.

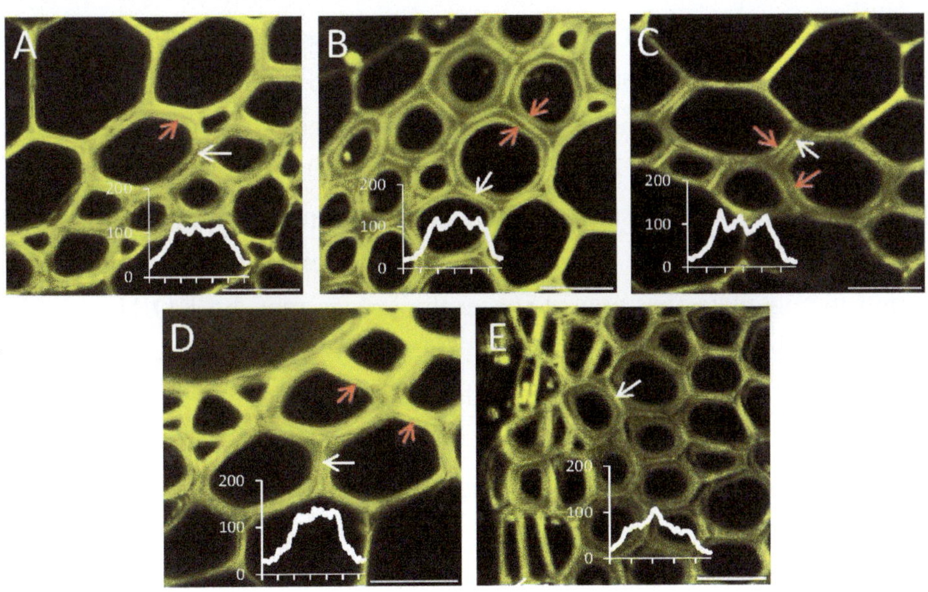

3. Discussion

The *in situ* distribution patterns of different glycoside hydrolase activities detected with the fluorogenic substrates in the sclerenchyma of *Arabidopsis* provide novel information about cell wall biochemistry in this tissue. Localized signals observed in cell walls most likely result from the activities of enzymes that are not free to diffuse. These enzymes could be either ionically bound to cell walls, as is known for some cellulases, pectate lyases, pectin methyl esterases or alpha-expansins, or be covalently-bound to the substrate, as is known for XET, or else be anchored in the plasma membrane by a transmembrane domain, like KOR1, or by a GPI anchor, like PMR6 [4,8].

In particular, the results indicate that developing secondary walls have an abundance of β-1,4-glucosidase activities (Figures 3 and 4; Supplemental Material 2). β-glucosidase activities were proposed to be present in growing primary walls [25] and have been extracted in the proteome of mature stems [21], but to our knowledge, this is the first report of the specific localization of these activities in the compound middle lamella and in developing secondary wall layers in the sclerenchyma cells. These observations imply that cell wall polysaccharide fragments released by various *endo*-glucanases or other glucosidases undergo trimming to release glucose to the apoplast, possibly followed by uptake and recycling.

The importance of β-1,4-galactosidase activity has been documented during cell wall formation in flax fibers [26,27]. These fibers represent a gelatinous fiber type (G-fibers), as opposed to the S-fibers of sclerenchyma composed of secondary walls rich in xylan [20]. β-1,4-galactosidase activity has been shown to be essential for the normal development of G-fibers, for cellulose crystallinity in the gelatinous layer and for fiber mechanical strength [26,27]. Our present data, revealing this activity in the sclerenchyma cells in *Arabidopsis* (Figures 2 and 4, Supplemental Material 5), imply a role for β-1,4-galactosidase also in S-fibers with xylan-type secondary walls. We observed galactosidase activity mainly in the compound middle lamella, where the majority of β-1,4-galactan is known to be located in S-fibers [20]. However, some signals were also detected in the innermost secondary wall layers. These signals may correspond to galactosidases being currently produced and delivered to cell walls for the modification of compound middle lamella.

Glc-Glc-β-Res may act as a substrate for both β-1,4-glucosidases and β-1,4-glucanases, whereas the non-cleavable internal thioglycosidic bond of Glc-*S*-Glc-β-Res confers activity specific to β-1,4-glucanases. The comparison of the signal distribution across Glc-β-Res, Glc-Glc-β-Res and Glc-*S*-Glc-β-Res (Figures 2–4; Supplemental Materials 2–4) allows us to conclude that both β-1,4-glucosidases and β-1,4-glucanases are active during secondary wall development in sclerenchyma cells, but that β-1,4-glucanases are less abundant or less active, at least against these artificial fluorogenic substrates. Interestingly, the β-1,4-glucanase-specific signal was highest in the innermost secondary cell wall layers for all activities tested (Figure 4). It is tempting to speculate that one of these glucanases may be KOR1, a membrane-bound cellulase implicated in cellulose biosynthesis in primary and secondary walls [28]. In secondary walls, KOR1 was found to reduce cellulose crystallinity, although the mechanisms are still not fully understood [28,29].

The most intriguing finding was the lack of xyloglucanase activity during sclerenchyma cell development at the primary-wall stage, when xyloglucan deposition is highest, while this activity was highest in mature sclerenchyma cells (Figures 3 and 4, Supplementary Material 6). We previously observed an absence of xyloglucanase activity in developing fibers [1]. Here, we have confirmed that finding and further observed that xyloglucanase activity was concentrated in the middle lamella of mature fibers (Figures 3 and 4; Supplemental Material 6). Currently, the only identified xyloglucanases are encoded by members of Group IIIa of the *XTH* gene family, a subgroup of glycoside hydrolase family 16 [6,30]. In some species with xyloglucan-rich seeds, such as nasturtium, xyloglucanase activity is detected in the storage cell walls during seed germination, when the polysaccharide reserve is mobilized [1]. In nasturtium, the corresponding Group IIIa enzymes, *Tm*NXG1 and *Tm*NXG2, have been biochemically and structurally characterized [30]. *Arabidopsis* has only two Group IIIa members, *XTH31* and *XTH32,* and *XTH31* is the main gene expressed in elongating roots [31]. *Arabidopsis* also lacks a mixed-function *endo*-xyloglucanase, ancestral to the *XTH* gene family in GH16, that is found in other dicots [32]. Phenotypic analysis of a *XTH31/XTH32* double mutant failed to identify a precise role of xyloglucanase activity in the growing roots [31]. Recently, the *XTH31* gene product was implicated in Al^{3+} resistance by interaction with the xyloglucan *endo*-transglycosylase XTH17 and modulation of the xyloglucan content of root cell walls [33,34]. It is possible that modulation of xyloglucan content in mature sclerenchyma may also occur. In support of this hypothesis, a decrease in xyloglucan was observed in mature xylem cells of aspen when *Ptt*XET34 was ectopically overexpressed [16]. Elucidation of the physiological role of such xyloglucan modulation is an exciting research topic for future studies.

4. Experimental Section

4.1. Substrates Synthesis

Resorufinyl glycosides were prepared from corresponding per-*O*-acetylated glycosyl bromides using glycosylation under phase-transfer catalysis and subsequent deacetylation, as described previously [1]. NMR and MS data of synthesized compounds were in full agreement with previously published data [1,23,35]. Hepta-*O*-acetyl-4-thio-4-*S*-cellobiosyl bromide for Glc-*S*-Glc-β-Res synthesis was prepared from methyl 2,3,6-tri-*O*-benzoyl-4-*S*-(2,3,4,6-tetra-*O*-acetyl-β-D-glucopyranosyl)-4-thio-β-D-glucopyranoside according to Orgeret *et al.* [36]. Novel NMR data for resorufinyl 2,3,6-tri-*O*-acetyl-4-*S*-(2,3,4,6-tetra-*O*-acetyl-β-D-glucopyranosyl)-4-thio-β-D-glucopyranoside and resorufinyl 4-*S*-(β-D-glucopyranosyl)-4-thio-β-D-glucopyranoside (Glc-*S*-Glc-β-Res) are given in Supplementary Material 7. NMR and MS data of synthesized compounds were in full agreement with previously published data [1,23,35].

4.2. Material and Growth Conditions

Seeds of *Arabidopsis* after stratification at 4 °C for 3 days in 0.1% agar were sown in pots filled with soil and placed in a room with long-day conditions (16 h day/8 h night) and a photon flux density of 120 to 150 $\mu E\ m^{-2} \cdot s^{-1}$. Plants were grown in 80% relative humidity and a constant temperature at 22 °C, for 6–7 weeks until Growth Stage 6.3 [37].

4.3. Optimization of the Enzymatic Reaction Conditions

Fresh hand cross-sections from the basal part of the inflorescence stems of *Arabidopsis* were placed on slides with 50 μL of the substrate, covered with a cover slip, the edges of which were sealed with nail polish. The fluorescent product of enzymatic activity (resorufin) was detected by confocal laser scanning microscopy (LSM-510, Carl Zeiss, Germany) at 567 nm excitation and a 580-nm detection wavelength. A transmitted light channel was used simultaneously to visualize the anatomical details. Experiments were repeated at least three times for each substrate, using low (10× objective lens ×8) and high (40× objective lens ×8) magnification. The profiles were made using ImageJ. Control sections were boiled 30 min at 95 °C in 25 mM MES buffer before incubation in the substrate solutions. Autofluorescence control sections were scanned without incubation in substrates with the same scanning confocal microscope settings as the experimental sections.

4.3.1. The Reaction Time Optimization

All tested substrates were dissolved in 25 mM 2-morpholino-ethanesulfonic acid (MES) buffer at pH 6.5 and a 4.7×10^{-4} M concentration. Sections immersed in substrates were scanned at five minute intervals, using the "time series" option.

4.3.2. pH Optimization

Substrates were dissolved in 25 mM MES at a 4.7×10^{-4} M concentration at different pH levels: 5.0, 5.5, 6.0, 6.5, 7.0. The incubation time for Glc-β-Res, Glc-Glc-β-Res and Gal-β-Res was 20 min, and for Glc-*S*-Glc-β-Res and XXXG-β-Res, it was 60 min.

4.3.3. Substrate Concentration Optimization

Each studied substrate was dissolved at the optimal pH to a final concentration of 9.3×10^{-4} M, 4.7×10^{-4} M, 9.7×10^{-5} M and 4.7×10^{-5} M with 25 mM MES buffer. Sections were incubated for 20 min for Glc-β-Res, Glc-Glc-β-Res and Gal-β-Res or 60 min for Glc-*S*-Glc-β-Res and XXXG-β-Res, chosen based on the observations of the time series.

5. Conclusions

In this work, we have demonstrated the presence of several glycoside hydrolases in sclerenchyma cell walls in *Arabidopsis* by direct, real-time observation of their activities using a library of fluorogenic probes. The results support the notion that cell walls are being constantly remodeled, not only during cell wall deposition, but also when cells are thought to be fully "mature." The physiological role of the remodeling is still not fully understood for all polysaccharides. It could involve the release of mechanical stress accrued during growth, recycling of carbohydrates and, relatedly, modulation of oligosaccharin signaling, or regulation of cell wall ionic content. In sclerenchyma cells, these modifications might also impact the lignification process.

The results described here demonstrate that the resorufin glycosides are effective for the identification of diverse hydrolytic activities at cellular resolution. Broader application of these substrates will facilitate a better understanding of the post-synthetic modification of the cell wall under different situations, including normal morphogenesis and in response to changing environmental conditions, e.g., pathogen attack and abiotic stress.

Acknowledgments

This work was supported by Formas, the Swedish Research Council (VR), the Swedish Governmental Agency for Innovation Systems (VINNOVA) and the Swedish Center for Biomimetic Fiber Engineering (funded by the Foundation for Strategic Research). Alicja Banasiak acknowledges the support of the European Cooperation in Science and Technology: COST-STSM-E50-03056. Harry Brumer is an Associate Member of the Department of Botany, University of British Columbia, and acknowledges funding from the Natural Sciences and Engineering Research Council of Canada (Discovery Grant).

Author Contributions

Alicja Banasiak and Ewa J. Mellerowicz performed microscopy observations and wrote the paper. Harry Brumer and Farid M. Ibatullin designed and synthesized the fluorogenic substrates and contributed to the manuscript preparation. All authors approved the final version of the paper.

Conflicts of Interest

The authors declare no conflict of interest.

References

1. Ibatullin, F.M.; Banasiak, A.; Baumann, M.J.; Greffe, L.; Takahashi, J.; Mellerowicz, E.J.; Brumer, H. A real-time fluorogenic assay for the visualization of glycoside hydrolase activity *in planta*. *Plant Physiol.* **2009**, *151*, 1741–1750.
2. Henrissat, B.; Coutinho, P.M.; Davies, G.J. A census of carbohydrate-active enzymes in the genome of *Arabidopsis thaliana*. *Plant Mol. Biol.* **2001**, *47*, 55–72.
3. Geisler-Lee, J.; Geisler, M.; Coutinho, P.M.; Segerman, B.; Nishikubo, N.; Takahashi, J.; Aspeborg, H.; Djerbi, S.; Master, E.; Andersson-Gunnerås, S.; *et al.* Poplar Carbohydrate-Active Enzymes (CAZymes). Gene identification and expression analyses. *Plant Physiol.* **2006**, *140*, 946–962.
4. Fry, S.C. Primary cell wall metabolism: Tracking the careers of wall polymers in living plant cells. *New Phytol.* **2004**, *161*, 641–675.
5. Del Bem, E.L.; Vincentz, M.G. Evolution of xyloglucan-related genes in green Plants. *BMC Evol. Biol.* **2010**, *10*, 341–388.
6. Eklöf, J.M.; Brumer, H. The XTH gene family: An update on enzyme structure, function, and phylogeny in xyloglucan remodelling. *Plant Physiol.* **2010**, *153*, 456–466.
7. Franková, L.; Fry, S.C. Biochemistry and physiological roles of enzymes that "cut and paste" plant cell-wall polysaccharides. *J. Exp. Bot.* **2013**, *64*, 3519–3550.
8. Cosgrove, D.J. Growth of the plant cell wall. *Nat. Rev. Mol. Cell Biol.* **2005**, *6*, 850–861.
9. Günl, M.; Neumetzler, L.; Kraemer, F.; de Souza, A.; Schultink, A.; Pena, M.; York, W.S.; Pauly, M. AXY8 encodes an α-fucosidase, underscoring the importance of apoplastic metabolism on the fine structure of *Arabidopsis* cell wall polysaccharides. *Plant Cell* **2011**, *23*, 4025–4040.
10. Günl, M.; Pauly, M. AXY3 encodes a α-xylosidase that impacts the structure and accessibility of the hemicellulose xyloglucan in *Arabidopsis* plant cell walls. *Planta* **2011**, *233*, 707–719.
11. Sampedro, J.; Pardo, B.; Gianzo, C.; Guitián, E.; Revilla, G.; Zarra, I. Lack of α-xylosidase activity in *Arabidopsis* alters xyloglucan composition and results in growth defects. *Plant Physiol.* **2010**, *3*, 1105–1115.
12. Sampedro, J.; Gianzo, C.; Iglesias, N.; Guitián, E.; Revilla, G.; Zarra, I. AtBGAL10 is the main xyloglucan β-galactosidase in *Arabidopsis*, and its absence results in unusual xyloglucan subunits and growth defects. *Plant Physiol.* **2012**, *3*, 1146–1157.
13. Brummell, D.A.; Cin, V.D.; Crisosto, C.H.; Labavitch, J.M. Cell wall metabolism during maturation, ripening and senescence of peach fruit. *J. Exp. Bot.* **2004**, *55*, 2029–2039.

14. Bourquin, V.; Nishikubo, N.; Abe, H.; Brumer, H.; Denman, S.; Eklund, M.; Christiernin, M.; Teeri, T.T.; Sundberg, B.; Mellerowicz, E.J. Xyloglucan endotransglycosylases have a function during the formation of secondary cell walls of vascular tissues. *Plant Cell* **2002**, *14*, 3073–3088.
15. Nishikubo, N.; Awano, T.; Banasiak, A.; Bourquin, V.; Ibatullin, F.; Funada, R.; Brumer, H.; Teeri, T.T.; Hayashi, T.; Sundberg, B.; *et al.* Xyloglucan *endo*-transglycosylase (XET) functions in gelatinous layers of tension wood fibers in poplar—A glimpse into the mechanism of the balancing act of trees. *Plant Cell Physiol.* **2007**, *48*, 843–855.
16. Nishikubo, N.; Takahashi, J.; Roos, A.A.; Derba-Maceluch, M.; Piens, K.; Brumer, H.; Teeri, T.T.; Stålbrand, H.; Mellerowicz, E.J. XET-mediated xyloglucan rearrangements in developing wood of hybrid aspen (*Populus tremula x tremuloides*). *Plant Physiol.* **2011**, *155*, 399–413.
17. Derba-Maceluch, M.; Awano, T.; Takahashi, J.; Lucenius, J.; Ratke, C.; Kontro, I.; Busse-Wicher, M.; Kosik, O.; Tanaka, R.; Winzéll, A.; *et al.* Suppression of xylan transglycosylase *Ptxt*Xyn10A affects cellulose microfibril angle in secondary wall in aspen wood. *New Phytol.* **2014**, in press.
18. Mellerowicz, E.J.; Immerzeel, P.; Hayashi, T. Xyloglucan—The molecular muscle of trees. *Ann. Bot.* **2008**, *101*, 659–665.
19. Baba, K.; Park, Y.W.; Kaku, T.; Kaida, R.; Takeuchi, M.; Yoshida, M.; Hosoo, Y.; Ojio, Y.; Okuyama, T.; Taniguchi, T.; *et al.* Xyloglucan for generating tensile stress to bend tree stem. *Mol. Plant* **2009**, *2*, 893–903.
20. Mellerowicz, E.J.; Gorshkova, T. Tensional stress generation in gelatinous fibres: A review and possible mechanism based on cell wall structure and composition. *J. Exp. Bot.* **2012**, *63*, 551–565.
21. Minic, Z.; Jamet, E.; Négroni, L.; der Garabedian, P.A.; Zivy, M.; Jouanin, L. A sub-proteome of *Arabidopsis* thaliana mature stems trapped on Concanavalin A is enriched in cell wall glycoside hydrolases. *J. Exp. Bot.* **2007**, *58*, 2503–2512.
22. Day, A.; Fénart, S.; Neutelings, G.; Hawkins, S.; Rolando, C.; Tokarski, C. Identification of cell wall proteins in the flax (*Linum usitatissimum*) stem. *Proteomics* **2013**, *13*, 812–825.
23. Coleman, D.J.; Studler, M.J.; Naleway, J.J. A long-wavelength fluorescent substrate for continuous fluorometric determination of cellulase activity: Resorufin-β-D-cellobioside. *Anal. Biochem.* **2007**, *371*, 146–153.
24. Tuomivaara, S.T.; Yaoi, K.; O'Neill, M.A.; York, W.S. Generation and structural validation of a library of diverse xyloglucan-derived oligosaccharides, including an update on xyloglucan nomenclature. *Carbohydr. Res.* **2014**, doi:10.1016/j.carres.2014.06.031.
25. Miyamoto, K.; Schopfer, P. Sugar release from maize coleoptiles during auxin-, fusicoccin- and acid-mediated elongation growth. *J. Plant Physiol.* **1997**, *150*, 309–316.
26. Mikshina, P.V.; Chemikosova, S.B.; Mokshina, N.E.; Ibragimova, N.N.; Gorshkova, T.A. Free galactose and galactosidase activity in the course of flax fiber development. *Russ. J. Plant Physiol.* **2009**, *56*, 58–67.
27. Roach, M.J.; Mokshina, N.Y.; Badhan, A.; Snegireva, A.V.; Hobson, N.; Deyholos, M.K.; Gorshkova, T.A. Development of cellulosic secondary walls in flax fibers requires b-galactosidase. *Plant Physiol.* **2011**, *156*, 1351–1363.
28. Takahashi, J.; Rudsander, U.J.; Hedenström, M.; Banasiak, A.; Harholt, J.; Amelot, N.; Immerzeel, P.; Ryden, P.; Endo, S.; Ibatullin, F.M.; *et al.* *KORRIGAN1* and its aspen homolog *PttCel9A1* decrease cellulose crystallinity in *Arabidopsis* stems. *Plant Cell Physiol.* **2009**, *50*, 1099–1115.

29. Maloney, V.J.; Mansfield, S.D. Characterization and varied expression of a membrane-bound endo-β-1,4-glucanase in hybrid poplar. *Plant Biotechnol. J.* **2010**, *8*, 294–307.
30. Baumann, M.J.; Eklöf, J.M.; Michel, G.; Kallas, Å.M.; Teeri, T.T.; Czjzek, M.; Brumer, H. Structural evidence for the evolution of xyloglucanase activity from xyloglucan endo-transglycosylases: Biological implications for cell wall metabolism. *Plant Cell* **2007**, *19*, 1947–1963.
31. Kaewthai, N.; Gendre, D.; Eklöf, J.M.; Ibatullin, F.M.; Ezcurra, I.; Bhalerao, R.P.; Brumer, H. Group III a XTH genes of *Arabidopsis* encode predominant xyloglucan endohydrolases that are dispensable for normal growth. *Plant Physiol.* **2013**, *161*, 440–454.
32. Eklöf, J.M.; Shojania, S.; Okon, M.; McIntosh, L.P.; Brumer, H. Structure-function analysis of a broad-specificity *Populus trichocarpa endo*-beta-glucanase reveals an evolutionary link between bacterial licheninases and plant *XTH* gene products. *J. Biol. Chem.* **2013**, *288*, 15786–15799.
33. Zhu, X.F.; Shi, Y.Z.; Lei, G.J.; Fry, S.C.; Zhang, B.C.; Zhou, Y.H.; Braam, J.; Jiang, T.; Xu, X.Y.; Mao, C.Z.; *et al.* XTH31, Encoding an *in vitro* XEH/XET-active enzyme, regulates aluminum sensitivity by modulating *in vivo* XET action, cell wall xyloglucan content, and aluminum binding capacity in *Arabidopsis*. *Plant Cell* **2012**, *24*, 4731–4747.
34. Zhu, X.F.; Wan, J.X.; Sun, Y.; Shi, Y.Z.; Braam, J.; Li, G.X.; Zheng, S.J. Xyloglucan ENDOTRANSGLUCOSYLASE-Hydrolase17 interacts with Xyloglucan Endotransglucosylase-Hydrolase31 to confer xyloglucan endotransglucosylase action and affect aluminum sensitivity in *Arabidopsis*. *Plant Physiol.* **2014**, *165*, 1566–1574.
35. Shoichi Tokutake, S.; Kasai, K.; Tomikura, T.; Yamaji, N.; Kato, M. Glycosides having chromophores as substrates for sensitive enzyme analysis. II. Synthesis of phenolindophenyl-β-D-glucopyranosides having an electron-withdrawing substituent as substrates for β-glucosidase. *Chem. Pharm. Bull.* **1990**, *38*, 3466–3470.
36. Orgeret, C.; Seillier, E.; Gautier, C.; Defaye, J.; Driguez, H. 4-Thiocellooligosaccharides. Their synthesis and use as ligands for the separation of cellobiohydrolases of *Trichoderma reesei* by affinity chromatography. *Carbohydr. Res.* **1992**, *224*, 29–40.
37. Boyes, D.C.; Zayed, A.M.; Ascenzi, R.; McCaskill, A.J.; Hoffman, N.E.; Davis, K.R.; Görlach, J. Growth stage-based phenotypic analysis of *Arabidopsis*: A model for high throughput functional genomics in plants. *Plant Cell* **2001**, *13*, 1499–1510.

10

Expression of the Arabidopsis Sigma Factor *SIG5* Is Photoreceptor and Photosynthesis Controlled

Marina Mellenthin [1], Ulrike Ellersiek [1], Anna Börger [2] and Margarete Baier [2,*]

[1] Plant Sciences, Heinrich-Heine-Universität, Universitätsstraße 1, Düsseldorf 40225, Germany; E-Mails: marina.mellenthin@uni-duesseldorf.de (M.M.); ulrike.ellersiek@uni-duesseldorf.de (U.E.)

[2] Plant Physiology, Freie Universität Berlin, Königin-Luise-Straße 12-16, Berlin 14195, Germany; E-Mail: anna.boerger@fu-berlin.de

* Author to whom correspondence should be addressed; E-Mail: margarete.baier@fu-berlin.de

Abstract: Two collections of Arabidopsis GAL4 enhancer trap lines were screened for light-intensity dependent reporter gene activation. Line N9313 was isolated for its strong light-intensity regulation. The T-DNA element trapped distant enhancers of the *SIG5* promoter, which drives expression of a sigma factor involved in regulation of chloroplast genes for photosystem II core proteins. The T-DNA insertion 715 bp upstream of the transcription initiation site splits the promoter in a distal and proximal part. Both parts are sensitive to blue and red light and depend on photosynthetic electron transport activity between photosystem II and the plastoquinone pool. The mainblue-light sensitivity is localized within a 196-bp sequence (–887 to –691 bp) in the proximal promoter region It is preferentially CRY1 and PHYB controlled. Type-I and type-II phytochromes mediate red-light sensitivity via various promoter elements spread over the proximal and distal upstream region. This work characterizes *SIG5* as an anterograde control factor of chloroplast gene expression, which is controlled by chloroplast signals in a retrograde manner.

Keywords: *SIG5*; sigma factor; phytochrome; cryptochrome; photosynthesis; plastoquinone; photosystem II; retrograde signal

Abbreviations

3-AT	3-amino-1,2,4-triazol
Act D	actinomycin D
ANOVA	analysis of variance
BL	blue light
BLRP	blue light responsive promoter
CDS	coding sequence
DCMU	3-(3,4-Dichlorophenyl)-1,1-dimethylurea
ET	enhancer trap FR far red
FRL	far-red light
GFP	green fluorescent protein
MS	Murashige and Skoog medium
PCR	polymerase chain reaction
PSII	photosystem II
qRT-PCR	quantitative real-time PCR
RL	red light
RUBISCO	ribulose-1,5-bisphosphate carboxylase/oxygenease
SD	synthetic dropout
SIG5	sigma factor 5
TAIL-PCR	thermal asymmetric interlaced PCR
UAS	upstream activation sequence
UTR	untranslated region
Y1H	yeast-one-hybrid

1. Introduction

Light provides the main source of energy for plant life. Furthermore, light is a key regulator of plant development and metabolism. Many light-responses include gene expression regulation [1,2]. *Cis*-acting regulatory elements control the combination, spacing and relative orientation of transcription factors, which in turn make the promoters sensitive to light intensities and light qualities in a gene-specific manner [2,3]. Most light responsive elements (LREs) have been identified by characterization of the promoter regions of photosynthesis-associated genes [4]. Motifs, such as G-boxes (CACGTG; [5]), Z-boxes (ATACGTGT; [6]), I-boxes (GATAAGR; [7]) and GATA-motifs (GATA; [8]) have mainly been identified in close vicinity (within 300 bp upstream) to the transcription initiation site. In the core-promoters, specific TATA-boxes/Inr-elements increase the light sensitivity [9]. More distant elements were identified by foot-printing and gel-retardation assays in screens for binding motifs of light-responsive transcription factors [1,10]. Despite these intensive gene-specific studies, knowledge on light-regulation of gene expression is still fragmentary [11,12].

Light regulates organellar and nuclear gene expression [13,14]. It controls germination, photomorphogenesis and plant development [15,16]. Light signals are sensed by photoreceptors, such as phytochromes, cryptochromes and phototropins, and the photosystems [17,18]. Some light signals are directly targeted to promoter element-bound transcription factors [10], others are interwoven in developmental and stress-regulated signaling cascades [14,19]. Light, besides water and nutrients, is the most important environmental regulator in plants.

To screen for light-regulated enhancers, irrespective of the function of the regulated gene and the position relative to the transcription initiation site, we analyzed *Arabidopsis thaliana GAL4-GFP* enhancer trap lines [20,21] for light-responsiveness. If the enhancer trap T-DNA is inserted in the vicinity of an enhancer element, the *GAL4-VP16* transcriptional activator is expressed. The GAL4-factor subsequently induces expression of an ER-targeted *GFP* (green fluorescent protein) (*mGFP5ER)* which can be monitored non-invasively [20,21].

Previously, enhancer trap lines have been used to identify regulatory sequences controlling developmental and organ-specific expression patterns, e.g., senescence [22], stomatal guard cell development [21] or lateral root development [23]. Here, we used *GAL4 GFP* enhancer trap populations (GAL4 *GFP* ET) to screen for light-responsive promoter elements irrespective of their location relative to the transcription start site. Lines with strong *GFP* expression in mesophyll cells were isolated and sub-selected for responsiveness to light intensity variation. A trap insertion in the distal part of the *SIG5* upstream region gave novel insights into the light-regulation of a nuclear-encoded regulator of chloroplast gene expression. *SIG5* (At5g24120) encodes the sigma factor, which activates expression of the D1 and D2 protein genes (*psbA* and *psbD*) inside chloroplasts [24]. It is required for light dependent regulation during the day [25] and, therefore, an essential factor in the nuclear control of chloroplast function. Here, we show that *SIG5* is regulated by the combined action of multiple blue- and red-light sensitive elements spread over the full promoter and by photosynthetic electron transport.

2. Results

2.1. Selection of Enhancer-Trap Lines

To identify distal promoter elements involved in light-regulation of gene expression in mesophyll cells, two collections of light grown *GAL4 GFP* enhancer trap lines [20,21] were screened by fluorescence microscopy for mesophyll activity of the enhancer at an age of 7–13 days. At this age, plants are fully shifted from heterotrophic lipid consumption to photoautotrophy and the first true leaves develop [26].

For each selected line, twenty four 10 day old seedlings were screened for light-intensity regulation by comparison of the *GFP*-activity at 10, 100 or 200 μmol·photons·m^{-2}·s^{-1} white light using a top reader fluorometer. The crude data were corrected by subtraction of the background fluorescence of wildtype C24 plants, which is caused e.g., by cell wall fluorescence. In several enhancer trap lines the *GFP* fluorescence positively correlated with the growth light intensity. The strongest gradual light-intensity regulation was observed for the lines N9266 and N9313 (Figure 1A,B). Smaller leaves in low light and larger ones at higher light intensities can mock light intensity regulation, while longer hypocotyls would partly mask it. Corrections were performed by standardization of the *GFP* activity on the leaf area. In N9266 and N9313, light intensity regulation was pronounced confirming light-intensity dependent regulation.

Figure 1. *GFP* fluorescence of *GAL4-GFP* enhancer trap lines. (**A** and **B**): Light-intensity regulation in N9266 and N9313. Seedlings were grown for 10 days in 10, 100 and 200 μmol·photons·m^{-2}·s^{-1} white light. One-way ANOVA was performed comparing all groups with Bonferroni's post-test. Statistical significance of difference is indicated as asterisks above bars ($p < 0.05$); (**C** and **D**): Transverse confocal section of spongy mesophyll of cotyledons. The fluorescence of ER-targeted *GFP* (m*GFP*5ER) is presented in green, Chlorophyll fluorescence in red.

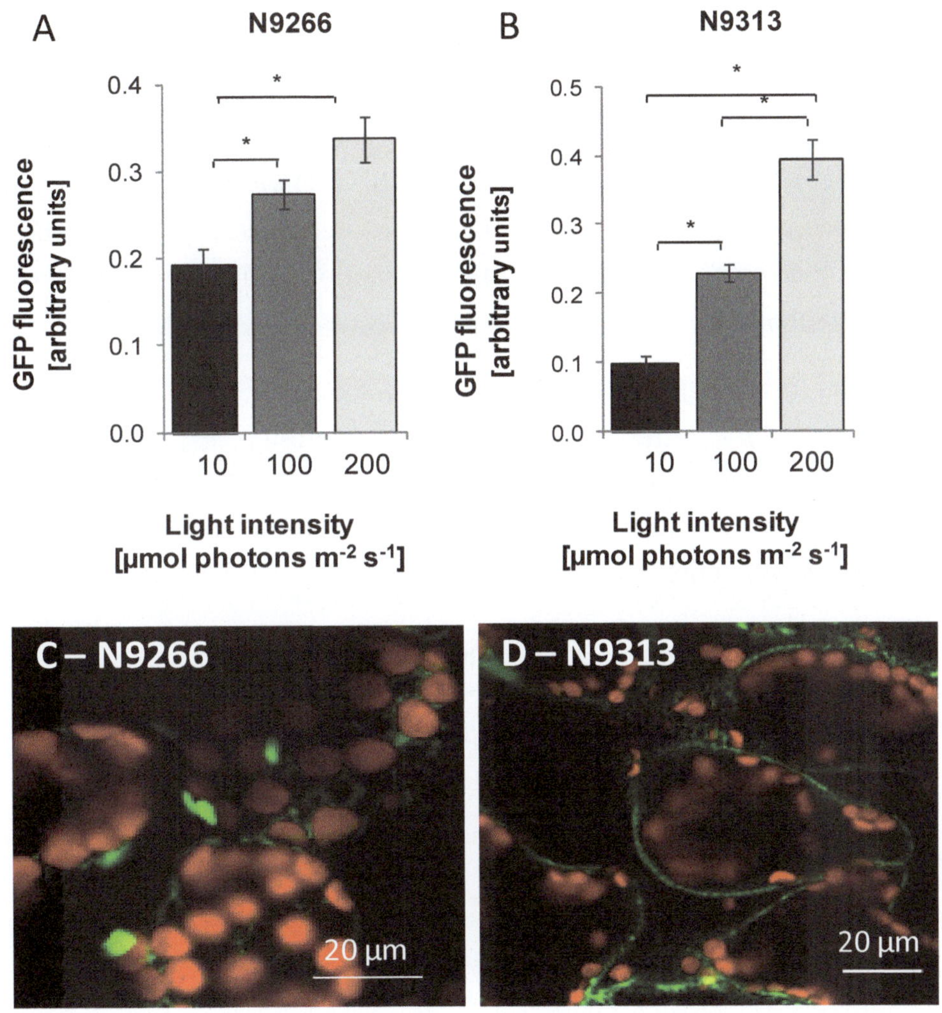

2.2. Identification of Enhancer Trapped Sequences and T-DNA Numbers

For line N9266, hybridization of Southern-Blots with probes against the GAL4 element gave two bands (Figure 2A). Since DNA was digested with *Bgl*I, which does not cut within the GAL4 element, two bands indicate two T-DNA insertions. In contrast, the hybridization pattern of line N9313 showed that the *GFP* fluorescence results from a single T-DNA insertion (Figure 2A).

To identify the T-DNA insertion site of N9313, 3-step thermal asymmetric interlaced PCR (TAIL-PCR; [27]) was performed with genomic DNA. The tertiary TAIL-PCR product was cloned into pJET1.2/blunt cloning vector (Fermentas, St. Leon-Rot, Germany) and sequenced using a vector specific primer. The sequencing product contained at one end the 35S-minimal promoter demonstrating that it is specific for the enhancer trap insertion site. Comparison of the sequence upstream of the 35S-minimal promoter with *Arabidopsis thaliana* genome sequence data revealed that the T-DNA of line N9313 is inserted 1198 bp upstream of the coding sequence of *SIG5* (At5g24120) (Figure 2B). The *SIG5* gene encodes a sigma factor which is involved in diurnal light regulation of plastid gene expression [24,25]. The T-DNA is also inserted in the vicinity (1302 bp upstream) of the coding sequence of a protein with unknown function (At5g24130) (Figure 2A). Publicly available microarray data (eFP browser; [28]; data not shown) suggest that it encodes an almost seed-specifically expressed gene. The direction of the inserted enhancer-trap construct matches with the direction of the *SIG5* gene, which is, like the reporter gene of N9313, leaf expressed and light-responsive [22,29,30]. Based on EST comparison, a 483 bp intron-containing 5'-UTR has been predicted for *SIG5*. Since it is only substantiated by two ESTs (and alternative transcription initiation cannot be excluded), we refer to the translation start site as position +1 for description of the *SIG5* promoter in this manuscript.

The T-DNA insertion site was confirmed by PCR with primers designed to anneal to the genomic DNA flanking the insertion site and to the right border of the T-DNA. The enhancer trap element is inserted 715 upstream of the predicted transcription initiation site of *SIG5* in N9313 and separates the *SIG5* promoter in a proximal part controlling *SIG5* transcription and a distal part regulating *GFP* expression (Figure 2B).

Figure 2. (**A**) Determination of the T-DNA insertion numbers by Southern blot hybridization with a DIG-labeled *GAL4* specific probe. Numbers of DIG-labeled fragments indicate numbers of T-DNA insertions in the different enhancer trap lines; (**B**) T-DNA insertion site and direction of the reporter element (*GFP*, green arrow) and the coding sequences of the flanking genes (red arrows) in the line N9313.

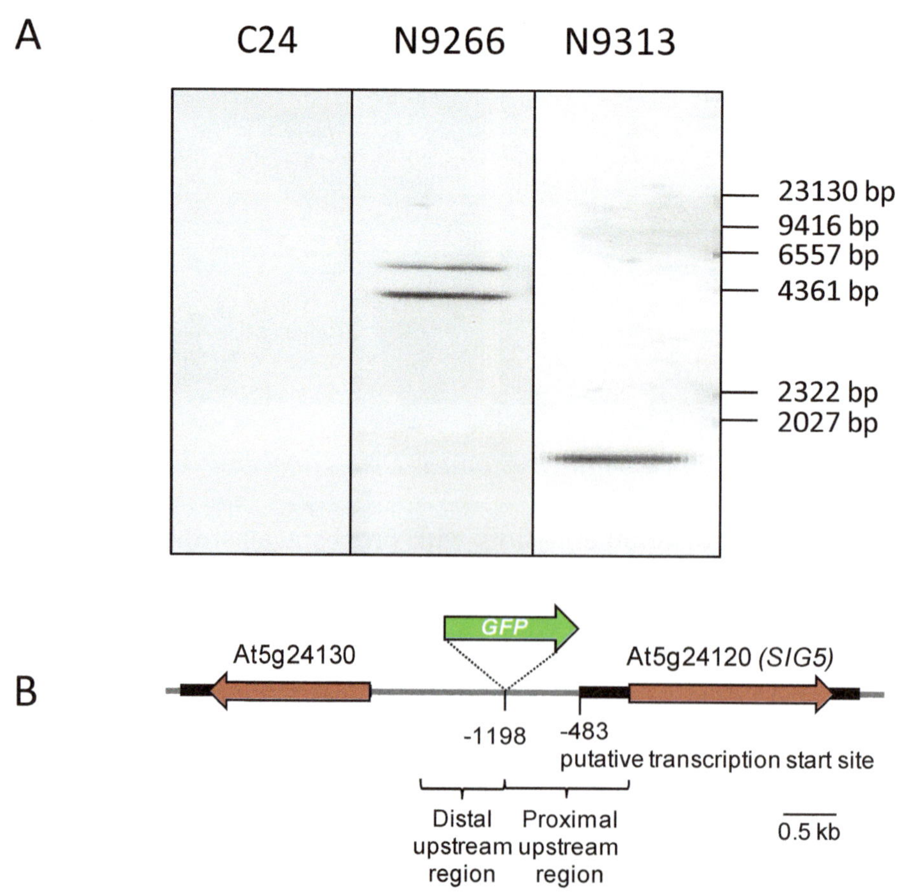

2.3. Light Quality Regulation of SIG5 by the Distal and Proximal Upstream Regions

SIG5 has previously been described to be blue-light regulated in a light-intensity-dependent manner [29–32]. To test whether and to which extent blue-light regulation is mediated by the trapped promoter elements, *GFP* fluorescence activity was tested in N9313 plants grown for 10 days in blue light (470 nm, 10 µmol photons m^{-2} s^{-1}). The reporter gene activity was only 1.25-fold higher in blue light than in darkness (Figure 3A) demonstrating that it is not sufficient to mediate the strong blue-light induction reported in the literature [30].

In parallel, seedlings were grown under red light (>600 nm). Originally, these plants were thought to function as controls, since it was reported that *SIG5* is not red-light regulated [31]. However, the *GFP* pattern showed that the distal upstream region is sensitive to red light (Figure 3A).

For further analysis of light quality sensing, *GFP* mRNA levels (resulting from activation of the enhancer trap by the distal upstream region) and the *SIG5* mRNA levels, which are under control of the proximal 715 bp of the upstream region and the *SIG5*-5'-UTR, were compared in the enhancer trap line N9313 and in C24, which is the wildtype background of N9313 and expresses *SIG5* under the full-length promoter. RNA-decay analysis (Figure A1) revealed that during a 24 h treatment 99% of *SIG5* mRNA and 97% of *GFP* mRNA can be expected to depend on *de-novo* synthesis. Here, plants were pre-cultivated for 12 days in 100–120 µmol·photons·m^{-2}·s^{-1} white light to minimize germination and early seedling development effects. Afterwards, they were transferred for 24 h into darkness for relaxation and exposed for 24 h to 100–120·µmol·photons·m^{-2}·s^{-1} monochromatic blue (471 nm), red (673 nm) or far-red light (745 nm), respectively. Controls were kept in the dark.

Figure 3. *GFP* and *SIG5* mRNA levels upon illumination with monochromatic light. (**A**) *GFP* fluorescence in 69–92 10 day old N9313 seedlings illuminated with 10 µmol·photons·m^{-2}·s^{-1} blue light or red/far-red light (R/FR); (**B–D**) *GFP* and *SIG5* transcript levels in N9313 and C24. Ten day old light grown seedlings were dark adapted for 24 h and then exposed to 100–120 µmol·photons·m^{-2}·s^{-1} monochromatic blue, red or far-red light for 24 h. Relative transcript levels were determined by qRT-PCR relative to *ACT2* transcript level and normalized on the transcript levels of dark adapted seedlings prior to the transfer to monochromatic light. Bars represent means (±SEM). * indicates significant differences from the dark adapted samples (Student's *t*-test, $p < 0.05$).

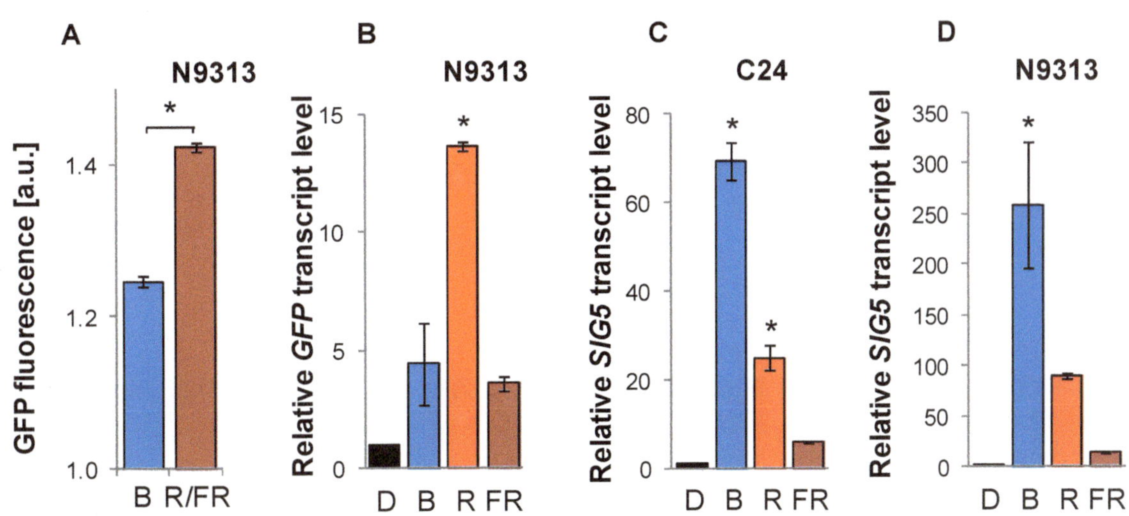

Blue light illumination resulted in low *GFP* mRNA levels in N9313 confirming that the distal upstream region is not blue-light sensitive. Consistent with the previous indication for red-light sensitivity of the distal upstream region, stronger *GFP* mRNA accumulation was observed in red-light illuminated N9313 seedlings (Figure 3B). Far-red light resulted in only very low *GFP* expression.

SIG5 transcript levels, which are under control of the proximal upstream region in N9313, were 260-fold higher in blue light in the enhancer trap line than in darkness showing that the previously reported blue-light sensitivity is mediated within the proximal upstream region. Red light resulted in almost 100-times the dark-level and far-red light gave around 15-times the dark mRNA level of *SIG5* in N9313 (Figure 3D).

In C24 (Figure 3C), the relative red-light and, especially, the far-red-light response were stronger compared to the blue-light reaction than in N9313 (Figure 3D) demonstrating that *SIG5* transcript abundance regulation in C24 combines the regulatory effects of proximal and distal upstream regulation in N9313 (Figure 3B,D).

2.4. Arabidopsis SIG5 Transcription Is Red Light Sensitive

Some of the sigma factors, e.g., *SIG2* and *SIG5*, are strongly induced in far-red light [33], while *SIG5* is the fastest blue-light induced sigma factor. Based on previous analyses, it was more or less excluded that *SIG5* is red-light sensitive [29–32]. Since most work was done on rosette leaves of 4 week old Arabidopsis plants of the accession *Landsberg erecta* (Ler) and *Columbia-0* (Col-0), while we observed red-light responses in up to 2 week old C24 seedlings, we first adjusted our growth conditions, illumination period and the duration of dark-adaptation and red-light illumination according to [31]. *SIG5* transcript levels were increased in 4 week N9313 plants after 3 h and after 24 h red-light illumination (Figure 4A). In a test for accession specific variation, *SIG5* mRNA levels were also increased in C24, Ler and Col-0 in response to red light (Figure 4B).

2.5. In Silico Analysis of the SIG5 Promoter for Putative Light Responsive Elements

In an *in silico* scan of the 2 kb upstream of the CDS of *SIG5* using the PlantCARE [34] and PLACE [35] databases for prediction of plant *cis*-acting regulatory DNA elements, 36 putative light-sensitive motifs were identified (Table 1): An AE-Box (−60) and two GATA-motifs (−71; −163) are located in the intron inserted in the 5'-UTR-region (−56 to −431 relative to the translation insertion site). Thirteen predicted elements map to the proximal upstream region (−483 to −1189). All other motifs were found in the distal upstream region which drives the enhancer trap. Eight predicted motifs cluster between −1428 and −1587. A second hot-spot for light regulated elements was observed between −1687 and −1794 bp upstream of the translation initiation site.

Figure 4. Red-light induction of *SIG5* transcripts. (**A**) Relative *SIG5* transcript levels in rosette leaves of 4 week old N9313 plants. N9313 was grown on soil under continuous white light (20 μmol·photons·m^{-2}·s^{-1}) for 4 weeks at 22 °C. The plants were dark adapted for 24 h and subsequently exposed to 50 μmol photons m^{-2} s^{-1} monochromatic red light (RL) at 22 °C for 3 h or 24 h; (**B**) *SIG5* transcript level of different Arabidopsis accessions. 10 day old light grown seedlings were dark adapted for 24 h and then exposed to 100–120 μmol·photons·m^{-2}·s^{-1} monochromatic red light for 24 h. * indicates significant differences from the dark adapted samples (Student's *t*-test, $p < 0.05$).

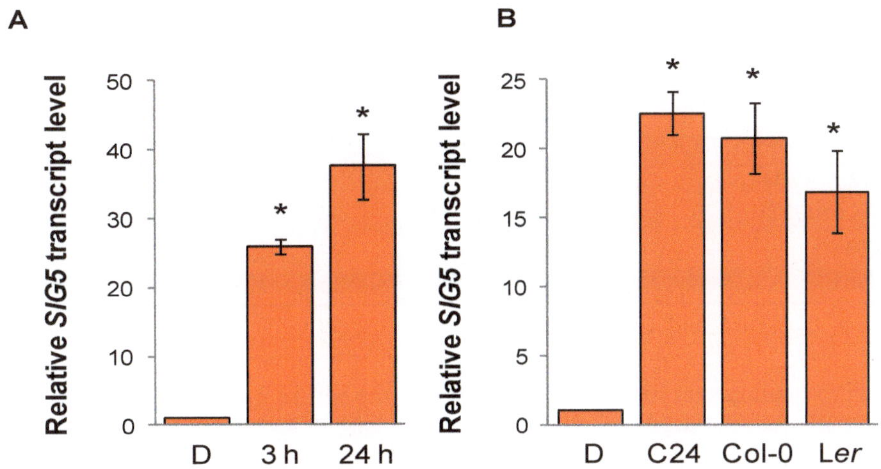

Table 1. Light responsive motifs identified in the 2 kb sequence upstream of the CDS of *SIG5* as predicted by PlantCARE and PLACE databases. A graphical overview is provided in Figure A2.

Motif	Sequence [a]	Position [b]
AE-box	AGAAACAT (+)	−60
	AGAAACAT (+)	−1021
GATA motifs	GATA (+)	−71
	GATA (−)	−163
	GATA (+)	−441
	GATA (+) [c]	−841
	GATA (−) [c]	−871
	GATA (+)	−914
	GATA (−)	−989
	GATA (−) [d]	−1593
	GATA (−) [d]	−1751
	GATA (−) [d]	−1944
	GATA (+) [d]	−1985
Box 4	ATTAAT (−)	−593
	ATTAAT (−)	−654
	ATTAAT (−)	−962
	ATTAAT (−)	−1428
GA motif	AAGGAAGA (−) [c]	−707
I-box	CACTTATGCT (−) [c]	−732
	aAGATAAGA (−)	−1587

Table 1. *Cont.*

Motif	Sequence [a]	Position [b]
ACE	AAAACGTTTA (−) [c]	−763
	CTAACGTATT (−) [d]	−1442
	ACGTGGA (−) [d]	−1795
Box I	TTTCAAA (−)	−917
	TTTCAAA (−)	−1751
	TTTCAAA (+)	−1901
3-AF1 binding site	AAGAGATATTT (−)	−922
ATCT-motif	AATGTAATCT (+)	−1205
	AATGTAATCT (+)	−1455
	AATGTAATCT (+)	−1460
L-box	AAATTAACCAAC (−)	−1426
TCT-motif	TCTTAC (+)	−1541
CATT-motif	GCATTC (+)	−1687
GAG-motif	GGAGATG (−)	−1699
GC-box	CACGTC (+)[d]	−1719
G-box	CACGTG (−)[d]	−1794

[a] (+) and (−) indicates the sense and complementary strand, respectively. [b] positions are relative to the translation start site of *SIG5*. [c] *cis*-acting elements within the 196-bp sequence mediating the main blue light response. [d] potential HY5 binding sites in the 0.8 kb upstream of the N9313 T-DNA.

2.6. Mapping Revealed Locally Restricted Blue-Light Sensitivity and Disperse Red-Light Sensitivity

To map the *SIG5* light sensitive promoter elements, the SALK-collection of T-DNA insertion lines [36] was screened for lines which interrupt the *SIG5* promoter at different positions. Five lines suitable for subsequent analysis were identified: SALK_015625 carries a T-DNA insertion at position −1618 relative to the *SIG5* CDS, SALK_077048 at position −1032, SALK_072457 at −887, SALK_019261 at −691 and SALK_133729 at −515. The T-DNA insertion sites were confirmed by PCR with T-DNA border primers and primers flanking the insertion site. Homozygous lines were isolated from segregating populations by PCR-based genotyping. For the light response tests, 10 day old seedlings grown in white light were dark-adapted for 24 h and treated with monochromatic blue, red and far-red light for 24 h prior to RNA isolation.

In blue-light, the *SIG5* transcript levels were similar in the three T-DNA lines with insertions at nucleotide positions −1618, −1032 and −887 in Col-0 (Figure 5A). Seedlings with T-DNA insertions at position −691 and −515 were strongly impaired in their blue light induction of *SIG5* transcription demonstrating that the 196-bp region between position −887 and position −691 mainly mediates blue-light regulation. Weaker blue-light sensitivity differences were observed between −515 and −691 in the proximal upstream region and upstream of −1618 in the distal region.

When the same lines were compared in red and far-red light, the strongest red-light difference was observed between the −887 and −1032 insertion lines (Figure 5B). Weaker, but still significantly strong positive elements are located between −515 and −691 and between −691 and −887. In the distal upstream region negative red-light regulation was observed if a SALK-line T-DNA was inserted upstream of the position of the enhancer trap insertion in N9313 (−1089), as if the T-DNA insertion destroyed the otherwise inducing element. The overall low far-red-light sensitivity increased gradually with the length of the upstream region (Figure 5C) indicating widely distributed regulation.

Figure 5. Mapping of the blue-light (BL; **A**), red-light (RL; **B**) and far-red-light (FRL; **C**) sensitive regions of the Arabidopsis *SIG5* promoter by qRT-PCR analysis of *SIG5* transcript levels in Col-0 wild-type and SALK T-DNA insertion lines exposed for 24 h to monochromatic light after 10 days growth in darkness. The numbers give the distances of the T-DNA insertions relative to the *SIG5* CDS. Values were normalized to the *SIG5* transcript level of 24 h dark-adapted seedlings. The data are means of 2–4 biological replicates (±SEM). * indicates significant differences from the dark adapted samples (Student's *t*-test, * $p < 0.05$). ℓ indicates significant differences from the line with the next longer upstream region (Student's *t*-test, $p < 0.05$).

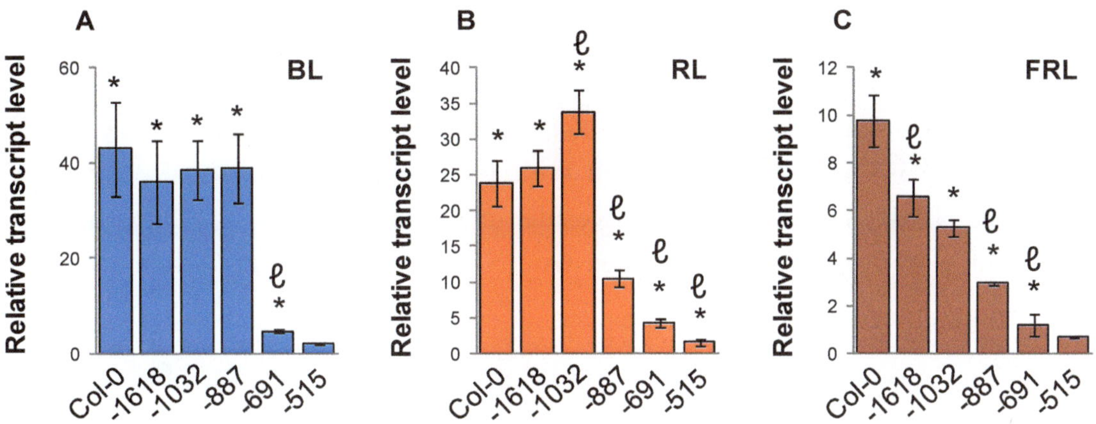

2.7. Blue- and Red-Light Regulation of the Proximal Upstream Region

As a first approach of signal transduction analysis, *SIG5* transcript levels were analyzed in photoreceptor mutant lines (Figure 6) grown next to their respective wild-type for 24 h in monochromatic light after 10 days pre-cultivation in white light and 24 h in darkness. In the *phot1phot2* and the *cry2* mutants, which are deficient in phototropins and cryptochrome 2, respectively [37,38], *SIG5* transcript levels were not significantly different from the respective wildtype lines. In *cry1* and *cry1cry2* mutants [39], it was strongly decreased (Figure 6A) demonstrating that CRY1 is involved in the blue light induction of *SIG5*. *SIG5* transcript levels were also decreased in the homozygous offspring of N9313 × cry1cry2 (Figure 6E) demonstrating that also the weak distal blue-light response is mediated by cryptochromes.

Blue light can alternatively be sensed via activation of the Soret absorption bands of phytochromes [40]. The comparison of *phyA*, *phyB* and the *phyAphyB* double mutants demonstrated that *phyB* mutants are impaired in the blue light induction of *SIG5* transcription, whereas *phyA* mutants are not (Figure 6B).

Comparison of *phyA*, *phyB* and *phyAphyB* mutants in red light showed that the red-light response of the *SIG5*-promoter is mediated by PHYB, but not by PHYA (Figure 6C). Consistent with a general low FR-response, the transcript levels of *SIG5* were low in *phyA*, *phyB* and *phyAphyB* mutants (Figure 6D).

2.8. Photoreceptor Control of the Distal SIG5 Upstream Region

For the analysis of the regulatory function of photoreceptors on the distal *SIG5* upstream region, the enhancer trap line N9313 was crossed with *phot1phot2* [37] and with *cry1* and *cry2* [39] mutants. Lines homozygous for the T-DNA and the respective mutant were selected and analyzed for *GFP* expression activity. Due to the severe differences in growth and greening between the mutants and wildtype plants, the reporter gene activities were standardized on chlorophyll-*a* levels after blue-light treatment (Figure 6F) and on fresh weight after red-light treatment (Figure 6G). The *GFP* fluorescence of N9313 x *phot1phot2* mutant plants was similar to that of N9313. The *GFP* values of N9313 × *cry1* were only slightly decreased, while the *GFP* fluorescence of *cry1cry2* double mutants were hardly detectable (Figure 7A), demonstrating that regulation of the distal upstream region depends on the availability of CRY2.

Figure 6. (A–E) Relative *SIG5* transcript levels in 10 day dark adapted old photoreceptor mutants, their background accessions and N9313 × *cry1cry2* in response to blue (BL), red (RL) and far-red light (FRL) after 24 h exposure to 100–120 μmol·photons·$m^{-2}·s^{-1}$ monochromatic light. The transcript levels were determined by qRT-PCR relative to *ACT2* transcript level. The transcript level of dark adapted seedlings was set to 1.0. (F + G) *GFP* fluorescence in N9313 and N9313 crossed with photoreceptor mutants. Relative *GFP* fluorescence of N9313 was set as 100%. Results are mean values of 3–8 measurements (±SEM). * indicates significant differences from the respective control line (Student's *t*-test, $p < 0.01$).

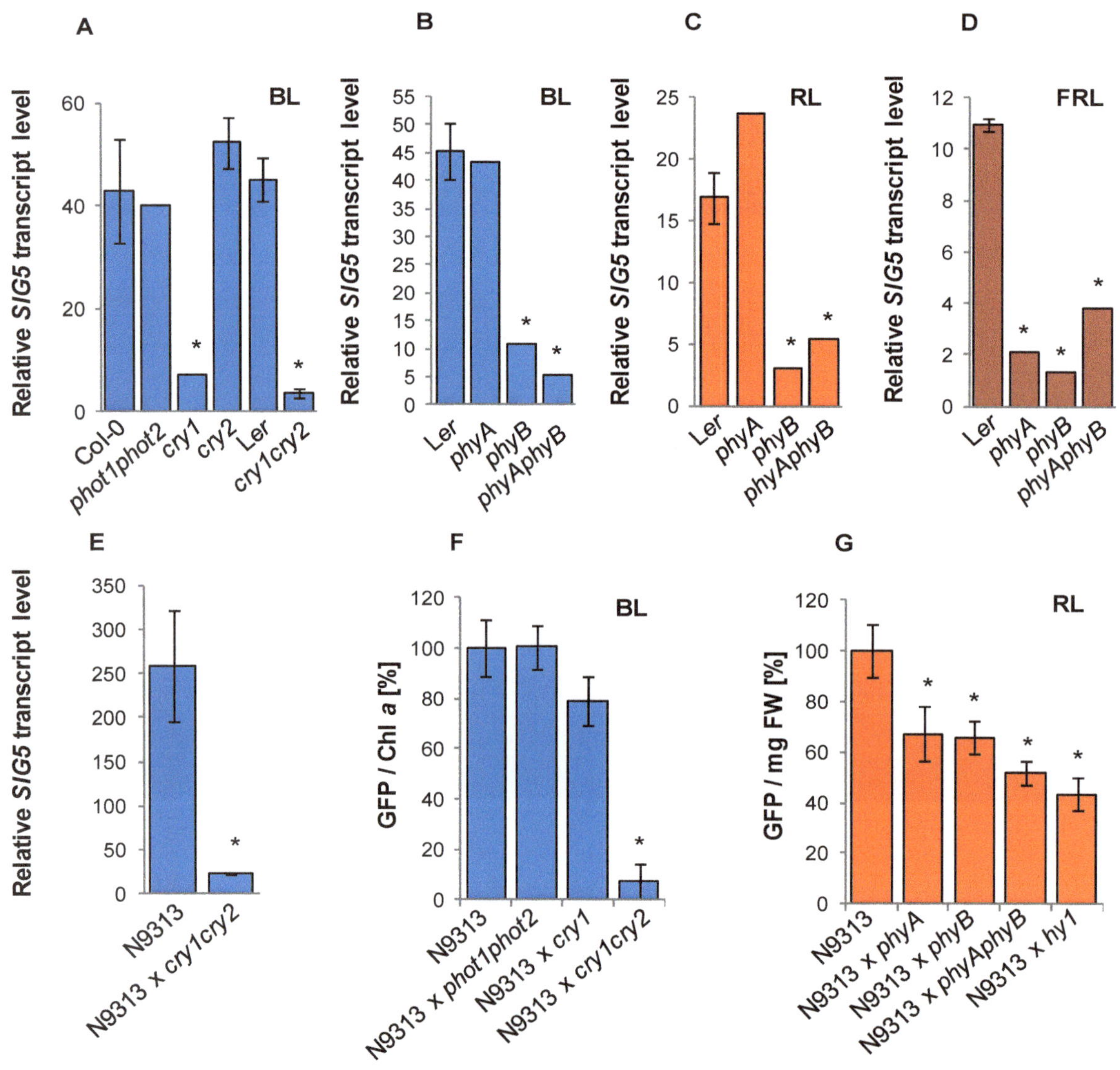

Figure 7. HY5 and COP1 regulation of the *SIG5* promoter in seedlings. (**A** and **B**) Relative *SIG5* transcript levels of old HY5 deficient hy5-1 and SALK_056405C and the respective wild-type in response to blue light or red light. Ten day old seedlings were dark adapted for 24 h and then exposed to 100–120 μmol photons m^{-2} s^{-1} monochromatic blue light or red light for 24 h. Relative transcript level of dark adapted seedlings prior to the transfer to monochromatic light was set to 1.0. (**C**) *GFP* fluorescence in N9313 and N9313 crossed with the HY5 deficient SALK_056405C (mean values of 4–8 samples with 10 seedlings each (±SEM). (**D**) Relative *SIG5* transcript abundances in Col-0 wild-type and in cop1 mutants as determined by qRT-PCR relative to ACT2 values. Light grown seedlings were dark-adapted for 24 h prior to RNA isolation. *SIG5* transcript levels of Col-0 were set to 1.0. * indicates significant differences from the N9313 or Col-0 value (Student's *t*-test, $p < 0.01$).

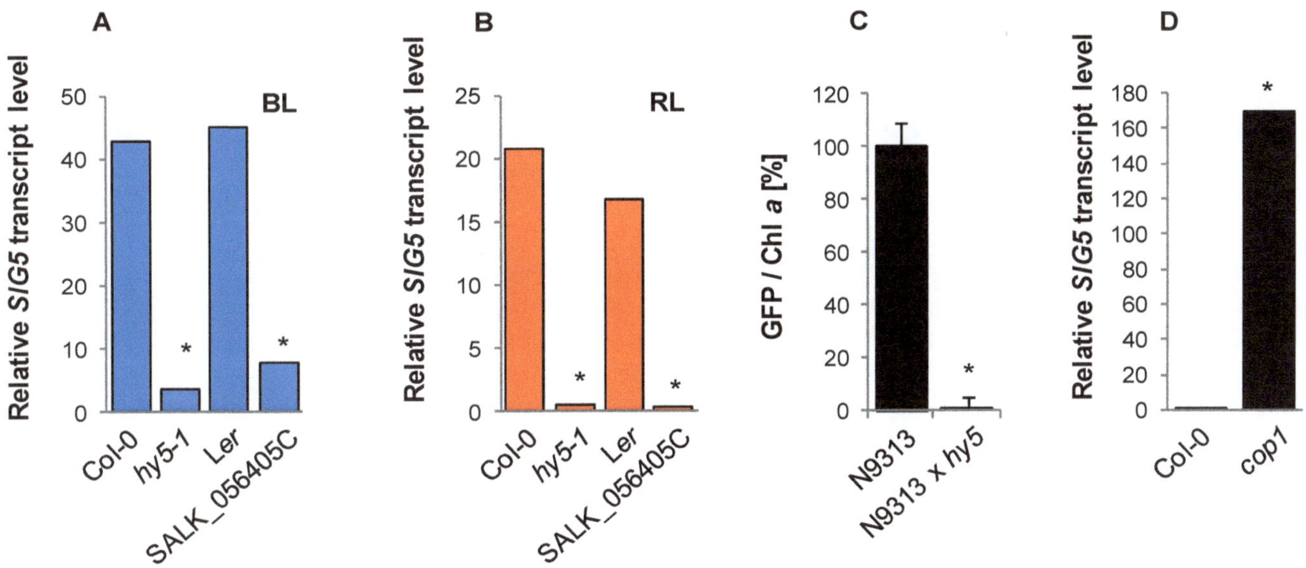

The reporter gene activity was also decreased in crosses of N9313 to *phyA* and *phyB* (Figure 6G). In the *phyAphyB* double mutant it was less than in each single mutant indicating a combined control of the promoter. Besides PHYB, Arabidopsis expresses three more type-II phytochromes. To investigate their overall impact, N9313 was crossed with the *hy1* mutant, which is deficient in phytochromobiline biosynthesis [41]. In response to the *hy1* mutation *GFP* activity was slightly more decreased than by *phyAphyB* (Figure 6G), demonstrating that the distal upstream region of *SIG5* is regulated by the combined action of various phytochromes, with strongest impact of PHYA and PHYB.

2.9. Regulation of the Distal SIG5 Upstream Region by HY5 and COP1

The bZIP transcription factor HY5 translates cryptochrome and phytochrome signals into gene expression regulation [42]. Its stability is under control of COP1 [43]. In the two HY5-deficient lines, *hy5-1* and SALK_056405C, *SIG5* transcript abundance was strongly decreased in red and in blue light (Figure 7A,B). qRT-PCR analysis showed a very strong accumulation of *SIG5* transcripts in dark-adapted *cop1-6* seedlings (Figure 8D), confirming that *SIG5* transcription is under control of HY5.

In the homozygous offspring of the cross of N9313 with the *hy5-1* mutant, *GFP* fluorescence was reduced to almost undetectable levels (Figure 7C) demonstrating that HY5 has strong impact on the distal upstream region.

Promoter motif analysis predicted HY5 target sites at the positions −1593 (GATA-motif), −1719 (GC-box) and −1794 (G-box), corresponding to 395, 521 and 596 bp upstream of the N9313 T-DNA insertion site (Table 2). To test whether the predicted motifs are involved in light quality regulation, they were amplified by PCR and fused to a truncated 35S promoter upstream of an open reading frame for *GFP-GUS*. In a PCR-based approach [44] the putative HY5 sites were mutagenized. Young tobacco leaves were transfected with mutagenized and non-mutagenized constructs. Expression activity was quantified from the GUS activity in leaf disks excised from the transfection site. As negative control, the minimal 35S-promoter (without regulatory upstream elements) was fused to *GFP-GUS*. The comparison of mutagenized and non-mutagenized constructs showed white- and red-light sensitivity for the G-box, but not for the other two predicted elements (Figure 8A).

Figure 8. Characterization of potential HY5 binding sites. (**A**) Transient GUS expression as regulated by distal 0.8 kb *SIG5* promoter fragments and variants with differently mutated putative HY5 target elements. Following infiltration, the leaves were illuminated for 5 days with 100–120 μmol photons $m^{-2} \cdot s^{-1}$ white light or red light. Statistical significance of difference (Bonferroni's *post-hoc* test) is indicated as asterisks above bars ($p < 0.05$); (**B**) Yeast-1-hybrid analysis of HY5 interaction with the *RBSC1A* promoter (p*RBCS1A*) and distal and proximal *SIG5* upstream regions (p$SIG5_{distal}$, p$SIG5_{proximal}$) in Y187. 5 μL of liquid culture with an OD_{600} of 0.01 were analyzed on SD/-H/-T (upper row) or SD/-H/-L/-T (lower row) supplemented with 3-AT. For control, the yeast transformants were cultivated on 3-AT-free SD/-T medium.

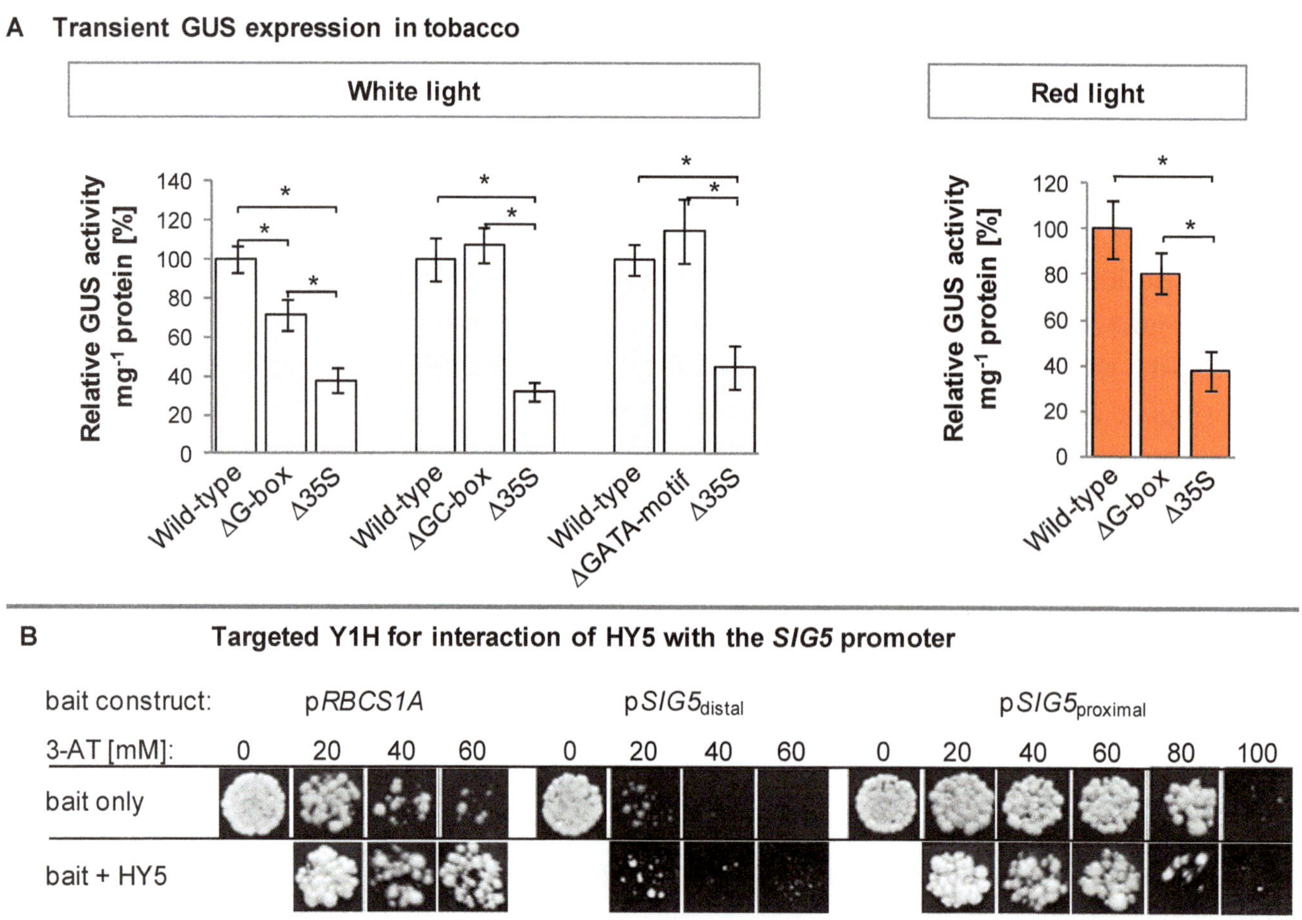

Table 2. HY5 binding sites identified in the distal 0.8 kb *SIG5* upstream region and introduced mutations.

cis-element	Sequence [a]	Position [b]	Introduced mutation [c]	Reference
G-box	CACGTG (+)	−1794	CAATTG	[45]
GC-box	GACGTG (+)	−1719	GAATTC	[45]
GATA-motif	GATAAG (+)	−1593	GTATTC	[46]

[a] (+) and (−) indicates the sense and complementary strand, respectively. [b] positions are relative to the translation start site of *SIG5*. [c] mutated nucleotides are highlighted in black.

2.10. Yeast-1-Hybrid Analysis of HY5 Binding to the SIG5 Promoter

To test for direct binding of HY5 to the *SIG5* promoter, yeast-one-hybrid assays were performed with *SIG5* promoter fragments as baits and HY5 as prey. The promoter of RUBISCO small subunit 1A (pRBCS1A) was used as positive control [43]. HY5 decreased auto-activation of the proximal *SIG5* upstream region in yeast demonstrating binding of the plant transcription factor (Figure 8B). Analysis of the distal *SIG5* upstream region gave no indication for direct HY5 interaction with the distal *SIG5* promoter (Figure 8B mid).

2.11. Impact of Photosynthetic Electron Transport on Light Regulation

Red light can also drive photosynthesis. Monochromatic red-light is predominantly absorbed by photosystem II [47]. During acclimation, expression of *psaAB*, which encodes the reaction center proteins of photosystem I, decreases and expression of the photosystem-II reaction center protein D1 (encoded by *psbA*) increases in order to optimize photosynthetic efficiency [48]. *SIG5* regulates transcription of the photoreaction center proteins D1 and D2 (encoded by *psbA* and *psbD*) [24,32]. Here, *SIG5* expression was shown to be light-intensity dependent (Figure 1) and red-light regulated (Figure 9).

To test the importance of photosynthetic electron transfer on *SIG5* expression, half of the 8 day old seedlings grown in white light were sprayed with 10 μM DCMU prior to shifting them for 24 h to red light. DCMU blocks the Q_B-site [49]. Electrons cannot be transferred from photosystem II to the plastoquinone pool. Therefore, DCMU should simulate over-excitation of photosystem II and function antagonistically to non-saturating doses of red light.

In C24, DCMU application decreased red-light dependent induction of the *SIG5* promoter (Figure 9A). In N9313 the *SIG5* transcript levels were hardly detectable after DCMU treatment (Figure 9B). The results demonstrated that regulation of the proximal upstream region depends strongly on photosynthetic electron transport. Red-light regulation of the distal upstream region is only partly dependent on electron transport activity as indicated from a weaker decrease in response to DCMU (Figure 9C).

Figure 9. (**A–C**) Relative *SIG5* and *GFP* transcript levels in C24 and N9313 in response to red and far-red light and red light 24 h after treatment with 10 μM DCMU and light exposure. The seedlings were pre-cultivated 10 days in 120 μmol photons m^{-2} s^{-1} white light followed by 24 h dark incubation. The transcript levels were standardized on ACT2 transcript levels and normalized to the transcript levels prior to the transfer to monocromatic light. Statistical significance of difference (Student T-Test) is indicated as asterisks above bars ($p < 0.05$); (**D–F**) *GFP* expression mediated by the 2 kb full-length *SIG5* promoter, the *SIG5* proximal and *SIG5* distal upstream regions fused to the Δ35S promoter and the Δ35S promoter after 4 days cultivation of the transfected tobacco leaves in monocromatic red and far-red light. In 1/3 of the plants the transfected leaves were sprayed with 1 mM DCMU prior to the exposure to the monochromatic light. The scale of the false color pictures was standardized on the same color scale.

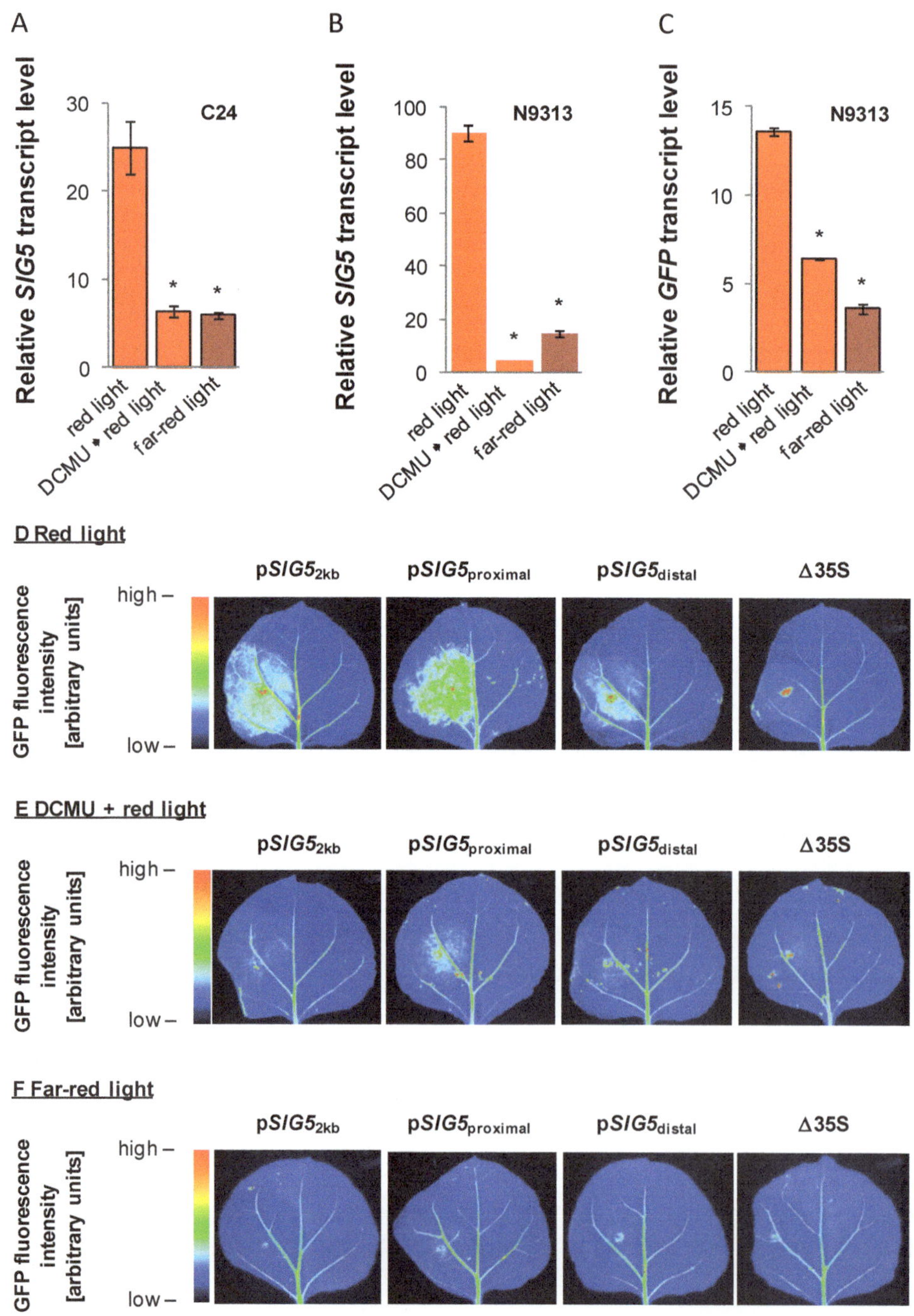

To test how strong the regulation by photosynthetic electron transport is manifested in the promoter, red-light- and DCMU-responses were analyzed in young tobacco leaves which were transfected with constructs expressing *GFP* under the control of the *SIG5* full length promoter (Figure 9D,E), the distal or proximal upstream regions (fused to the Δ35S minimal promoter), or the Δ35S minimal promoter. Similar as in seedlings, DCMU antagonized the red-light dependent induction of the full-length, the distal and the proximal *SIG5* upstream regions.

In non-transfected plant material, the redox state of the plastoquinone pool (1-qP) and the quantum yield of photosystem II (F_V/F_M) were determined (Figure 10). The redox state of the plastoquinone pool was highly oxidized in red light in the presence and absence of DCMU. In red light the quantum yield of photosystem II was slightly higher than in white light in tobacco and in the range of white light treated plants in Arabidopsis in absences of DCMU and decreased after DCMU treatment reflecting PSII damage.

Figure 10. Photosynthetic performance upon light and DCMU treatment. (**A**) Quantum yield of photosystem II following dark adaptation; (**B**) Redox state of the plastoquinone pool. Arabidopsis was grown for 10 days in 100–120 µmol photons m^{-2} s^{-1} white light. Following 24 h dark relaxation, they were transferred to red or white light for 24 h. Half of the plant transferred to red light were treated with DCMU prior to the light shift. DCMU-treated and non-treated tobacco leaves were incubated for 4 days in 100–120 µmol photons m^{-2} s^{-1} white light or red light. Statistical significance of difference (Student's *t*-test) is indicated as asterisks above bars ($p < 0.05$).

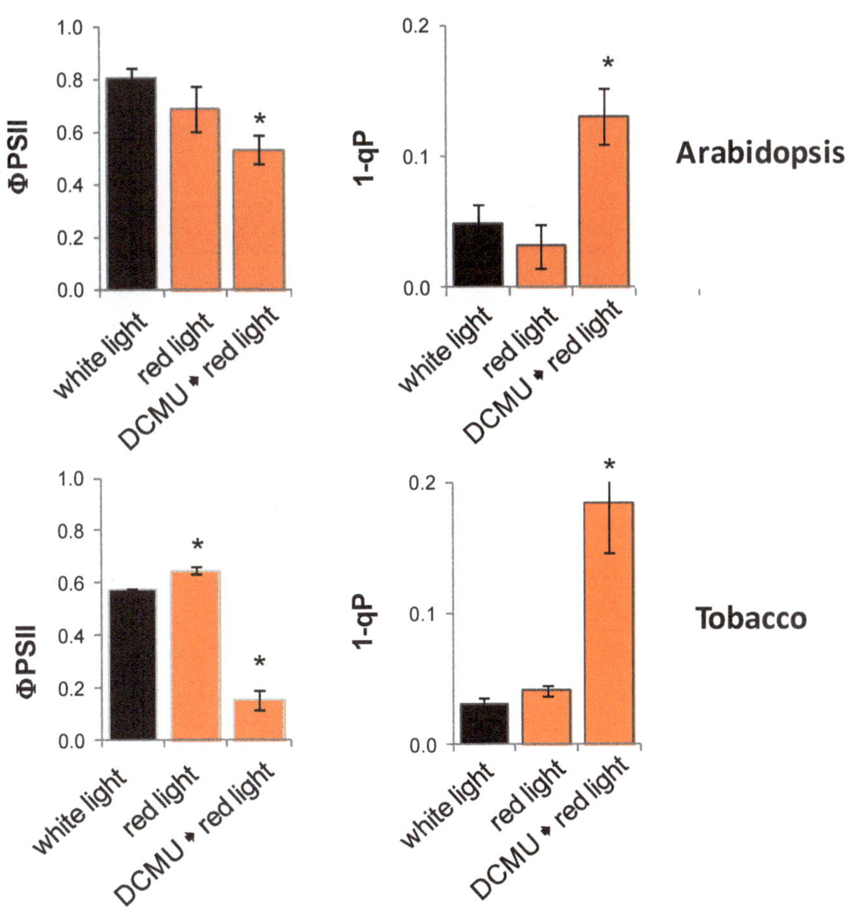

2.12. Sugar Effect on SIG5 Transcription

Intensity dependent light responsiveness can be mediated by photosynthate signatures. *Vice versa*, high levels of carbohydrates can indicate high photosynthetic activity. As part of the feed-back mechanisms, they inhibit the activity of the Calvin-Benson-Cycle and, therefore, the main sink for electrons transported in the photosynthetic light reaction. The plastoquinone pool is more reduced and the quantum yield on photosystem II decreased [50].

The impact of carbohydrates on the light-regulated *SIG5* promoter was tested by application of sucrose. The osmotic side effects of sucrose were simulated by sorbitol. Forty eight hours after application of sucrose or sorbitol to 10 day old light-grown seedlings, the *SIG5* mRNA and *GFP* mRNA levels were quantified in C24 and N9313. Sucrose resulted in decreased *GFP*-levels in N9313. *SIG5* transcript amounts were decreased in C24, but not in N9313 (Figure 11) demonstrating that only the distal upstream region is sucrose-responsive. Sorbitol did not alter *SIG5* expression significantly.

Figure 11. Relative *SIG5* and *GFP* transcript levels in C24 and N9313 in response to sucrose and sorbitol. The transcript levels were standardized on ACT2 and normalized on the transcript levels determined on 0% sucrose. Statistical significance of difference (Student's *t*-test) is indicated as asterisks above bars ($p < 0.05$).

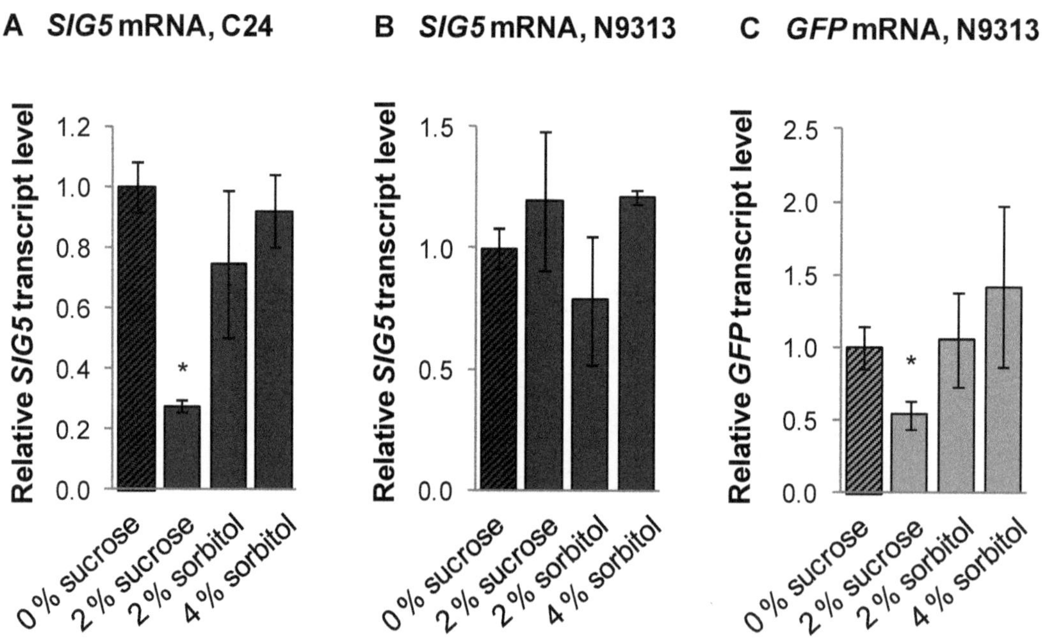

3. Experimental

3.1. Plant Material and Growth Conditions

Seeds of all SALK-lines, Arabidopsis wildtype accessions (C24, Col-0 and L*er*), the *GAL4 GFP* enhancer trap population, and the mutants *cry2*, *hy1-1* and *hy5-1* were obtained from the Nottingham Arabidopsis Stock Centre (NASC, Nottingham, UK). Other mutants were kindly provided by Prof. A. Batschauer (Phillips-Universität, Marburg, Germany).

For sterile growth, Arabidopsis seeds were surface sterilized for 1 min in 80% (v/v) ethanol followed by 8 min in 20% (v/v) household bleach. After five washing steps in sterile water, they were transferred on sterile MS medium (pH 5.7) (Duchefa, The Netherlands) supplemented with 2.5 g L^{-1} Phytagel (Sigma, Taufkirchen, Germany) and 0.5% (w/v) sucrose. Unless otherwise described, experiments were performed with seedlings grown under short day conditions (10 h light, 22 °C/14 h dark, 18 °C) at 120 μmol photons m^{-2} s^{-1} white light (Philips F17TS/TL741 ALTO) after two days of stratification. Alternative light regimes were performed in FloraLED systems (CLF PlantClimatics, Wertingen, Germany) or growth cabinets in blue light (471 nm), red light (673 nm) and far-red light (745 nm).

For sucrose and sorbitol effector studies, 12 day old seedlings grown on sucrose-free media in white light (Philips F17TS/TL741 ALTO; 100 μmol photons m^{-2} s^{-1}) were transferred to media supplemented with 2% sucrose and 2% and 4% sorbitol, respectively. The plants were harvested after 24 illumination with white light.

Nicotiana benthamiana plants were grown on soil for four to five weeks in a day/night regime of 16 h light and 8 h darkness. Light intenstity and light quality studies were performed in FloraLED systems or growth cabinets in blue light (471 nm), red light (673 nm), far-red light (745 nm) or white light (Philips F17TS/TL741 ALTO, 17 watt).

3.2. Extraction of Genomic DNA, TAIL-PCR and Sequence Analysis

For small scale isolation of genomic DNA, plant material was homogenized in 200 μL 50 mM Tris/HCl pH 8.0, 25 mM EDTA, 250 mM NaCl and 0.5% (w/v) SDS. After extraction with 200 μL phenol-chloroform-isoamyl alcohol (25:24:1), the samples were centrifuged at 20,000 ×g for 5 min at room temperature. The DNA was precipitated for 1 h at −20 °C after transfer of the upper phase to 200 μL isopropanol. Following 15 min centrifugation at 20,000 ×g, the precipitated DNA was resuspended in 100 μL water.

TAIL-PCR was performed with 2 μL DNA solution per 20 μL total volume as described in [27] with the primers listed in Table 3. PCR products were cloned into pJET1.2/blunt (Fermentas/Thermo-Fischer, St. Leon-Rot, Germany). Sanger sequencing was performed by GATC (Konstanz, Germany). Database searches were performed using the BLAST program on the NCBI server [51]. The insertion sites were confirmed by PCR, designed to anneal the genomic DNA flanking the mapped insertion sites and T-DNA primer TAIL-TR3.2.

3.3. Southern Blot Analysis

DNA was extracted from plant material using the DNeasy Plant Mini Kit (Qiagen, Hilden, Germany). Three μg of genomic DNA was digested with *Bgl*II (Fermentas) and purified by sodium acetate precipitation prior to separation on 1% (w/v) TAE agarose gels and capillary transfer onto Hybond™-N membranes (GE-Healthcare, Freiburg, Germany). A 504 bp digoxigenin (DIG) labeled *GAL4* probe was amplified from genomic DNA of N9313 with the DIG Probe Synthesis Kit (Roche, Germany) by PCR using primers mPPR1-5 (5'-CGGCAAGCTTGGATCCAACAATG-3') and mPPR1-3 (5'-CCCGG AGCTCGTCCCCCAGGCTG-3'). After crosslinking of the DNA to the membrane by UV-light, the blot was hybridized under high stringency conditions using DIG Easy Hyb Granules (Roche, Mannheim, Germany). Following two washing steps at 68 °C in 75 mM NaCl, 7.5 mM sodium citrate (pH 7.0) with 0.1% (w/v) SDS and two in 15 mM NaCl, 1.5 mM sodium citrate (pH 7.0) with 0.1% (w/v)

SDS, the bound probes were detected in a Luminescent Image Analyzer LAS-4000 (GE-Healthcare) using the CDP-Star, ready to use Kit (Roche).

3.4. Identification and Isolation of Homozygous T-DNA Insertion Lines by PCR

T-DNA insertion lines were tested for the T-DNA insertion site and homozygosity by PCR with T-DNA border primer LBb1.3 (5'-ATTTTGCCGATTTCGGAAC-3') and specific primers (Table 3) binding upstream and downstream of the predicted T-DNA insertion site. For confirmation, the PCRs were repeated with 10 seedlings of the progeny.

Table 3. Primers and annealing temperatures used for PCR.

PCR reaction/ product	T_A [°C]	Forward primer/reverse primer (5'→3')
TAIL-PCR, T-DNA primers	73.1	CACTTGGCGCACTTCGGCTTC
	67.5	AGCTTCTTGAGGCGGCAGA
	63.3	GGAGCTTCATTGTTGGATCC
TAIL-PCR, AD primers	46.0	NTCGA(G/C)T(A/T)T(G/C)G(A/T)GTT
	46.8	NGTCGA(G/C)(A/T)GANA(A/T)GAA
	34.8	(A/T)GTGNAG(A/T)ANCANAGA
Verification of N9266 T-DNA	58	CGTATCACGCGGCGC
Verification of N9313 T-DNA	54	CTCCGACTCTTGCGATAT
qRT-PCR, *ACT2*	60	TCTTCCGCTCTTTCTTTCCAAGC ACCATTGTCACACACGATTGGTTG
qRT-PCR, *SIG5*	60	TGGAGCTAATAACAGCAGACAGC TCGGCTTCAATGAATCGAGCAC
qRT-PCR, *GFP*	60	CCATTACCTGTCCACACAATC GTTCATCCATGCCATGTG
qRT-PCR, *HY5*	60	AGAACAAGCGGCTGAAGAGGTTG TCCTCTCTCTTGCTTGCTGAGCTG
Construction of p*SIG5*$_{wildtype}$	45	ACTAGTTTTTCTGCAGGTAACTCCGACTCTTGCG GCTTGAGAGATTACATTATT
Construction of Δ35S	62	TTCGCAAGACCCTTCCTCTATATAAGG GGGTACCGGTCGCCACC
Analysis of SALK_015625	50	CAATCATGGTTTAATTCGT [a] GATCCACAACCACAAGCC
Analysis of SALK_077048 and 072457	45	GTTATTGATCTGTACCTAGC [a] AAATACGATAGATGTGTTG
Analysis of SALK_019261	45	ATCACAATCTTAAGGCTCAAAA AAATACGATAGATGTGTTG [a]
Analysis of SALK_133729	45	ATCACAATCTTAAGGCTCAAAA [a] AAATACGATAGATGTGTTG
Identification of *cry2-1* allele	58	CAGTTTTATCCTGGAAGAGCCTC CTTCTCCTTTACGGTATGGTCC
Identification of *hy1-1* allele	56	GGAATTAGCAGAGAAGGATCC TATCCGCTCTGCCACCTG
Identification of *phyA-201* allele	55	CCTTAAATGAAGTGTTGACTGC GCAAGATGCACAGAACG

Table 3. *Cont.*

PCR reaction/ product	T_A [°C]	Forward primer/reverse primer (5'→3')
Identification of *phyB-5* allele	55	GTTGTGGAGTGGTTGCTTG CATAGCCGCCTCAGATTC
Identification of *phot1-5* allele	58	CCACTTGCAACCTATGCG CTCTTTCACTGCGGTTTCTTC
Identification of *phot2-1* allele	54	CTCTGCCTCACAATAAGGAG CTGCCAGTATCACCAGAGC
Identification of SALK_056405C	58	GCGGTAGCCAGAGTAATCTATTCC TCCTCTCTCTTGCTTGCTGAGCTG ATTTTGCCGATTTCGGAAC [b]
Construction of p$SIG5_{\Delta GATA}$-1a	60	GGTGGCGACCGGTACC CATCTTTTACTGAATACTTTGAGTTATTTGCACATATAG
Construction of p$SIG5_{\Delta GATA}$-1b	55	CTATATGTGCAAATAACTCAAAGTATTCAGTAAAAGATG GTAACTCCGACTCTTGCGAT
Construction of p$SIG5_{\Delta G}$-1a	60	GGTGGCGACCGGTACC CTGAGAAGACCATCCAATTGTATAATTCCTGATC
Construction of p$SIG5_{\Delta G}$-1b	55	GATCAGGAATTATACAATTGGATGGTCTTCTCAG GTAACTCCGACTCTTGCGAT
Construction of p$SIG5_{\Delta GC}$-1a	60	GGTGGCGACCGGTACC CTATAAATTGGCCAATTCGTCTCTCTCTC
Construction of p$SIG5_{\Delta GC}$-1b	55	GAGAGAGAGACGAATTGGCCAATTTATAG GTAACTCCGACTCTTGCGAT
Construction of p*SIG5* fragments-2	63	GGTGGCGACCGGTACC GTAACTCCGACTCTTGCGAT
Y1H-construction of p*RBCS1A*	55	TTTTTGAGCTCGATTTTGAGTGTGGATATGTGT TTTTTGAATTCCCAGGCAAGTAAAATGAGCAAG
Y1H-construction of HY5 CDS		TTTTTGGATCC**TA**CAGGAACAAGCGACTAGCTC [c] TTTTTCTCGAGTCAAAGGCTTGCATCAGC
Y1H-construction of p$SIG5_{distal}$	57	TTTTTGAGCTCCACAATCTTAAGGCTCAAAAATTG TTTTTGGGCCCTCGGATGCTTTACATGGTG
Y1H-construction of p$SIG5_{proximal}$	60	TTTTTGAATTCGTAACTCCGACTCTTGCG TTTTTGAGCTCGCTTGAGAGATTACATTATT

[a] Leading to amplification of a T-DNA specific PCR product when combined with primer LBb1.3 annealing with the left border of the SALK T-DNA. [b] LBb1.3, annealing with the left border of the SALK T-DNA. LBb1.3 leads to a PCR product in SALK_056405C if combined with the reverse primer annealing the HY5 coding sequence. [c] two nucleotides added to HY5 CDS are marked in bold.

3.5. Determination of Reporter Gene Activity

GFP activity screens were performed in 96-well microtiter plates on 100 µL MS medium per well supplemented with 0.5% (w/v) sucrose in a Fluoroskan Ascent FL fluorometer (Thermo Fisher Scientific, St. Leon-Rot, Germany) in the top-reader modus with 500 ms integration time, 485 nm excitation at 527 nm. Each well was covered with nine measurement points to scan all plants equally.

Background fluorescence was subtracted from the means by analyzing the autofluorescence at 527 nm of parallel grown C24 seedlings. Normalization was performed based on area determination using the ImageJ software package [52].

In crosses of N9313 with the photoreceptor mutants, the *GFP* activity was quantified in triplicates from 100 µL extracts because of the high variability in seedling morphology. Plant material (10–20 mg) was homogenized in 500 µL sodium phosphate, pH 7.0. After 2 min centrifugation at 20,000 ×g *GFP* fluorescence was quantified in a Mithras LB 940 top reader fluorometer (Berthold Technologies GmbH & Co. KG, Bad Wildbad, Germany) with 100 ms counting time (excitation filter 460/10 and emission filter F510). Chlorophyll contents were determined according to [53].

GUS expression was quantified in 15–30 mg leaf material and standardized on protein levels as described in [54].

3.6. Transcript Abundance Analysis

RNA was isolated from 10–15 frozen seedlings in Precellys® 24 (Peqlab, Erlangen, Germany) using the RNeasy Plant Mini Kit. The purity of the RNA was determined spectrophotometrically from the A_{260}/A_{280} ratio. cDNA was synthesized with the High-Capacity cDNA Reverse Transcription Kit (Applied Biosystems, Darmstadt, Germany). Primers for qRT-PCR were designed by the QuantPrime software [55,56] (Table 3). The primer specificity was checked by assaying that the melting curves display only a single peak for each PCR product of interest, and by gel electrophoresis. Real-time amplification was performed according to the MIQE standards [57] using the Brilliant II SYBR® Green Master Mix (Agilent Technologies, Böblingen, Germany) in a Stratagene MX3005P Cycler (Agilent Technologies, Böblingen, Germany). No-template-controls were performed for each gene. Each biological replicate represents an independent RNA isolation. Levels of each transcript relative to the constitutively expressed *ACT2* control gene [58] were quantified as described by Pfaffl [59].

3.7. mRNA Decay Analysis

Transcription was inhibited by transferring 10 day old N9313 seedlings to liquid MS medium containing 200 µM Actinomycin D (Act D) (Sigma-Aldrich, Taufkirchen, Germany) at 120 µmol photons m^{-2} s^{-1}. Control plants were incubated in MS medium. The mRNA half-life was quantified for *SIG5* and *GFP* transcripts from the C_t values determined by *ACT2*-standardized qRT-PCR in samples treated for 1, 2, 4 and 8 h. Since mRNA decay generally obeys first-order kinetics [60,61], an exponential regression model ($A = 1e^{-kt}$) was applied for determination of the decay coefficient (k_{decay}). The mRNA half-life was then calculated using the following equation: $t½ = \ln(2)/k_{decay}$.

3.8. Reporter Gene Construct Design, Site-Directed Mutagenesis and Tobacco Transfection

SIG5 promoter fragments were amplified from genomic DNA of the accession C24 by PCR with specific primers containing *Pst*I and a *Bcu*I restriction sites (Table 3) and cloned into the TA cloning site of the pCR8/GW/TOPO vector (Invitrogen, Darmstadt, Germany). The truncated (–48) 35S minimal promoter from cauliflower mosaic virus (CaMV) which was amplified by PCR from

T-DNA of the enhancer trap line and cloned into the restriction sites. The promoter fragments were transferred into the binary vector pHGWFS7.0 [62] with LR Clonase II enzyme mix (Invitrogen, Darmstadt, Germany). *Agrobacterium tumefaciens* strain GV3101 [63] was transformed using the freeze-thaw method [64].

For transfection of tobacco leaves [65], transformed *Agrobacteria* were cultivated in 10 mL YEB medium at 28 °C and 180 rpm to OD_{600} 0.5. For co-infiltrations, GV3101(pMP90) and the respective *SIG5*-promoter strain were mixed in a 2:3-ratio. Cells were collected by 8 min centrifugation at 3000 ×*g* at room temperature and resuspended in 100 mM MES pH 5.6 plus 10 mM $CaCl_2$ supplemented with 150 μM acetosyringone. After 2 h incubation, the agrobacteria suspensions were infiltrated into tobacco leaves according to [66]. Site-directed mutagenesis was performed according to [44].

3.9. *In Silico Analysis of Promoter Sequences*

Promoter motif searches were performed *in silico* using PLACE [35,67] and PlantCARE [34,68].

3.10. *Yeast One-Hybrid Analysis*

SIG5 and *RBCS1A* promoter fragments and *HY5* cDNA were generated by PCR using genomic DNA of the ecotype C24 as template. The *HY5* cDNA was cloned into the *Bam*HI/*Xho*I site of pACT2 (Clontech, Montain View, CA, USA). *SIG5* promoter fragments and a 196-bp fragment of the Arabidopsis *RBCS1A* promoter were cloned into the *Sac*I/*Eco*RI site of pHIS2 (Clontech, Mountain View, CA, USA) using the LigaFast Rapid DNA Ligation System (Promega, Fitchburg, Madison, WI, USA). The *Saccharomyces cerevisiae* strain Y187 (*MATα, ura3-52, his3-200, ade2-101, trp1-901, leu2-3, 112, gal4Δ, met-, gal80Δ, URA3:GAL1*$_{UAS}$*GAL1*$_{TATA}$*-lacZ, MEL1*, with reporter genes *HIS3* and *lacZ*) was transformed with lithium acetate (LiAc) [69]. Transformants were selected on minimal synthetic dropout (SD) medium (Clontech, Montain View, CA, USA) lacking the amino acids corresponding to the autotrophy markers of the plasmids. For yeast colony-PCR, a single colony was resuspended in 50 μL water supplemented with 60 U mL^{-1} lyticase and incubated at 30 °C for 30 min. After cell lysis at 95 °C (10 min) and 2 min centrifugation at 20,000 ×*g*, 2 μL of the supernatant was used for PCR in a total volume of 20 μL.

For the interaction tests, 5 μL of overnight cultures of double-transformed yeast stains (adjusted to OD_{600} = 0.01) were dropped onto SD/-His/-Trp plates containing 10–100 μM 3-amino-1,2,4-triazol (3-AT) to increase the stringency. Plates were incubated 2–3 days at 30 °C.

3.11. *Crossing Arabidopsis thaliana Plants and Mutant Selection*

Arabidopsis lines were cross-pollinated. From the progenies, homozygous *cry1*, *phyA* or *phyB* mutants were selected according to their hypocotyl length in blue, red or white light, respectively [18]. Plants carrying the homozygous *phot1* allele were identified by measuring the phototropic response [70]. The presence of the *GFP*-enhancer element was tested by PCR.

3.12. Fluorescence Microscopy

Tissue-specific *GFP* expression was analyzed by confocal laser scanning microscopy using a Zeiss LSM 510 Meta with a multiline argon ion laser (Carl Zeiss, Jena, Germany). Cells were examined with a 40X Zeiss oil-immersion objective (1.3 numerical aperture). *GFP* was excited at 488 nm and the emission was recorded through the meta-channel at 497–550 nm. Fluorescence images were analyzed with LSM Image Browser software [71].

3.13. Chlorophyll-a Fluorescence Analysis

The quantum yield of photosystem II and the redox state of the plastoquinone pool were determined with a Dual-PAM-100 (Heinz Walz, Effeltrich, Germany) at room temperature after 20 min dark incubation with saturating light flashes of 6000 µmol photons m^{-2} s^{-1} as F_V/F_M [72]. The redox state of the plastoquinone pool was calculated from the steady state photochemical quench (qP) in 100 µmol photons m^{-2} s^{-1} red light (600 nm) as (1-qP) with light flashes of 6000 µmol photons m^{-2} s^{-1}.

4. Discussion

4.1. Transcriptional Regulation of SIG5 Promoter

SIG5 is one of six sigma factors of Arabidopsis [73,74] controlling plastid gene expression [75]. It activates *psbD* and *psbA* transcription upon dark-light shifts and controls diurnally chloroplast gene expression in a light-dependent manner [24,25,32]. Whereas the structure of the *psbD* promoter was a subject of intensive investigation [76,77], the structure of the *SIG5* promoter has been barely investigated. From transcript abundance analysis [24,29,30] and GUS-reporter gene assays [32], *SIG5* transcription was shown to be strongly blue-light responsive and stress-sensitive. Here, we showed that it is also red-light responsive in a photoreceptor- and photosynthesis-dependent manner (Figures 6, 7 and 9).

In blue light, *SIG5* is the fastest transcriptionally activated sigma factor [30]. *SIG5* binds to the −948 upstream region of the *psbD* promoter, which drives the *psbD-psbC-psbZ* operon encoding core-proteins of photosystem II, in response to blue and white light and various stress conditions [32,76]. *SIG5* is hardly expressed in far-red light (Figure 3C and [33]) and, therefore, can be assumed to be less important than other sigma factors, especially *SIG2* and *SIG6*, in shade avoidance reactions, such as hypocotyl elongation, and activation of greening upon germination [78,79]. As shown by promoter-GUS analysis [32], *SIG5* transcription activity quickly decreases three and four days after onset of germination and becomes activated again in rosette leaves. *SIG5* transcript abundance increases in response to high light, NaCl, mannitol and cold in rosette leaves [32]. Analysis of *SIG5*-deficient T-DNA insertion lines demonstrated that *SIG5* is essential for maintaining photosynthetic activity upon stress [32].

In contrast to the blue/far-red type of sigma factors, e.g., *SIG2* and *SIG6*, *SIG5* is induced by red and blue light (Figures 3 and 5). The regulatory elements are spread widely over the proximal and distal upstream region. Comparison of distal and proximal regulatory activities (Figure 3) and stepwise deleted promoters (Figure 5) show complex combinational regulation. Most of the red-light induction, part of the photosynthetic control and the sugar regulation take place in the distal upstream region

(Figures 3, 9 and 11), while the proximal upstream region mediates the main blue-light sensitivity (Figure 5).

4.2. Blue-light Regulation

The 196 bp blue-light responsive region, located 208–404 bp upstream of the predicted transcription initiation site (Figure 5A), is under control by CRY1 and PHYB (Figure 6A,B). CRY2 and the phototropins PHOT1 and PHOT2 are of negligible importance (Figure 6A).

The CRY2-mediated blue-light responsive enhancers of the distal upstream region (Figures 3A,B and 5A) may support CRY1 regulation of the distal upstream region (Figure 7A) under low light, such as upon shading or during germination. In the rosette stage, Arabidopsis leaves are frequently exposed to high light intensities. Under these conditions, CRY2 can be expected to be light inactivated [80,81].

4.3. Red-light Sensitivity

SIG5 transcription and transcript abundance were regulated by various phytochromes via widely dispersed red-light sensitive promoter elements (Figures 5 and 6). The region −887 to −1032 in the proximal upstream region showed preference toward PHYB-mediated regulation (Figure 6B), which is also the most abundant phytochrome in light-grown *Arabidopsis* [82].

In blue-light around 78% and in red-light 59% of the *SIG5* transcripts depended on PHYB after 24 h (Figure 6B,C). In a similar experiment, in which light-grown Arabidopsis plants were transferred back to light after 30 h dark treatment, *SIG5* transcript levels were wild-type-like in *phyA* and *phyB* mutants after 1.5 h of illumination [32]. Differences can be observed after 6 h. At this time, the overall *SIG5* expression was low in the analysis [32]. The comfortable access to a wide range of transcript abundance studies from various labs via tools like the eFP browser [28] showed strong diurnal regulation of *SIG5* expression. Therefore, we compared transcript levels exactly after 24 h (except Figure 4A; 3 h data) and harvested plant material at the time of highest transcript abundance 2–4 h after onset of light. This approach demonstrated that *SIG5* expression is also strongly red-light sensitive. Light signals are transmitted mainly via CRY1, PHYA and PHYB in a HY5 and COP1 dependent manner. Consistent with ChIP-based prediction of HY5-binding sites in the *SIG5* promoter, [83] and previously shown HY5-dependent transcript abundance regulation [32], HY5 binds to the proximal upstream region (Figure 8B) and a HY5-binding motif, a G-box, was shown to transmit white- and red-light signals (Figure 8A).

4.4. Antero- and Retrograde Signaling Connecting Chloroplast and Nuclear Gene Expression

SIG5 expression is promoted in red- and blue-light (Figure 3) and in a light intensity dependent manner (Figure 1). Depending on *SIG5* regulation, chloroplast gene expression and photoprotection are regulated [24,32] indicating that its main function of *SIG5* is anterograde light acclimation, such as compensation of photosynthetic imbalances due to suboptimal photoreaction center innervation.

In addition to the anterograde function [32], *SIG5* is regulated by retrograde signals (Figure 9): Red light preferentially activates photosystem II [84]. Blocking the Q_B-site of photosystem II with DCMU [49], decreased activation of *SIG5* expression in red light (Figure 9). In red light and after DCMU treatment, the plastoquinone pool was oxidized as determined by chlorophyll-a fluorescence (1-qP; Figure 10). This result excludes the possibility of regulation by the redox state of the plastoquinone pool and showed that full *SIG5* expression depends on photosynthetic electron transfer from photosystem II to the plastoquinone-pool.

psbA and *psaAB* expression respond to light-quality changes in order to adjust photoreaction center I and II availability to the relative innervation of the photosystems [84]. Transcription of *psaAB* is under control of the redox state of the plastoquinone pool, but not *psbA* [85]. *SIG5* controls expression of *psbD* and, less specifically, *psbA* [32]. It is the anterograde control factor [32] which can adjust plastid gene expression to light signals due to its complex light quality sensitivity (Figures 1, 3 and 5).

Excess carbohydrates inhibit photosynthetic electron transport [50,86]. Like DCMU, sucrose suppresses *SIG5* expression (Figure 11). If *SIG5* would be regulated depending on the redox state of the plastoquinone pool, sucrose application should have resulted in a higher reduction status. The sucrose data give additional evidence for independence from plastoquinone redox regulation, which controls state transition [87], expression of photosystem I core proteins [84] and STN7-dependent regulation of nuclear gene expression [88]. Regulation of *SIG5* expression by red light and DCMU (Figures 9 and 11) demonstrates that transcription of the nuclear encoded *SIG5* gene responds to photosynthetic signals.

The antagonism of the sucrose effect to the red-light effect excludes carbohydrates as direct signaling molecules in chloroplast-to-nucleus transmission of the photosynthetic electron transport signal. Insensitivity to similar concentrations of sorbitol (Figure 11B,C) and induction by higher concentrations of sorbitol (Figure 11C) and mannitol [32] excludes osmotic signaling. The comparison of the red-light and DCMU response (Figures 3 and 9) and the general stress-induction reported before [32], which decrease photosystem II activity [89], let assume that the signal emerges at PSII in response to high excitation pressure. A link between generation of signaling by D1 damage or the D1 repair cycle, demonstrates that *SIG5* controls a perfect regulatory circuitry: In chloroplasts, D1 damage reflects over-excitation of photosystem II [90]. In response to retrograde signals, *SIG5* drives expression of D1 and its closest protein partners D2, *PsbC* and *PsbZ* [24,32] and stabilizes photosystem II activity (Figure 12).

5. Conclusions

SIG5 expression is regulated by multiple light-responsive elements spread over the 2 kb upstream region (Figure 5). Signal transduction is controlled by cryptochromes, especially CRY1, and phytochromes (Figure 6) in a HY5 and COP1-dependent manner (Figure 7) and by photosynthetic signals (Figure 9). We conclude that the chloroplast sigma factor *SIG5* is a retrograde and light controlled regulator of chloroplast function. It combines intrinsic and extrinisic information important in adjusting nuclear and plastid gene expression upon light acclimation processes and links the signaling and regulation potential of the eukaryotic-type extra-plastidic gene expression system with the chloroplast regulatory system.

Figure 12. *SIG5* expression control in the light-signaling pathway regulating the expression of the chloroplast genes *psbD* and *psbA* for the photoreaction center II core proteins D1 and D2 in response to light intensity and quality. Light signals regulate *SIG5* expression by photoreceptors in a HY5- and COP1-controlled manner via various promoter elements. *SIG5* is post-translationally targeted to chloroplasts, where the sigma factor is involved in tuning expression of *psbD* and *psbA*. Photoreaction center expression is involved in the PSII repair cycle. Retrograde signals transmit information correlating with the photosystem II excitation status and damage, respectively, to the nucleus to adjust the intensity of *SIG5* transcription.

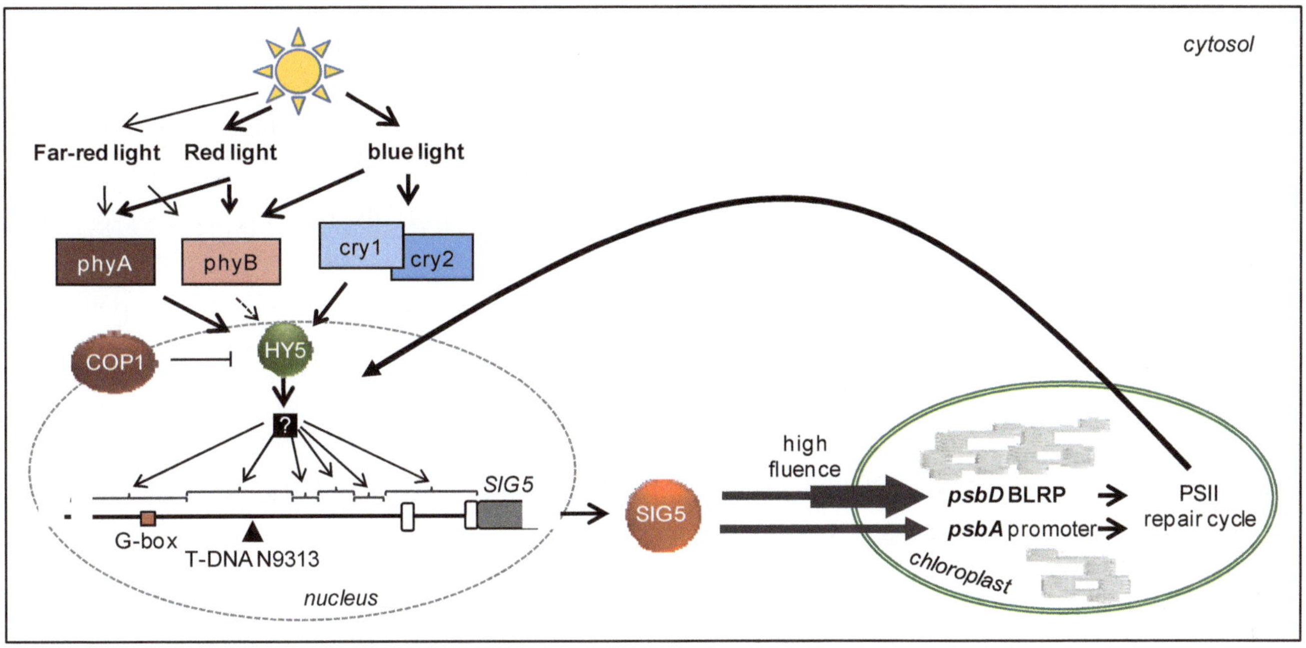

Acknowledgements

This work was supported by the German Research Foundation (Grant Ba2011/5-1), the Heinrich-Heine-Universität Düsseldorf and the Freie Universität Berlin. We thank Alfred Batschauer for providing seed stocks of *cry2*, *phot1phot2*, *phyA*, *phyB phyAphyB* and *cop1*. Special thanks to Rüdiger Simon for hosting M.M. for the final phase of her project, to our former lab neighbor, Peter Jahns, for providing his equipment for chlorophyll—a fluorescence analysis after we had moved ours to Berlin, and to Rainer Bode for critical reading.

Author Contributions

Marina Mellenthin (90%), Ulrike Ellersiek (8%) and Anna Börger (2%) performed the experiments under supervision of Margarete Baier. Marina Mellenthin and Margarete Baier drafted the manuscript.

Conflicts of Interest

The authors declare no conflict of interest.

References

1. Terzaghi, W.B.; Cashmore, A.R. Light-regulated transcription. *Annu. Rev. Plant Physiol. Plant Mol. Biol.* **1995**, *46*, 445–474.
2. Troxler, R.F.; Zhang, F.; Hu, J.; Bogorad, L. Evidence that sigma factors are components of chloroplast RNA polymerase. *Plant Physiol.* **1994**, *104*, 753–759.
3. Block, A.; Dangl, J.L.; Hahlbrock, K.; Schulze-Lefert, P. Functional borders, genetic fine-structure, and distance requirements of cis elements mediating light responsiveness of the parsley chalcone cynthase promoter. *Proc. Natl. Acad. Sci. USA* **1990**, *87*, 5387–5391.
4. Gilmartin, P.M.; Sarokin, L.; Memelink, J.; Chua, N.-H. Molecular light switches for plant genes. *Plant Cell* **1990**, *2*, 369–378.
5. Giuliano, G.; Pichersky, E.; Malik, V.S.; Timko, M.P.; Scolnik, P.A.; Cashmore, A.R. An evolutionarily conserved protein-binding sequence upstream of a plant light-regulated gene. *Proc. Natl Acad. Sci. USA* **1988**, *85*, 7089–7093.
6. Ha, S.B.; An, G.H. Identification of upstream regulatory elements involved in the developmental expression of the *Arabidopsis thaliana cab1* gene. *Proc. Natl. Acad. Sci. USA* **1988**, *85*, 8017–8021.
7. Borello, U.; Ceccarelli, E.; Giuliano, G. Constitutive, light-responsive and circadian clock-responsive factors compete for the different L-box elements in plant light-regulated promoters. *Plant J.* **1993**, *4*, 611–619.
8. Grob, U.; Stuber, K. Discrimination of phytochrome dependent light inducible from non-light inducible plant genes—Prediction of a common light-responsive element (Lre) in phytochrome dependent light inducible plant genes. *Nucleic Acids Res.* **1987**, *15*, 9957–9973.
9. Srivastava, R.; Rai, K.M.; Srivastava, M.; Kumar, V.; Pandey, B.; Singh, S.P.; Bag, S.K.; Singh, B.D.; Tuli, R.; Sawant, S.V. Distinct role of core promoter architecture in regulation of light-mediated responses in plant genes. *Mol. Plant* **2014**, *7*, 626–641.
10. Martinez-Garcia, J.F.; Huq, E.; Quail, P.H. Direct targeting of light signals to a promoter element-bound transcription factor. *Science* **2000**, *288*, 859–863.
11. Rüdiger, W.; Oster, U. Intracellular signaling from plastids to the nucleus. In *Photosynthesis*, 34th ed.; Eaton-Rye, J.J., Tripathy, B.C., Sharkey, T.D., Eds.; Springer: Dortrecht, The Netherlands, 2012; pp. 175–190.
12. Chory, J. Light signal transduction: An infinite spectrum of possibilities. *Plant J.* **2010**, *61*, 982–991.
13. Nelson, T.; Harpster, M.H.; Mayfield, S.P.; Taylor, W.C. Light-regulated gene expression during maize leaf development. *J. Cell Biol.* **1984**, *98*, 558–564.
14. Lorrain, S.; Genould, T.; Fankhauser, C. Let there be light in the nucleus! *Curr. Opin. Plant Biol.* **2006**, *9*, 509–514.
15. Millar, A.J. Input signals to the plant circadian clock. *J. Exp. Bot.* **2004**, *55*, 277–283.
16. Sullivan, J.A.; Deng, X.W. From seed to seed: The role of photoreceptors in *Arabidopsis* development. *Dev. Biol.* **2003**, *260*, 289–297.
17. Larkin, R.M.; Ruckle, M.E. Integration of light and plastid signals. *Curr. Opin. Plant Biol.* **2008**, *11*, 593–599.
18. Neff, M.M.; Chory, J. Genetic Interactions between phytochrome A, phytochrome B, and cryptochrome 1 during *Arabidopsis* development. *Plant Physiol.* **1998**, *118*, 27–35.

19. Chen, F.; Li, B.; Li, G.; Charron, J.B.; Dai, M.; Shi, X.; Deng, X.W. Arabidopsis phytochrome A directly targets numerous promoters for individualized modulation of genes in a wide range of pathways. *Plant Cell* **2014**, *26*, 1949–1966.
20. Haseloff, J. *GFP* variants for multispectral imaging of living cells. *Methods Cell Biol.* **1999**, *58*, 139–151.
21. Laplaze, L.; Parizot, B.; Baker, A.; Ricaud, L.; Martiniére, A.; Auguy, F.; Franche, C.; Nussaume, L.; Bogusz, D.; Haseloff, J. GAL4-*GFP* enhancer trap lines for genetic manipulation of lateral root development in *Arabidopsis thaliana*. *J. Exp. Bot.* **2005**, *56*, 2433–2442.
22. He, Y.H.; Tang, W.N.; Swain, J.D.; Green, A.L.; Jack, T.P.; Gan, S.S. Networking senescence-regulating pathways by using Arabidopsis enhancer trap lines. *Plant Physiol.* **2001**, *126*, 707–716.
23. Gardner, M.J.; Baker, A.J.; Assie, J.M.; Poethig, R.S.; Haseloff, J.P.; Webb, A.A.R. GAL4 *GFP* enhancer trap lines for analysis of stomatal guard cell development and gene expression. *J. Exp. Bot.* **2009**, *60*, 213–226.
24. Tsunoyama, Y.; Ishizaki, Y.; Morikawa, K.; Kobori, M.; Nakahira, Y.; Takeba, G.; Toyoshima, Y.; Shiina, T. Blue light-induced transcription of plastid-encoded *psbD* gene is mediated by a nuclear-encoded transcription initiation factor, At*SIG5*. *Proc. Natl. Acad. Sci. USA* **2004**, *101*, 3304–3309.
25. Noordally, Z.B.; Ishii, K.; Atkins, K.A.; Wetherill, S.J.; Kusakina, J.; Walton, E.J.; Kato, M.; Azuma, M.; Tanaka, K.; Hanaoka, M.; *et al.* Circadian control of chloroplast transcription by a nuclear-encoded timing signal. *Science* **2013**, *339*, 1316–1319.
26. Pena-Ahumada, A.; Kahmann, U.; Dietz, K.J.; Baier, M. Regulation of peroxiredoxin expression versus expression of Halliwell-Asada-Cycle enzymes during early seedling development of *Arabidopsis thaliana*. *Photosyn. Res.* **2006**, *89*, 99–112.
27. Liu, Y.G.; Mitsukawa, N.; Oosumi, T.; Whittier, R.F. Efficient isolation and mapping of *Arabidopsis thaliana* T-DNA insert junctions by thermal asymmetric interlaced PCR. *Plant J.* **1995**, *8*, 457–463.
28. Winter, D.; Vinegar, B.; Nahal, H.; Ammar, R.; Wilson, G.V.; Provart, N.J. An "Electronic fluorescence pistograph" browser for exploring and analyzing large-scale data sets. *PLoS One* **2007**, *2*, e178.
29. Mochizuki, T.; Onda, Y.; Fujiwara, E.; Wada, M.; Toyoshima, Y. Two independent light signals cooperate in the activation of the plastid *psbD* blue light-responsive promoter in *Arabidopsis*. *FEBS Lett.* **2004**, *571*, 26–30.
30. Onda, Y.; Yagi, Y.; Saito, Y.; Takenaka, N.; Toyoshima, Y. Light induction of Arabidopsis *SIG1* and *SIG5* transcripts in mature leaves: Differential roles of cryptochrome 1 and cryptochrome 2 and dual function of *SIG5* in the recognition of plastid promoters. *Plant J.* **2008**, *55*, 968–978.
31. Tsunoyama, Y.; Morikawa, K.; Shiina, T.; Toyoshima, Y. Blue light specific and differential expression of a plastid sigma factor, *SIG5* in *Arabidopsis thaliana*. *FEBS Lett.* **2002**, *516*, 225–228.
32. Nagashima, A.; Hanaoka, M.; Shikanai, T.; Fujiwara, M.; Kanamaru, K.; Takahashi, H.; Tanaka, K. The multiple-stress responsive plastid sigma factor, *SIG5*, directs activation of the *psbD* blue light-responsive promoter (BLRP) in *Arabidopsis thaliana*. *Plant Cell Physiol.* **2004**, *45*, 357–368.

33. Oh, S.; Montgomery, B.L. Phytochrome-dependent coordinate control of distinct aspects of nuclear and plastid gene expression during anterograde signaling and photomorphogenesis. *Front. Plant Sci.* **2014**, *5*, e171.
34. Rombauts, S.; Déhais, P.; van Montagu, M.; Rouzé, P. PlantCARE, a plant *cis*-acting regulatory element database. *Nucleic Acids Res.* **1999**, *27*, 295–296.
35. Higo, K.; Ugawa, Y.; Iwamoto, M.; Korenaga, T. Plant *cis*-acting regulatory DNA elements (PLACE) database. *Nucleic Acids Res.* **1999**, *27*, 297–300.
36. Alonso, J.M.; Stepanova, A.N.; Leisse, T.J.; Kim, C.J.; Chen, H.M.; Shinn, P.; Stevenson, D.K.; Zimmerman, J.; Barajas, P.; Cheuk, R.; *et al.* Genome-wide insertional mutagenesis of *Arabidopsis thaliana*. *Science* **2003**, *301*, 653–657.
37. Kagawa, T.; Kasahara, M.; Abe, T.; Yoshida, S.; Wada, M. Function analysis of phototropin2 using fern mutants deficient in blue light-induced chloroplast avoidance movement. *Plant Cell Physiol.* **2004**, *45*, 416–426.
38. Guo, H.; Duong, H.; Ma, N.; Lin, C. The Arabidopsis blue light receptor cryptochrome 2 is a nuclear protein regulated by a blue light-dependent post-transcriptional mechanism. *Plant J.* **1999**, *19*, 279–287.
39. Lopez-Juez, E.; Dillon, E.; Magyar, Z.; Khan, S.; Hazeldine, S.; de Jager, S.M.; Murray, J.A.H.; Beemster, G.T.S.; Bogre, L.; Shanahan, H. Distinct light-initiated gene expression and cell cycle programs in the shoot apex and cotyledons of Arabidopsis. *Plant Cell* **2008**, *20*, 947–968.
40. Schäfer, E.; Haupt, W. Blue light effects in phytocrome-mediated responses. In *Encyclopedia of Plant Physiology*; Shropshire, W., Jr., Mors, H., Eds.; Springer: New York, NY, USA, 1983; p. 16A.
41. Muramoto, T.; Kohchi, T.; Yokota, A.; Hwang, I.H.; Goodman, H.M. The *Arabidopsis* photomorphogenic mutant *hy1* is deficient in phytochrome chromophore biosynthesis as a result of a mutation in a plastid heme oxygenase. *Plant Cell* **1999**, *11*, 335–347.
42. Oyama, T.; Shimura, Y.; Okada, K. The *Arabidopsis* HY5 gene encodes a bZIP protein that regulates stimulus-induced development of root and hypocotyl. *Genes Dev.* **1997**, *11*, 2983–2995.
43. Chattopadhyay, S.; Puente, P.; Deng, X.W.; Wei, N. Combinatorial interaction of light-responsive elements plays a critical role in determining the response characteristics of light-regulated promoters in *Arabidopsis*. *Plant J.* **1998**, *15*, 69–77.
44. Montemartini, M.; Kalisz, H.M.; Hecht, H.J.; Steinert, P.; Flohe, L. Activation of active-site cysteine residues in the peroxiredoxin-type tryparedoxin peroxidase of *Crithidia fasciculata*. *Eur. J. Biochem.* **1999**, *264*, 516–524.
45. McKendree, W.L., Jr.; Ferl, R.J. Functional elements of the *Arabidopsis* Adh promoter include the G-box. *Plant Mol. Biol.* **1992**, *19*, 859–862.
46. Donald, R.G.; Cashmore, A.R. Mutation of either G box or I box sequences profoundly affects expression from the Arabidopsis rbcS-1A promoter. *EMBO J.* **1990**, *9*, 1717–1726.
47. Laisk, A.; Oja, V.; Eichelmann, H.; Dall'Osto, L. Action spectra of photosystems II and I and quantum yield of photosynthesis in leaves in State 1. *Biochim. Biophys. Acta* **2014**, *1837*, 315–325.
48. Pfannschmidt, T.; Nilsson, A.; Allen, J.F. Photosysnthetic control of chloroplast gene expression. *Nature* **1999**, *397*, 625–628.
49. Trebst, A.; Depka, B.; Kraft, B.; Johanningmeier, U. The Q_B site modulates the conformation of the photosystem II reaction center polypeptides. *Photosyn. Res.* **1988**, *18*, 163–177.

50. Rook, F.; Corke, F.; Baier, M.; Holman, R.; May, A.G.; Bevan, M.W. *Impaired sucrose induction1* encodes a conserved plant-specific protein that couples carbohydrate availability to gene expression and plant growth. *Plant J.* **2006**, *46*, 1045–1058.
51. NCBI-Server. Available online: http://www.ncbi.nlm.nih.gov (accessed on 18 April 2010).
52. Sheffield, J.B. ImageJ, a useful tool for biological image processing and analysis. *Microsc. Microanal.* **2007**, *13*, 200–201.
53. Porra, R.J.; Thompson, W.A.; Kriedemann, P.E. Determination of accurate extinction coefficients and simultaneous equations for assaying chlorophyll-a and chlorophyll-b extracted with four different solvents: Verification of the concentration of chlorophyll standards by atomic absorption spectroscopy. *Biochim. Biophys. Acta* **1989**, *975*, 384–394.
54. Baier, M.; Ströher, E.; Dietz, K.J. The acceptor availability at photosystem I and ABA control nuclear expression of 2-Cys peroxiredoxin-A in *Arabidopsis thaliana. Plant Cell Physiol.* **2004**, *45*, 997–1006.
55. Arvidsson, S.; Kwasniewski, M.; Riano-Pachon, D.M.; Mueller-Roeber, B. QuantPrime—A flexible tool for reliable high-throughput primer design for quantitative PCR. *BMC Bioinform.* **2008**, *9*, e465.
56. Quantprime. Avilable online: http://www.quantprime.mpimp-golm.mpg.de (accessed on 28 January 2010).
57. Bustin, S.A.; Benes, V.; Garson, J.A.; Hellemans, J.; Huggett, J.; Kubista, M.; Mueller, R.; Nolan, T.; Pfaffl, M.W.; Shipley, G.L.; *et al.* The MIQE guidelines: Minimum information for publication of quantitative real-time PCR experiments. *Clin. Chem.* **2009**, *55*, 611–622.
58. An, Y.Q.; McDowell, J.M.; Huang, S.; McKinney, E.C.; Chambliss, S.; Meagher, R.B. Strong, constitutive expression of the *Arabidopsis* ACT2/ACT8 actin subclass in vegetative tissues. *Plant J.* **1996**, *10*, 107–121.
59. Pfaffl, M.W. A new mathematical model for relative quantification in real-time RT-PCR. *Nucleic Acids Res.* **2001**, *29*, e45.
60. Ross, J. Messenger-RNA stability in mammalian cells. *Microbiol. Rev.* **1995**, *59*, 423–450.
61. Gutierrez, R.A.; Ewing, R.M.; Cherry, J.M.; Green, P.J. Identification of unstable transcripts in *Arabidopsis* by cDNA microarray analysis: Rapid decay is associated with a group of touch- and specific clock-controlled genes. *Proc. Natl. Acad. Sci. USA* **2002**, *99*, 11513–11518.
62. Karimi, M.; Inze, D.; Depicker, A. GATEWAY((TM)) vectors for *Agrobacterium*-mediated plant transformation. *TIPS* **2002**, *7*, 193–195.
63. Koncz, C.; Schell, J. The promoter of TL-DNA gene *5* controls the tissue-specific expression of chimaeric genes carried by a novel type of *Agrobacterium* binary vector. *Mol. Gen. Genet.* **1986**, *204*, 383–396.
64. Weigel, D.; Glazebrook, J. *Arabidopsis: A Laboratory Manual*; Cold Spring Harbor Laboratory Press: New York, NY, USA, 2002.
65. Voinnet, O.; Rivas, S.; Mestre, P.; Baulcombe, D. An enhanced transient expression system in plants based on suppression of gene silencing by the p19 protein of tomato bushy stunt virus. *Plant J.* **2003**, *33*, 949–956.

66. English, J.J.; Davenport, G.F.; Elmayan, T.; Vaucheret, H.; Baulcombe, D.C. Requirement of sense transcription for homology-dependent virus resistance and trans-inactivation. *Plant J.* **1997**, *12*, 597–603.
67. PLACE. Available online: http://dna.affrc.go.jp/PLACE/signal-scan.html (accessed on 8 May 2011).
68. PlantCARE. Available online: http://bioinformatics.psb.ugent.be/web-tools/plantcare/html (accessed on 8 May 2011).
69. Gietz, R.D.; Schiestl, R.H. High-efficiency yeast transformation using the LiAc/SS carrier DNA/PEG method. *Nat. Protoc.* **2007**, *2*, 31–34.
70. Lascéve, G.; Leymarie, J.; Olney, M.A.; Liscum, E.; Christie, J.M.; Vavasseur, A.; Briggs, W.R. *Arabidopsis* contains at least four independent blue-light-activated signal transduction pathways. *Plant Physiol.* **1999**, *120*, 605–614.
71. *LSM Image Browser Software*. Carl Zeiss: Jena, Germany, 2009.
72. Schreiber, U.; Bilger, W. Progress in chlorophyll fluorescence research: Major developments during the past years in retrospect. *Prog. Bot.* **1993**, *54*, 151–173.
73. Tanaka, K.; Tozawa, Y.; Mochizuki, N.; Shinozaki, K.; Nagatani, A.; Wakasa, K.; Takahashi, H. Characterization of three cDNA species encoding plastid RNA polymerase sigma factors in *Arabidopsis thaliana*: Evidence for the sigma factor heterogeneity in higher plant plastids. *FEBS Lett.* **1997**, *413*, 309–313.
74. Fujiwara, M.; Nagashima, A.; Kanamaru, K.; Tanaka, K.; Takahashi, H. Three new nuclear genes, *sigD*, *sigE* and *sigF*, encoding putative plastid RNA polymerase [sigma] factors in *Arabidopsis thaliana*. *FEBS Lett.* **2000**, *481*, 47–52.
75. Sugiura, M. The Chloroplast Chromosomes in Land Plants. *Annu. Rev. Cell. Biol.* **1989**, *5*, 51–70.
76. Hoffer, P.H.; Christopher, D.A. Structure and blue-light-responsive transcription of a chloroplast *psbD* promoter from *Arabidopsis thaliana*. *Plant Physiol.* **1997**, *115*, 213–222.
77. Thum, K.E.; Kim, M.; Morishige, D.T.; Eibl, C.; Koop, H.U.; Mullet, J.E. Analysis of barley chloroplast *psbD* light-responsive promoter elements in transplastomic tobacco. *Plant Mol. Biol.* **2001**, *47*, 353–366.
78. Kanamaru, K.; Nagashima, A.; Fujiwara, M.; Shimada, H.; Shirano, Y.; Nakabayashi, K.; Shibata, D.; Tanaka, K.; Takahashi, H. An *Arabidopsis* sigma factor (*SIG2*)-dependent expression of plastid-encoded tRNAs in chloroplasts. *Plant Cell Physiol.* **2001**, *42*, 1034–1043.
79. Ishizaki, Y.; Ozono, K.; Takenaka, C.; Tsunoyama, Y.; Nakahira, Y.; Tanaka, K.; Kanamaru, K.; Hanaoka, M.; Shiina, T. An *Arabidopsis* double mutant *sig2sig6* exhibits albino phenotype. *Plant Cell Physiol.* **2006**, *47*, S187.
80. Shalitin, D.; Yu, X.; Maymon, M.; Mockler, T.; Lin, C. Blue light-dependent *in vivo* and *in vitro* phosphorylation of *Arabidopsis* cryptochrome 1. *Plant Cell* **2003**, *15*, 2421–2429.
81. Shalitin, D.; Yang, H.; Mockler, T.C.; Maymon, M.; Guo, H.; Whitelam, G.C.; Lin, C. Regulation of *Arabidopsis* cryptochrome 2 by blue-light-dependent phosphorylation. *Nature* **2002**, *417*, 763–767.
82. Sharrock, R.A.; Clack, T. Patterns of expression and normalized levels of the five *Arabidopsis* phytochromes. *Plant Physiol.* **2002**, *130*, 442–456.

83. Lee, J.; He, K.; Stolc, V.; Lee, H.; Figueroa, P.; Gao, Y.; Tongprasit, W.; Zhao, H.; Lee, I.; Deng, X.W. Analysis of transcription factor HY5 genomic binding sites revealed its hierarchical role in light regulation of development. *Plant Cell* **2007**, 19, 731–749.
84. Pfannschmidt, T.; Nilsson, A.; Tullberg, A.; Link, G.; Allen, J.F. Direct transcriptional control of the chloroplast genes *psbA* and *psaAB* adjusts photosynthesis to light energy distribution in plants. *IUBMB Life* **1999**, *48*, 271–276.
85. Tullberg, A.; Alexciev, K.; Pfannschmidt, T.; Allen, J.F. Photosynthetic electron flow regulates transcription of the *psaB* gene in pea (*Pisum sativum* L.) chloroplasts through the redox state of the plastoquinone pool. *Plant Cell Physiol.* **2000**, *41*, 1045–1054.
86. Koch, K.E. Carbohydrate-modulated gene expression in plants. *Annu. Rev. Plant Physiol. Plant Mol. Biol.* **1999**, *47*, 509–540.
87. Allen, J.F.; Pfannschmidt, T. Balancing the two photosystems: Photosynthetic electron transfer governs transcription of reaction centre genes in chloroplasts. *Philos. Trans. R. Soc. Lond. B* **2000**, *355*, 1351–1357.
88. Tikkanen, M.; Gollan, P.J.; Suorsa, M.; Kangasjarvi, S.; Aro, E.M. STN7 operates in retrograde signaling through controlling redox balance in the electron transfer chain. *Front. Plant Sci.* **2012**, *3*, e277.
89. Aro, E.M.; Virgin, I.; Andersson, B. Photoinhibition of Photosystem II. Inactivation, protein damage and turnover. *Biochim. Biophys. Acta* **1993**, *1143*, 113–134.
90. Keren, N.; Berg, A.; van Kan, P.J.; Levanon, H.; Ohad, I. Mechanism of photosystem II photoinactivation and D1 protein degradation at low light: The role of back electron flow. *Proc. Natl. Acad. Sci. USA* **1997**, *94*, 1579–1584.

Appendix

Figure A1. RNA decay analysis. 10 day old seedlings were treated with 200 μM Actinomycin-D in liquid MS-medium for 8 h. The amounts of *SIG5* and *GFP* transcripts were quantified by qRT-PCR and standardized on ACT2.

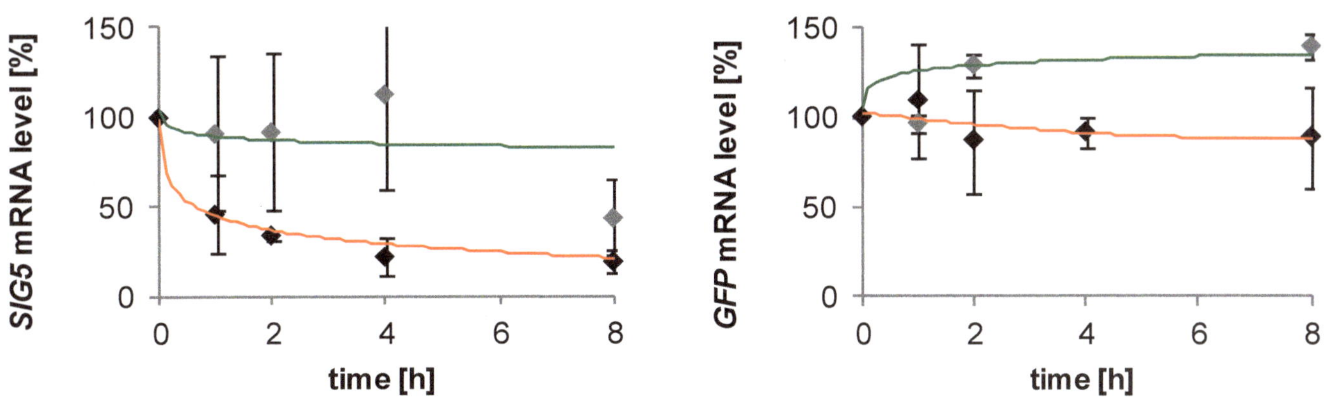

Figure A2. Graphical overview of the distribution of motifs in the *SIG5* upstream region. Information on the precise positions of the motifs is listed in Table 1.

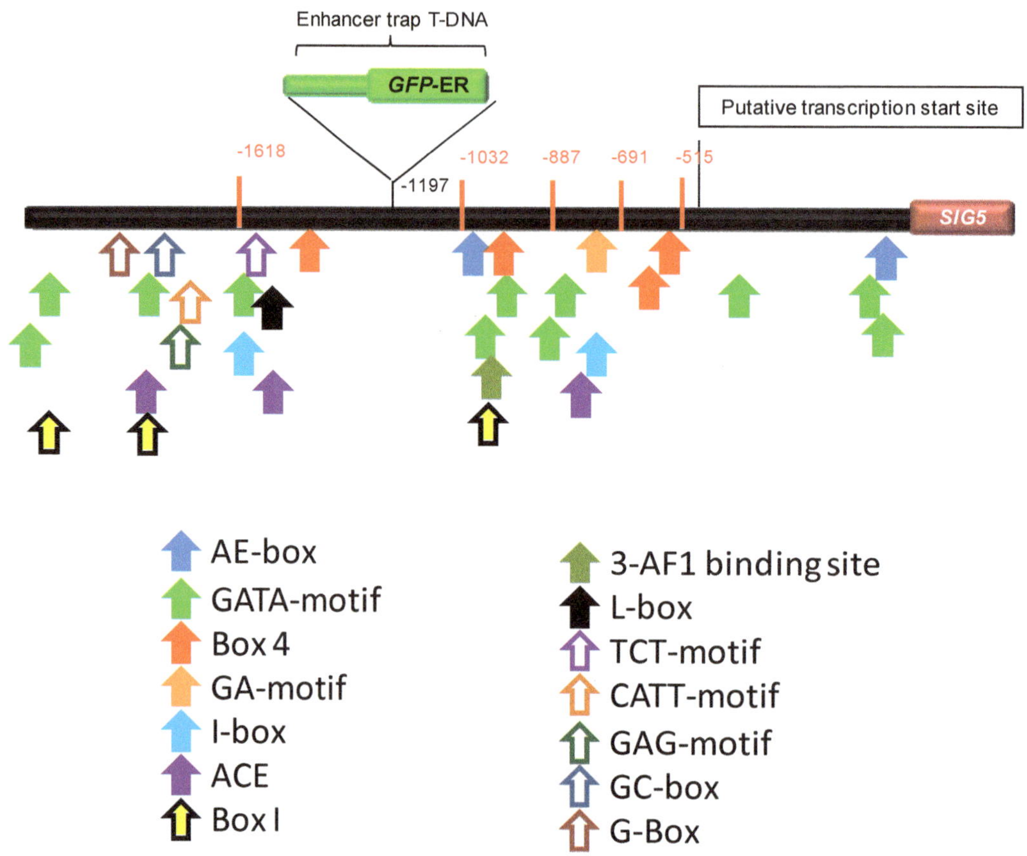

11

Functions of Calcium-Dependent Protein Kinases in Plant Innate Immunity

Xiquan Gao [1,*] Kevin L. Cox Jr. [2] and Ping He [3,*]

[1] State Key Laboratory of Crop Genetics and Germplasm Enhancement, College of Agriculture, Nanjing Agricultural University, Nanjing 210095, China

[2] Department of Plant Pathology and Microbiology, Institute for Plant Genomics and Biotechnology, Texas A&M University, College Station, TX 77843, USA

[3] Department of Biochemistry and Biophysics, Institute for Plant Genomics and Biotechnology, Texas A&M University, College Station, TX 77843, USA

* Authors to whom correspondence should be addressed; E-Mails: xgao@njau.edu.cn (X.G.); pinghe@tamu.edu (P.H.)

Abstract: An increase of cytosolic Ca^{2+} is generated by diverse physiological stimuli and stresses, including pathogen attack. Plants have evolved two branches of the immune system to defend against pathogen infections. The primary innate immune response is triggered by the detection of evolutionarily conserved pathogen-associated molecular pattern (PAMP), which is called PAMP-triggered immunity (PTI). The second branch of plant innate immunity is triggered by the recognition of specific pathogen effector proteins and known as effector-triggered immunity (ETI). Calcium (Ca^{2+}) signaling is essential in both plant PTI and ETI responses. Calcium-dependent protein kinases (CDPKs) have emerged as important Ca^{2+} sensor proteins in transducing differential Ca^{2+} signatures, triggered by PAMPs or effectors and activating complex downstream responses. CDPKs directly transmit calcium signals by calcium binding to the elongation factor (EF)-hand domain at the *C*-terminus and substrate phosphorylation by the catalytic kinase domain at the *N*-terminus. Emerging evidence suggests that specific and overlapping CDPKs phosphorylate distinct substrates in PTI and ETI to regulate diverse plant immune responses, including production of reactive oxygen species, transcriptional reprogramming of immune genes, and the hypersensitive response.

Keywords: calcium-dependent protein kinase; PAMP-triggered immunity; effector-triggered immunity; phosphorylation

1. Introduction

Calcium ions (Ca^{2+}) are one of the most important cellular ionic species that function as second messengers in many aspects of biological processes in plants and animals. When the host plants encounter a diverse number of physiological stimuli or stresses, including cold, drought, salinity stresses, and pathogen or herbivory attacks, an increase in cytosolic free Ca^{2+} levels ((Ca^{2+})cyt) is rapidly triggered [1]. The Ca^{2+} signals are subsequently sensed and translated into intracellular responses, mainly via different Ca^{2+} sensor proteins, some of which are highly conserved in all eukaryotic organisms, thereby triggering complex downstream signaling in response to developmental and environmental cues. The temporal and spatial changes of Ca^{2+} binding to the sensor proteins are monitored by the conformational changes of sensor proteins in a Ca^{2+}-dependent manner [2–4].

2. Calcium and Calcium Sensor Proteins

There are two types of Ca^{2+} sensor proteins, one of which relays calcium signals, such as calcineurin B-like/CBL-interacting protein kinases (CBL/CIPKs) and calmodulin proteins (CaMs), while the other constitute sensor protein kinases, such as calcium-dependent protein kinases (CDPKs) and calmodulin-dependent protein kinases (CaMKs) [5,6]. CaMs are found ubiquitously in all eukaryotes, generally being defined as CaM, CaM-like (CML), and CaM-related proteins; all of which typically contain four elongation factor (EF)-hand domains for calcium binding [7–9]. EF-hands are found in pairs in calcium sensor proteins; each of which is considered the basic functional unit to stabilize the protein and facilitate high-affinity binding to Ca^{2+} [9]. The binding of calcium to the EF-hands subsequently leads to the conformational change of the CaM globular structure, which allows the interaction of CaMs with their target proteins [10,11]. Multiple target proteins have been identified and the interaction of CaM/CML with the target proteins is proposed to function in diverse cellular responses, including transcriptional reprogramming, activation of the phosphorylation cascade, and accumulation of secondary metabolites [1,12,13]. For example, CaMs/CMLs interact with no pollen germination 1 (NPG1) [14], the kinesin-like calmodulin binding protein (KCBP) [15], and the *Arabidopsis thaliana* Na^+/H^+ antiporter 1 (AtNHX1) [16] to modulate development and growth processes. CaM/CMLs also target the barley recessive resistance protein MLO [17], *Arabidopsis thaliana* cyclic nucleotide-gated ion channel 2 (AtCNGC2) [18], nicotinamide adenine dinucleotide (NAD^+) kinases (NADKs) [19], the NAC and WRKY family of transcription factors [20,21], and *Arabidopsis* calmodulin binding protein *60* (CBP60) [22] to regulate defense responses to pathogen attack.

Although CBLs contain three EF-hand motifs, unlike CaMs, CBLs were only found in plants and some protists [2]. Substitutions within the Ca^{2+}-binding loops of CBLs indicate that the Ca^{2+}-binding activity of CBLs is different from the EF-hands of CaM, and different CBLs display differential Ca^{2+} affinities to Ca^{2+}, supporting the complexity and flexibility of CBLs in sensing Ca^{2+} signals [2]. CBL-interacting protein kinases (CIPKs), the direct interacting substrate proteins of CBLs, belong to a type of kinase that contain domains related to the sucrose non-fermenting-like kinase 1 (SNF1) and AMP-dependent kinases (AMPKs) from yeast and mammals, respectively [23]. CBLs interact with CIPKs via the conserved NAF domain at the CIPK *C*-terminus, subsequently releasing the kinase from auto-inhibition [24]. It seems that the competitive recruitment of different CIPKs by various CBLs, and *vice versa*, at distinct subcellular compartments dictates not only the specificity, but also the versatility of the CBLs/CIPKs complex in decoding Ca^{2+} signals, thereby leading to subsequent response outputs [1,23].

3. Structure, Distribution, and Physiological Functions of CDPKs

Distinctly different from CaM and CBLs, which do not exert their function via possessing any intrinsic activity, but transduce calcium signals to target proteins, CDPKs and CaMKs can directly transmit the calcium signals by calcium binding to the EF-hands at their *C*-terminus and by substrate phosphorylation via the catalytic kinase domain at their *N*-terminus. Upstream of the kinase domain, there is a variable *N*-terminal domain ranging in length from 40–180 amino acids that has very low homology among different family members [5]. Many CDPKs contain potential myristoylation and palmitoylation sites at the beginning of their highly variable *N*-terminal domains, implicating the association of CDPKs with the plasma membrane. In addition to plasma membrane localization, CDPKs are also proposed to be localized in other subcellular organelles, including the cytosol, endoplasmic reticulum, and peroxisomes [6,25–27]. Furthermore, a conserved PEST (proline, glutamine, serine and threonine) domain, which exerts rapid proteolytic degradation, has also been found in the *N*-terminus of some CDPKs [25]. The conserved protein kinase domain contains serine/threonine residues and the activation loop has acidic residues to prevent the loop from being unnecessarily phosphorylated [5,28,29]. The protein kinase domain links to a calmodulin-like domain through a pseudosubstrate-containing auto-inhibitory junction domain, which is capable of inhibiting kinase activity via interaction with the active site in the absence of Ca^{2+} and keeping CDPKs in a state of low activity [28,29]. Upon calcium influx into the cell, the intramolecular interaction between the CaM-like domain and the auto-inhibitory domain causes the conformational change of the protein, and leads to the activation of the kinase domain [28–33].

CDPKs have been found widely in flowering plants, algae, and certain apicomplexa protists [30]. The *Arabidopsis* and rice genomes encode a large gene family with 34 and 29 members, respectively [6,34]. The genome-wide analysis of *Arabidopsis* revealed the occurance of gene duplication and subsequent divergence of CDPKs with distinct functions. The large gene family of CDPKs consist of at least 12 subfamilies that are proposed to be derived from a common ancestor of monocots and eudicots [28]. In addition to the model plants *Arabidopsis* and rice, the diversity in the multigene family of CDPKs also exists in other plant species, including soybean *(Glycine max*), tomato (*Lycopersicon esculentum*), and maize (*Zea mays*) [6,28].

In plants, CDPKs and related kinases (CRKs) have been reported to function in various developmental processes [35], including pollen tube growth [36], root development [37], stem elongation and vascular development [38], as well as cell division and differentiation [5]. Moreover, CDPKs are also involved in plant responses to diverse abiotic stresses. For instance, transient expression of *Arabidopsis thaliana* calcium-dependent protein kinase 10 (AtCPK10) or AtCPK30 could strongly activate the expression of a stress-regulated promoter that responds to multiple stresses, including cold, drought, and salt [27]. Accumulating evidence suggests that plant CDPKs are also involved in phytohormone-mediated signaling pathways. Using a combination of genetic and biochemical approaches, it has been shown that *Arabidopsis* CPK4 and CPK11 could phosphorylate abscisic acid (ABA)-responsive element-binding factor 1 (ABF1) and ABF4, to regulate ABA signaling during seed germination, seedling growth, guard cell opening, and the insensitivity to salt stress [39]. A recent study showed that *Arabidopsis* CPK6 positively regulated the activation of ion channels in ABA signaling, and activated the stomatal closure by yeast elicitor (YEL) in *Arabidopsis* [40]. Mutation of *Arabidopsis* CPK6 impaired the methyl jasmonate (MeJA)-induced stomatal closure, implying that CPK6 also functions as a positive regulator in MeJA signaling in *Arabidopsis* guard cells [40,41].

4. Two Branches of Plant Innate Immunity

Similar to animals, plants possess a surveillance system mediated by pattern-recognition receptors (PRRs) to sense invading microorganisms and trigger potent immune responses via detection of conserved pathogen or microbe-associated molecular patterns (PAMPs/MAMPs) (Figure 1) [42,43]. Plants utilize receptor-like kinases (RLKs) localized on the cell surface to perceive PAMPs, thereby activating PAMP-triggered immunity (PTI) [44,45]. The well-studied PAMPs include flg22 derived from the bacterial flagellin and elf18 from the elongation factor Tu (EF-Tu), peptidoglycan (PGN) [46,47], lipopolysaccharides (LPS) [48], as well as chitin from fungal cell wall polymers [49]. Different PAMPs likely trigger distinct, yet convergent, downstream signaling pathways. For instance, perception of flg22 by the PRR protein flagellin sensing 2 (FLS2) in *Arabidopsis* involves a complex formation of FLS2 with the brassinosteroid (BR)-insensitive 1 (BRI1)-associated receptor kinase 1(BAK1) [50], which appears to be a shared component of the signaling complex activated by multiple PAMPs. A cytoplasmic receptor kinase *Botrytis*-induced kinase 1 (BIK1) was found to constitutively associate with FLS2 and BAK1. Phosphorylation of BIK1 by BAK1 results in the activation of downstream PTI signaling, whereas phosphorylation of BIK1 by BRI1 negatively regulates plant hormone BR signaling [51–53]. Moreover, two E3-ubiquitin ligases, PUB12 and PUB13, function as negative regulators via BAK1 phosphorylation-mediated interaction and polyubiquitination of FLS2, which leads to the degradation of FLS2 to down-regulate flagellin signaling [54].

To encounter PTI, successful pathogens often secrete various virulence proteins, so-called effectors, through the bacterial type III secretion system or fungal haustorium. The effectors are in turn recognized by a type of intracellular nucleotide-binding domain leucine-rich repeat (NB-LRR or plant NLR) proteins [55–57], structurally similar to mammalian nucleotide-binding oligomerization domain receptors (NOD-like receptor or NLR) that sense intracellular PAMPs [58]. Recognition of effectors by plant NLRs eventually initiates effector-triggered immunity (ETI) [45,59–61]. Plant NLRs are also known as resistance (R) proteins, which have been widely used in agriculture for the improvement of resistance to a wide range of pathogens [62,63]. Plant NLR proteins often trigger diverse immune responses, including apoptosis-like programmed cell death (PCD), also called hypersensitive response (HR), to restrict the growth and spread of biotrophic pathogens, production of reactive oxygen species (ROS), as well as transcriptional reprogramming of immune genes [61,64]. The *Arabidopsis* NLR protein resistance to *Pseuodomonas syringae* 2 (RPS2) initiates the resistance to *Pseudomonas syringae* carrying AvrRpt2, whereas resistance to *Pseudodomonas syringae pv. maculicola* 1 (RPM1) recognizes two unrelated *P. syringae* effectors: AvrRpm1 and AvrB [65–67]. AvrRpt2, AvrRpm1, and AvrB activate ETI signaling by modification of host target RPM1-interacting 4 (RIN4), which is sensed by the corresponding RPS2 and RPM1 proteins [68]. RPM1 has been reported to localize and function on the plasma membrane [69]. It is likely that different NLR proteins deploy distinct mechanisms through recognizing and interacting with the effectors and/or downstream components in multiple subcellular compartments to trigger complex downstream immune responses.

Figure 1. Model of calcium-dependent protein kinase (CDPK)-mediated plant innate immune signaling. Plants employ cell surface pattern-recognition receptors (PRRs) or intracellular nucleotide-binding domain leucine-rich repeat (NB-LRR or plant NLR) proteins to perceive pathogen-associated molecular patterns (PAMPs) or effector proteins, respectively. The signatures of increased cytosolic calcium (Ca^{2+}) levels triggered by interaction of PAMP-PRR or effector-NLR are sensed by specific CDPKs, which trigger subsequent activation of distinct immune events. CDPKs could act synergistically or independently with the mitogen-activated protein kinase (MAPK) cascade, which consititutes three tiered kinases, MAPK (MPK), MAPK kinase (MKK) and MKK kinase (MKKK), in regulating reactive oxygen species (ROS) production and transcriptional reprogramming in PAMP-triggered immunity (PTI) signaling. On the other hand, CDPKs phosphorylate distinct substrates in regulating bifurcate effector-triggered immunity (ETI) signaling, including phosphorylation of WRKY8/28/48 transcription factors (TF) for immune gene expression and respiratory burst oxidase homolog D/F (RbohD/F) for ROS production.

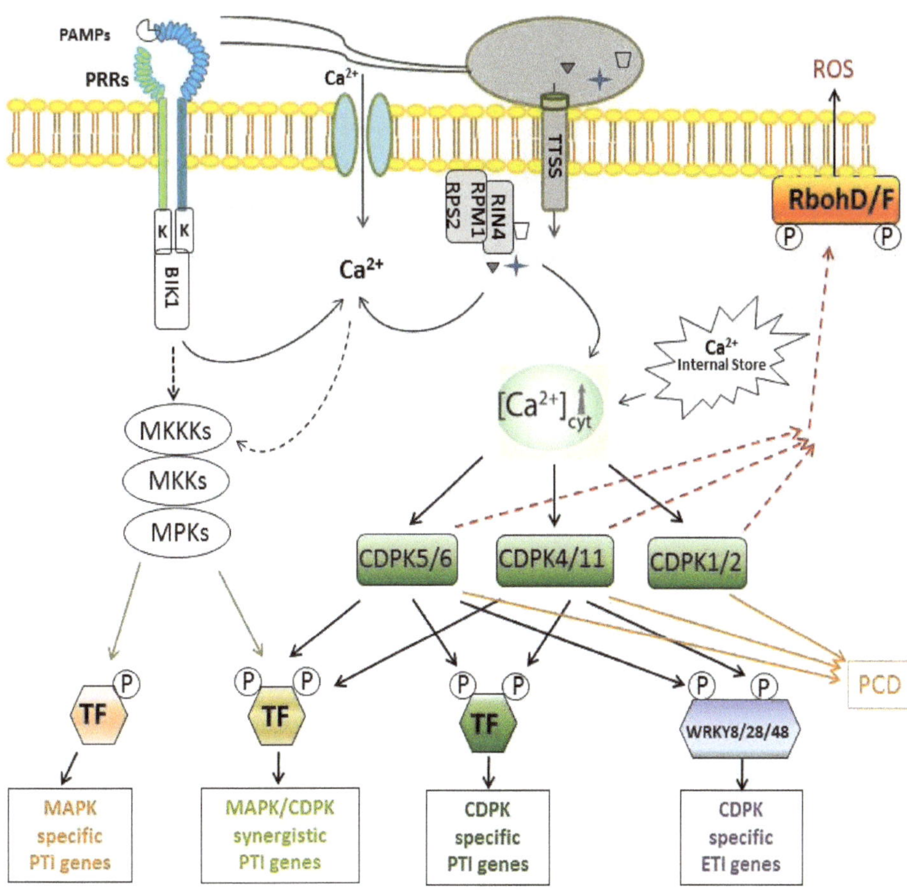

Ca^{2+} influx is one of the earliest responses upon recognition of pathogens in plant innate immunity, and distinct calcium signatures have been observed in plant PTI and ETI signaling. For example, different PAMPs or elicitors, such as flg22 and elf18, triggered the rapid and transient increase of cytosolic Ca^{2+} concentration in distinct cellular compartments, including the cytosol, nucleus, mitochondria, or chloroplasts [70]. While challenging of plants with bacteria carrying *avrRpm1* or *avrB* induced a much prolonged and sustained increase of cytosolic Ca^{2+} in *Arabidopsis* [71], calcium-channel blockers completely suppressed AvrRpm1-meditated cell death. The data suggest that Ca^{2+} signaling plays essential roles in modulating diverse aspects of innate immunity through differential Ca^{2+} sensory machineries [18,71].

5. Functions of CDPKs in PTI Signaling

5.1. CDPK-Regulated Transcriptional Reprogramming in PTI

Accumulating evidence has suggested the essential roles of CDPKs in plant basal resistance mediated by PTI signaling. For instance, *Triticum aestivum* calcium-dependent protein kinase 2 (TaCDPK2), a wheat CDPK, was found to be required for powdery mildew resistance, and the overexpression of TaCDPK2 in rice resulted in an enhanced resistance to bacterial blight (*Xanthomonas oryzae pv. oryzae*, Xoo) [72]. In *Arabidopsis*, overexpression of AtCPK1 confers broad-spectrum resistance to both bacteria and fungi [73]. CDPKs function in innate immunity by controlling transcriptional reprogramming of immune genes. In *Arabidopsis*, for example, a functional genomic screen using a flg22-responsive reporter gene, *NHL10-LUC* (*NDR1/HIN1-like10-luciferase*), as a marker enabled the identification of a group of closely related CDPKs, AtCPK4, AtCPK5, AtCPK6 and AtCPK11, as early transcriptional regulators in flg22 signaling [74]. While the Ca^{2+} channel blockers La^{3+} and Gd^{3+} eliminated the activation of *NHL10-LUC* by flg22, active AtCPK4, AtCPK5, AtCPK6 and AtCPK11 induced *NHL10-LUC* to a level comparable to that by flg22. Moreover, three genes (*PROPEP1*, *PROPEP2,* and *PROPEP3*) encoding endogenous defense peptides were also quickly activated by flg22 and active AtCPK5 and AtCPK11. More importantly, ectopic expression of active AtCPK5 or AtCPK11 induced a set of largely overlapping genes with flg22 and other PAMP-induced early genes [74], suggesting that CDPKs are the convergent point of PTI signaling triggered by diverse PAMPs. Similarly, transient expression of constitutively active *Nicotiana benthamiana* calcium-dependent protein kinase 2 (NtCDPK2) triggered the accumulation of jasmonic acid (JA) and ethylene (ET), but not salicyclic acid (SA), accompanied by the activation of JA- and ET-regulated genes [75], indicating the complexity of CDPKs in regulating innate immunity and defense-related phytohormone signaling. Future identification of transcription factors that function downstream of CDPKs will enable the elucidation of how CDPKs regulate the transcriptional reprogramming of defense genes in PTI signaling.

5.2. CDPK-Regulated ROS Production in PTI

Two potato CDPKs, *Solanum tuberosum* calcium-dependent protein kinase 4 (StCDPK4) and StCDPK5, were reported to directly phosphorylate *Solanum tuberosum* respiratory burst oxidase homologue B (StRbohB) to activate ROS production. The *N*-terminal variable domains seem to be required for phosphorylation of StRbohB and ROS production, as mutations of *N*-terminal myristoylation and palmitoylation sites in the V-domain resulted in the incapability of StCDPK5 to phosphorylate StRbohB [76]. In line with these findings, a recent report also showed that activation of *Arabidopsis* AtCPK5 resulted in enhanced SA-mediated resistance to *Pseudomonas syringae* pv. *tomato* strain DC3000 (*Pst* DC3000), defense gene expression, and ROS production via direct phosphorylation of RbohD, further confirming the potential functions of CDPKs in PTI signaling via ROS production [77]. The phosphopeptide analysis, based on a targeted mass spectrometry (MS) approach, selected reaction monitoring (SRM), supported that AtCPK5-regulated ROS production is flg22 ligand-dependent [77]. On the other hand, rice CDPK12 seems to play a negative role in ROS production as well as the immunity to both virulent and avirulent blast fungus [78]. The complexity of CDPKs in regulating ROS production may be mediated by differential phosphorylation of Rboh family members.

Although it has been suggested that substrate itself, Ca^{2+} affinities of CDPKs, the kinetic parameters of certain substrate and the phosphorylation sites contribute to the specificity, how CDPKs determine specific substrates remains enigmatic [2,79]. It has been proposed that substrate, specificity of kinases, largely relies on the interaction between the activation domain of kinases and the sequences surrounding the phosphorylation sites of substrates [26]. Until now, multiple phosphorylation motifs of CDPKs have been identified through in-depth analysis, including basic-XX-S/T and S-X-basic, which have been extensively and effectively used to search plant proteomes for predicting potential CDPK substrates [30]. For instance, StCDPKs were found to phosphorylate StRbohB at residues Serine (Ser)-82 and Ser-97 [76], while AtCPK5 showed strong phosphorylation activity on AtRbohD at Ser-39, Ser-148, Ser-163, and Ser-347 [77].

5.3. Crosstalk of MAPKs and CDPKs

The mitogen-activated protein kinase (MAPK) cascade is a prevalent component in PTI signaling [61]. While CDPKs have been identified at different cellular compartments [6,25–27], MAPKs were reported to be mainly localized in the cytosol and the nucleus [80]. Despite this, crosstalk between MAPKs and CDPKs has been implicated in plant innate immune responses. For instance, both MAPKs and CDPKs could regulate 1-aminocyclopropane-1-carboxylic acid synthase (ACS) activity, likely through phosphorylation by both MAPKs and CDPKs [81]. Moreover, CDPK signaling compromised stress-induced MAPK activation in tobacco, depending on ethylene synthesis and perception [75]. In addition, global gene expression analysis revealed that CDPKs and MAPKs function either synergistically or independently in PTI responses. In this study, different sets of flg22-inducible genes in protoplasts expressing the active form of AtCPK5 and AtMKK4 were identified to be either MAPK specific (*FRK1*), MAPK dominant (*CYP81F2*), CDPK specific (*PHI-1*), or CDPK/MAPK synergistic (*NHL10*) [74]. However, overexpression of active AtCPK5 in mesophyll protoplasts did not activate AtMPK3 and AtMPK6 [74]. This data indicates that a concerted activation of both CDPK and MAPK pathways controls the specific yet overlapping downstream immune signaling events in PTI (Figure 1).

6. Functions of CDPKs in ETI Signaling

6.1. CDPK-Regulated Transcriptional Reprogramming in ETI

So far, there are limited data showing that CDPKs regulate immune gene expression in ETI signaling. A previous study showed that overexpression of AtCPK1 resulted in an increase of SA levels through the activation of SA biosynthesis and regulatory genes, such as phytoalexin-deficient 4 (*PAD4*) and SA induction deficient 2/isochorismate synthase 1 (*SID2/ICS1*) [73]. Using a functional genomic screen, in combination with a genetic study, we found recently that constitutive expression of active form of AtCPK4, AtCPK5, AtCPK6, or AtCPK11 moderately inducedthe expression of the ETI marker gene *WRKY46*. This activation could be synergistically enhanced with the co-expression of a group of specific WRKY transcription factors, including WRKY8, WRKY28, and WRKY48, from subgroup IIc in *Arabidopsis*. Supporting this, bioinformatic analyses suggested that the amino acid sequence surrounding Threonine 247 (T247) and T248 of WRKY48 [basic-X-TT-X-X-X-X-hydrophobic (h)-basic] closely matches an optimal phosphorylation substrate target of CDPKs (basic-h-X-basic-X-XS/T-X-X-X-h-basic) [6,82], indicating that CDPKs control the transcriptional reprogramming likely through phosphorylating WRKYs in ETI signaling. Indeed, the experimental *in vitro* phosphorylation assay demonstrated that AtCPK4, AtCPK5, AtCPK6, or AtCPK11 strongly phosphorylated WRKY8, WRKY28, and WRKY48, resulting in enhanced binding of WRKYs to the cis-element in the promoter region of immune responsive gene *WRKY46* [82]. Future study will focus on the identification of CDPK-mediated WRKY targeted genes and elucidation their importance in ETI signaling.

6.2. CDPK-Regulated ROS Production in ETI

PAMP-induced ROS production occurs immediately, usually between 2–30 min upon PAMP elicitation or pathogen attack [74]. In contrast to this rapid and transient ROS burst in PTI signaling, effector-induced ROS production seems to be more prolonged and robust [71], which is correlated with the sustained increase of Ca^{2+} levels upon elicitation in effector-NLR interaction. The involvement of CDPKs in the production of ROS in ETI signaling was supported by *Arabidopsis cpk1/2* mutants showing reduced ROS accumulation on the leaves challenged with *Pst* DC3000 carrying *avrRmp1* or *avrRpt2* with a diaminobenzidine (DAB) staining assay. An immunocomplex kinase assay further demonstrated that AtCPK1 and AtCPK2, as well as AtCPK4 and AtCPK11, strongly phosphorylated AtRbohD and AtRbohF which are responsible for the NLR-triggered ROS production [82]. While StCDPKs phosphorylated StRbohB at residues Ser-82 and Ser-97 [76], mutation of Ser-148 in AtRbohD which corresponds to Ser-97 in StRbohB, but not Ser-133 in AtRbohD which corresponds to Ser-82 in StRbohB, resulted in the reduced phosphorylation of AtRbohD by AtCPK2, AtCPK4, and AtCPK11 [82]. Comparing to AtCPK1, AtCPK5 and AtCPK6 only weakly phosphorylated AtRbohD and AtRbohF, suggesting the specificity of CDPKs in phosphorylating Rbohs. A recent report also suggested that AtCPK5 phosphorylated AtRbohD with the peptide of AtRbohD amino acid 143–152 as a substrate and full length AtCPK5 protein as a kinase and this phosphorylation associated with ROS production in PTI signaling [76]. Together, these findings suggest the specificity and diversity of CDPK substrate phosphorylation in regulating ROS production in both PTI and ETI signaling pathways.

6.3. CDPK-Regulated HR in ETI

Distinct from the rapid and transient increase of cytosolic Ca^{2+} concentration induced by PAMPs [83–85], inoculation with bacteria carrying *avrRpm1*, *avrB* or *avrRpt2* triggered a prolonged and sustained increase of cytosolic Ca^{2+} concentration, which is associated with PCD in *Arabidopsis* leaves [71,86]. Furthermore, expression of AvrRpm1, AvrB, or AvrRpt2 in *Arabidopsis* protoplasts triggered distinct kinetics of PCD as detected by Evan's blue staining with an earlier PCD occurring as early as 2 hours post-transfection (hpt) for AvrRpm1 or AvrB, and a later PCD observed at 16 hpt for AvrRpt2 [82]. The specificity and redundancy of individual CDPKs that control PCD in ETI signaling have been demonstrated recently. *Arabidopsis* CPK1, CPK2, CPK5, CPK6 and probably CPK4 and CPK11 function synergistically in controlling PCD, as the *cpk1/2/5/6* quadruple mutants, but not double mutants, displayed a reduced HR [82]. In line with this finding, silencing of NtCDPK2 subfamily resulted in the compromised induction of Avr9/Cf-9-dependent HR-like PCD symptom in *Nicotiana benthamiana* [87], while transgenic potato plants carrying *StCDPK5* showed enhanced HR-like PCD and resistance to the hemibiotrophic pathogen *Phytophthora infestans* [76]. In addition, transient expression of a constitutively active form of barley *Hordeum vulgare* calcium-dependent protein kinase 4 (HvCDPK4) in *Nicotiana benthamiana* triggered cell death in tobacco mesophyll cells [88], whereas virus-induced gene silencing (VIGS) of the *NtCDPK2* and *NtCDPK3* gene family eliminated the HR elicited by race-specific pathogens [87]. Identification of CDPK substrates that control PCD will shed light on how CDPKs regulate PCD in ETI signaling.

6.4. Nuclear Dynamics of CDPKs in Immune Signaling

Although it has been widely known that CDPKs contain myristoylation and palmitoylation sites in the *N*-terminal variable domain indicating their potential plasma membrane localization, it has been suggested that a second post-translational modification, such as phosphorylation, affects the translocation of CDPKs from the membranes to other organelles, such as the cytosol or the nucleus, upon activation [89,90]. In line with this hypothesis, the ice plant (*Mesembryanthemum crystallinum*) McCPK1 and the groundnut (*Arachis hypogaea*) AhCPK2 were reported to accumulate largely in the nucleus upon environmental stress [90,91]. A similar study showed that calcium/calmodulin-dependent protein kinase (CCaMK) interacts with and phosphorylates the nuclear localized coiled-coil protein CYCLOPS required for both arbuscular mycorrhizal (AM) and rhizobial infection [92,93].We also showed that a significant portion of AtCPK5 re-localized to the nucleus where it could interact and phosphorylate WRKY8, WRKY28 and WRKY48 upon eliciting the *Arabidopsis* cells with effector protein AvrRpt2 [82]. Even though that several transcription factors, including WRKYs, REPRESSION OF SHOOT GROWTH (RSG), and ABF4 were identified as CDPK substrates [82,94], it is tempting to speculate that the translocation of CDPKs to the nucleus upon elicitation by effectors is essential to relay downstream signaling in ETI. Previous reports showed that several R proteins required effector-induced nuclear translocation for full downstream immune signaling [95–97]. Using the recombinant effector protein AvrRpt2 tagged with Nuclear Export Sequence (NES) or the mutated sequence of NES (nes) at its *C*-terminus, we found that AvrRpt2-NES did not reduce the ability to activate *WRKY46::LUC* compared to AvrRpt2 or AvrRpt2-nes [82,98]. More importantly, AtCPK5-NES was no longer able to display synergistic effect with WRKY8 or WRKY48 to activate *WRKY46::LUC* compared to AtCPK5 itself. Therefore, these data together demonstrated that the nuclear localization and subcellular dynamics of CDPKs are essential to orchestrating bifurcate ETI signaling via distinct substrate specificity and subcellular dynamics, for instance, to interact with and phosphorylate transcription factors in the nucleus to regulate transcriptional reprogramming crucial for restriction of pathogen growth.

7. Conclusions and Perspectives

Extensive research efforts made in the past decade using integrative approaches have significantly advanced our understanding of how plant CDPKs function as the versatile Ca^{2+}-sensor proteins in translating Ca^{2+} signatures into downstream responses, such as phosphorylation events and transcriptional reprogramming of immune genes upon pathogen challenges.With the unique binding motif EF-hands, as well as the catalytic kinase domain at the *N*-terminus distinct from other Ca^{2+}-sensor proteins, such as CaMs, CMLs, and CBL/CIPK complexes, CDPKs exert their functions via not only direct calcium ion binding but also through the phosphorylation of their substrate proteins. Interaction of CDPKs with their target proteins or substrates consecutively leads to the regulation of diverse immune responses, including ROS production and regulation of immune gene expression in both PTI and ETI, as well as HR-like PCD in ETI (Figure 1). Apparently, the subcellular dynamics of CDPKs and the specificity, as well as the diversity of substrate phosphorylation, are essential for orchestrating various signaling events, allowing a fine tune of the plant innate immune responses.

Presently, a major challenge for studying CDPK functions is to determine and distinguish substrate phosphorylation events responsible for distinct signaling pathways in PTI and ETI. To achieve this goal, a combination of diverse biochemical approaches have been deployed in identifying potential phosphorylation sites in plants. Proteome-based surveys of candidate substrate proteins followed by mass spectrometry, as well as constructing custom-designed peptide libraries, provide a strategy to rapidly test a large number of mapped phosphorylation sites for their potential to be phosphorylated by a specific kinase [94]. Using SRM for example, several serine residues, including Ser-39, Ser-148, Ser-163, and Ser-347, in AtRbohD were identified as the phosphorylation sites of AtCPK5 [77]. Another challenge is to elucidate the convergent points and mechanisms for the crosstalk between MAPK and CDPK signaling pathways, especially how differential signaling components of CDPKs and MAPKs are integrated in different subcellular compartments. Finally, it would be interesting to investigate whether, and how, the CDPK family coordinate with other Ca^{2+}-sensor proteins, such as CaMs, CMLs, and CBL/CIPK, in translating the Ca^{2+} signals from the cytosol into the nucleus to subsequently activate the corresponding physiological responses when challenged by pathogens.

Acknowledgements

This work was supported by NIH grant #R01GM092893 (P.H.), and by the Start-up funding and the Fundamental Research Funds for the Central Universities from Nanjing Agricultural University (X.G.).

Author Contributions

X.G and P.H prepared the outline of the manuscript. X.G. wrote the draft, and K.C and P.H revised and edited the manuscript.

Conflicts of Interest

The authors declare no conflicts of interest.

References

1. Hashimoto, K.; Kudla, J. Calcium decoding mechanisms in plants. *Biochimie* **2011**, *93*, 2054–2059.
2. DeFalco, T.A.; Bender, K.W.; Snedden, W.A. Breaking the code: Ca^{2+} sensors in plant signalling. *Biochem. J.* **2010**, *425*, 27–40.
3. Sanders, D.; Pelloux, J.; Brownlee, C.; Harper, J.F. Calcium at the crossroads of signaling. *Plant Cell* **2002**, *14*, S401–S417.
4. Sanders, D.; Brownlee, C.; Harper, J.F. Communicating with calcium. *Plant Cell* **1999**, *11*, 691–706.
5. Harmon, A.C.; Gribskov, M.; Harper, J.F. CDPKs-a kinase for every Ca^{2+} signal? *Trends Plant Sci.* **2000**, *5*, 154–159.
6. Cheng, S.H.; Willmann, M.R.; Chen, H.C.; Sheen, J. Calcium signaling through protein kinases. The *Arabidopsis* calcium-dependent protein kinase gene family. *Plant Physiol.* **2002**, *129*, 469–485.
7. Zielinski, R.E. Calmodulin and calmodulin-binding proteins in plants. *Annu. Rev. Plant Physiol. Plant Mol. Biol.* **1998**, *49*, 697–725.
8. Bouche, N.; Yellin, A.; Snedden, W.A.; Fromm, H. Plant-specific calmodulin-binding proteins. *Annu. Rev. Plant Biol.* **2005**, *56*, 435–466.
9. Luan, S.; Kudla, J.; Rodriguez-Concepcion, M.; Yalovsky, S.; Gruissem, W. Calmodulins and calcineurin B-like proteins: Calcium sensors for specific signal response coupling in plants. *Plant Cell* **2002**, *14*, S389–S400.
10. Yamniuk, A.P.; Vogel, H.J. Structural investigation into the differential target enzyme regulation displayed by plant calmodulin isoforms. *Biochemistry* **2005**, *44*, 3101–3111.
11. Lee, S.H.; Johnson, J.D.; Walsh, M.P.; van Lierop, J.E.; Sutherland, C.; Xu, A.; Snedden, W.A.; Kosk-Kosicka, D.; Fromm, H.; Naravanan, N.; *et al.* Differential regulation of Ca^{2+}/calmodulin-dependent enzymes by plant calmodulin isoforms and free Ca^{2+} concentration. *Biochem. J.* **2000**, *350*, 299–306.
12. Batistic, O.; Kudla, J. Integration and channeling of calcium signaling through the CBL calcium sensor/CIPK protein kinase network. *Planta* **2004**, *219*, 915–924.
13. Batistic, O.; Kudla, J. Analysis of calcium signaling pathways in plants. *Biochim. Biophys. Acta* **2012**, *1820*, 1283–1293.
14. Golovkin, M.; Reddy, A.S. A calmodulin-binding protein from *Arabidopsis* has an essential role in pollen germination. *Proc. Natl. Acad. Sci. USA* **2003**, *100*, 10558–10563.

15. Vos, J.W.; Safadi, F.; Reddy, A.S.; Hepler, P.K. The kinesin-like calmodulin binding protein is differentially involved in cell division. *Plant Cell* **2000**, *12*, 979–990.
16. Yamaguchi, T.; Aharon, G.S.; Sottosanto, J.B.; Blumwald, E. Vacuolar Na^+/H^+ antiporter cation selectivity is regulated by calmodulin from within the vacuole in a Ca^{2+}- and pH-dependent manner. *Proc. Natl. Acad. Sci. USA* **2005**, *102*, 16107–16112.
17. Kim, M.C.; Panstruga, R.; Elliott, C.; Muller, J.; Devoto, A.; Yoon, H.W.; Park, H.C.; Cho, M.J.; Schulze-Lefert, P. Calmodulin interacts with MLO protein to regulate defence against mildew in barley. *Nature* **2002**, *416*, 447–451.
18. Ali, R.; Ma, W.; Lemtiri-Chlieh, F.; Tsaltas, D.; Leng, Q.; Bodman, S.; Berkowitz, G.A. Death don't have no mercy and neither does calcium: *Arabidopsis* CYCLIC NUCLEOTIDE GATED CHANNEL2 and innate immunity. *Plant Cell* **2007**, *19*, 1081–1095.
19. Harding, S.A.; Oh, S.H.; Roberts, D.M. Transgenic tobacco expressing a foreign calmodulin gene shows an enhanced production of active oxygen species. *EMBO J.* **1997**, *16*, 1137–1144.
20. Reddy, A.S.; Ali, G.S.; Celesnik, H.; Day, I.S. Coping with stresses: Roles of calcium- and calcium/calmodulin-regulated gene expression. *Plant Cell* **2011**, *23*, 2010–2032.
21. Popescu, S.C.; Popescu, G.V.; Bachan, S.; Zhang, Z.; Seay, M.; Gerstein, M.; Snyder, M.; Dinesh-Kumar, S.P. Differential binding of calmodulin-related proteins to their targets revealed through high-density *Arabidopsis* protein microarrays. *Proc. Natl. Acad. Sci. USA* **2007**, *104*, 4730–4735.
22. Wang, L.; Tsuda, K.; Sato, M.; Cohen, J.D.; Katagiri, F.; Glazebrook, J. *Arabidopsis* CaM binding protein CBP60g contributes to MAMP-induced SA accumulation and is involved in disease resistance against *Pseudomonas syringae*. *PLoS Pathog.* **2009**, *5*, e1000301.
23. Kolukisaoglu, U.; Weinl, S.; Blazevic, D.; Batistic, O.; Kudla, J. Calcium sensors and their interacting protein kinases: Genomics of the *Arabidopsis* and rice CBL-CIPK signaling networks. *Plant Physiol.* **2004**, *134*, 43–58.
24. Albrecht, V.; Ritz, O.; Linder, S.; Harter, K.; Kudla, J. The NAF domain defines a novel protein-protein interaction module conserved in Ca^{2+}-regulated kinases. *EMBO J.* **2001**, *20*, 1051–1063.
25. Boudsocq, M.; Sheen, J. CDPKs in immune and stress signaling. *Trends Plant Sci.* **2013**, *18*, 30–40.
26. Asai, S.; Ichikawa, T.; Nomura, H.; Kobayashi, M.; Kamiyoshihara, Y.; Mori, H.; Kadota, Y.; Zipfel, C.; Jones, J.D.; Yoshioka, H. The variable domain of a plant Calcium-dependent Protein Kinase (CDPK) confers subcellular localization and substrate recognition for NADPH oxidase. *J. Biol. Chem.* **2013**, *288*, 14332–14340.
27. Sheen, J. Ca^{2+}-dependent protein kinases and stress signal transduction in plants. *Science* **1996**, *274*, 1900–1902.
28. Harper, J.F.; Breton, G.; Harmon, A. Decoding Ca(2+) signals through plant protein kinases. *Annu. Rev. Plant Biol.* **2004**, *55*, 263–288.
29. Harper, J.F.; Sussman, M.R.; Schaller, G.E.; Putnam-Evans, C.; Charbonneau, H.; Harmon, A.C. A calcium-dependent protein kinase with a regulatory domain similar to calmodulin. *Science* **1991**, *252*, 951–954.

30. Harper, J.F.; Harmon, A. Plants, symbiosis and parasites: A calcium signalling connection. *Nat. Rev. Mol. Cell Biol.* **2005**, *6*, 555–566.
31. Wernimont, A.K.; Amani, M.; Qiu, W.; Pizarro, J.C.; Artz, J.D.; Lin, Y.H.; Lew, J.; Hutchinson, A.; Hui, R. Structures of parasitic CDPK domains point to a common mechanism of activation. *Proteins* **2011**, *79*, 803–820.
32. Wernimont, A.K.; Artz, J.D.; Finerty, P., Jr.; Lin, Y.H.; Amani, M.; Allali-Hassani, A.; Senisterra, G.; Vedadi, M.; Tempel, W.; Mackenzie, F.; *et al.* Structures of apicomplexan calcium-dependent protein kinases reveal mechanism of activation by calcium. *Nat. Struct. Mol. Biol.* **2010**, *17*, 596–601.
33. Jeong, J.C.; Shin, D.; Lee, J.; Kang, C.H.; Baek, D.; Cho, M.J.; Kim, M.C.; Yun, D.J. Isolation and characterization of a novel calcium/calmodulin-dependent protein kinase, AtCK, from arabidopsis. *Mol. Cells* **2007**, *24*, 276–282.
34. Asano, T.; Tanaka, N.; Yang, G.; Hayashi, N.; Komatsu, S. Genome-wide identification of the rice calcium-dependent protein kinase and its closely related kinase gene families: Comprehensive analysis of the CDPKs gene family in rice. *Plant Cell Physiol.* **2005**, *46*, 356–366.
35. Klimecka, M.; Muszynska, G. Structure and functions of plant calcium-dependent protein kinases. *Acta Biochim. Pol.* **2007**, *54*, 219–233.
36. Myers, C.; Romanowsky, S.M.; Barron, Y.D.; Garg, S.; Azuse, C.L.; Curran, A.; Davis, R.M.; Hatton, J.; Harmon, A.C.; Harper, J.F. Calcium-dependent protein kinases regulate polarized tip growth in pollen tubes. *Plant J.* **2009**, *59*, 528–539.
37. Ivashuta, S.; Liu, J.; Liu, J.; Lohar, D.P.; Haridas, S.; Bucciarelli, B.; VandenBosch, K.A.; Vance, C.P.; Harrison, M.J.; Gantt, J.S. RNA interference identifies a calcium-dependent protein kinase involved in *Medicago truncatula* root development. *Plant Cell* **2005**, *17*, 2911–2921.
38. Matschi, S.; Werner, S.; Schulze, W.X.; Legen, J.; Hilger, H.H.; Romeis, T. Function of calcium-dependent protein kinase CPK28 of *Arabidopsis thaliana* in plant stem elongation and vascular development. *Plant J.* **2013**, *73*, 883–896.
39. Zhu, S.Y.; Yu, X.C.; Wang, X.J.; Zhao, R.; Li, Y.; Fan, R.C.; Shang, Y.; Du, S.Y.; Wang, X.F.; Wu, F.Q.; *et al.* Two calcium-dependent protein kinases, CPK4 and CPK11, regulate abscisic acid signal transduction in *Arabidopsis*. *Plant Cell* **2007**, *19*, 3019–3036.
40. Ye, W.; Muroyama, D.; Munemasa, S.; Nakamura, Y.; Mori, I.C.; Murata, Y. Calcium-dependent protein kinase, CPK6, positively functions in induction by YEL of stomatal closure and inhibition by YEL of light-induced stomatal opening in *Arabidopsis*. *Plant Physiol.* **2013**, *163*, 591–599.
41. Munemasa, S.; Hossain, M.A.; Nakamura, Y.; Mori, I.C.; Murata, Y. The *Arabidopsis* calcium-dependent protein kinase, CPK6, functions as a positive regulator of methyl jasmonate signaling in guard cells. *Plant Physiol.* **2011**, *155*, 553–561.
42. Takeuchi, O.; Akira, S. Pattern recognition receptors and inflammation. *Cell* **2010**, *140*, 805–820.
43. Ausubel, F.M. Are innate immune signaling pathways in plants and animals conserved? *Nat. Immunol.* **2005**, *6*, 973–979.
44. Boller, T.; Felix, G. A renaissance of elicitors: Perception of microbe-associated molecular patterns and danger signals by pattern-recognition receptors. *Annu. Rev. Plant Biol.* **2009**, *60*, 379–406.

45. Chisholm, S.T.; Coaker, G.; Day, B.; Staskawicz, B.J. Host-microbe interactions: Shaping the evolution of the plant immune response. *Cell* **2006**, *124*, 803–814.
46. Erbs, G.; Silipo, A.; Aslam, S.; de Castro, C.; Liparoti, V.; Flagiello, A.; Pucci, P.; Lanzetta, R.; Parrilli, M.; Molinaro, A.; *et al.* Peptidoglycan and muropeptides from pathogens *Agrobacterium* and *Xanthomonas* elicit plant innate immunity: Structure and activity. *Chem. Biol.* **2008**, *15*, 438–448.
47. Zipfel, C. Pattern-recognition receptors in plant innate immunity. *Curr. Opin. Immunol.* **2008**, *20*, 10–16.
48. Erbs, G.; Newman, M.A. The role of lipopolysaccharide and peptidoglycan, two glycosylated bacterial microbe-associated molecular patterns (MAMPs), in plant innate immunity. *Mol. Plant Pathol.* **2012**, *13*, 95–104.
49. Miya, A.; Albert, P.; Shinya, T.; Desaki, Y.; Ichimura, K.; Shirasu, K.; Narusaka, Y.; Kawakami, N.; Kaku, H.; Shibuya, N. CERK1, a LysM receptor kinase, is essential for chitin elicitor signaling in *Arabidopsis*. *Proc. Natl. Acad. Sci. USA* **2007**, *104*, 19613–19618.
50. Chinchilla, D.; Zipfel, C.; Robatzek, S.; Kemmerling, B.; Nurnberger, T.; Jones, J.D.G.; Felix, G.; Boller, T. A flagellin-induced complex of the receptor FLS2 and BAK1 initiates plant defence. *Nature* **2007**, *448*, 497–500.
51. Zhang, J.; Zhou, J.M. Plant immunity triggered by microbial molecular signatures. *Mol. Plant* **2010**, *3*, 783–793.
52. Lu, D.; Wu, S.; Gao, X.; Zhang, Y.; Shan, L.; He, P. A receptor-like cytoplasmic kinase, BIK1, associates with a flagellin receptor complex to initiate plant innate immunity. *Proc. Natl. Acad. Sci. USA* **2010**, *107*, 496–501.
53. Lin, W.; Lu, D.; Gao, X.; Jiang, S.; Ma, X.; Wang, Z.; Mengiste, T.; He, P.; Shan, L. Inverse modulation of plant immune and brassinosteroid signaling pathways by the receptor-like cytoplasmic kinase BIK1. *Proc. Natl. Acad. Sci. USA* **2013**, *110*, 12114–12119.
54. Lu, D.; Lin, W.; Gao, X.; Wu, S.; Cheng, C.; Avila, J.; Heese, A.; Devarenne, T.P.; He, P.; Shan, L. Direct ubiquitination of pattern recognition receptor FLS2 attenuates plant innate immunity. *Science* **2011**, *332*, 1439–1442.
55. Eitas, T.K.; Dangl, J.L. NB-LRR proteins: Pairs, pieces, perception, partners, and pathways. *Curr. Opin. Plant Biol.* **2010**, *13*, 472–477.
56. Elmore, J.M.; Lin, Z.J.; Coaker, G. Plant NB-LRR signaling: Upstreams and downstreams. *Curr. Opin. Plant Biol.* **2011**, *14*, 365–371.
57. DeYoung, B.J.; Innes, R.W. Plant NBS-LRR proteins in pathogen sensing and host defense. *Nat. Immunol.* **2006**, *7*, 1243–1249.
58. Davis, B.K.; Wen, H.; Ting, J.P. The inflammasome NLRs in immunity, inflammation, and associated diseases. *Annu. Rev. Immunol.* **2011**, *29*, 707–735.
59. Dodds, P.N.; Rathjen, J.P. Plant immunity: Towards an integrated view of plant-pathogen interactions. *Nat. Rev. Genet.* **2010**, *11*, 539–548.
60. Boller, T.; He, S.Y. Innate immunity in plants: An arms race between pattern recognition receptors in plants and effectors in microbial pathogens. *Science* **2009**, *324*, 742–744.
61. Jones, J.D.; Dangl, J.L. The plant immune system. *Nature* **2006**, *444*, 323–329.

62. Gawehns, F.; Cornelissen, B.J.; Takken, F.L. The potential of effector-target genes in breeding for plant innate immunity. *Microb. Biotechnol.* **2013**, *6*, 223–229.
63. Dangl, J.L.; Horvath, D.M.; Staskawicz, B.J. Pivoting the plant immune system from dissection to deployment. *Science* **2013**, *341*, 746–751.
64. Nimchuk, Z.; Eulgem, T.; Holt, B.F., 3rd; Dangl, J.L. Recognition and response in the plant immune system. *Annu. Rev. Genet.* **2003**, *37*, 579–609.
65. Kunkel, B.N.; Bent, A.F.; Dahlbeck, D.; Innes, R.W.; Staskawicz, B.J. RPS2, an *Arabidopsis* disease resistance locus specifying recognition of *Pseudomonas syringae* strains expressing the avirulence gene avrRpt2. *Plant Cell* **1993**, *5*, 865–875.
66. Mindrinos, M.; Katagiri, F.; Yu, G.L.; Ausubel, F.M. The *A. thaliana* disease resistance gene RPS2 encodes a protein containing a nucleotide-binding site and leucine-rich repeats. *Cell* **1994**, *78*, 1089–1099.
67. Grant, M.R.; Godiard, L.; Straube, E.; Ashfield, T.; Lewald, J.; Sattler, A.; Innes, R.W.; Dangl, J.L. Structure of the *Arabidopsis* RPM1 gene enabling dual specificity disease resistance. *Science* **1995**, *269*, 843–846.
68. Axtell, M.J.; Staskawicz, B.J. Initiation of RPS2-specified disease resistance in *Arabidopsis* is coupled to the AvrRpt2-directed elimination of RIN4. *Cell* **2003**, *112*, 369–377.
69. Gao, Z.; Chung, E.H.; Eitas, TK.; Dangl, J.L. Plant intracellular innate immune receptor Resistance to *Pseudomonas syringae* pv. *maculicola* 1 (RPM1) is activated at, and functions on, the plasma membrane. *Proc. Natl. Acad. Sci. USA* **2011**, *108*, 7619–7624.
70. Ranf, S.; Eschen-Lippold, L.; Pecher, P.; Lee, J.; Scheel, D. Interplay between calcium signalling and early signalling elements during defence responses to microbe- or damage-associated molecular patterns. *Plant J.* **2011**, *68*, 100–113.
71. Grant, M.; Brown, I.; Adams, S.; Knight, M.; Ainslie, A.; Mansfield, J. The RPM1 plant disease resistance gene facilitates a rapid and sustained increase in cytosolic calcium that is necessary for the oxidative burst and hypersensitive cell death. *Plant J.* **2000**, *23*, 441–450.
72. Geng, S.; Li, A.; Tang, L.; Yin, L.; Wu, L.; Lei, C.; Guo, X.; Zhang, X.; Jiang, G.; Zhai, W.; *et al.* TaCPK2-A, a calcium-dependent protein kinase gene that is required for wheat powdery mildew resistance enhances bacterial blight resistance in transgenic rice. *J. Exp. Bot.* **2013**, *64*, 3125–3136.
73. Coca, M.; San Segundo, B. AtCPK1 calcium-dependent protein kinase mediates pathogen resistance in *Arabidopsis*. *Plant J.* **2010**, *63*, 526–540.
74. Boudsocq, M.; Willmann, M.R.; McCormack, M.; Lee, H.; Shan, L.; He, P.; Bush, J.; Cheng, S.; Sheen, J. Differential innate immune signalling via Ca(2+) sensor protein kinases. *Nature* **2010**, *464*, 418–422.
75. Ludwig, A.A.; Saitoh, H.; Felix, G.; Freymark, G.; Miersch, O.; Wasternack, C.; Boller, T.; Jones, J.D.; Romeis, T. Ethylene-mediated cross-talk between calcium-dependent protein kinase and MAPK signaling controls stress responses in plants. *Proc. Natl. Acad. Sci. USA* **2005**, *102*, 10736–10741.
76. Kobayashi, M.; Ohura, I.; Kawakita, K.; Yokota, N.; Fujiwara, M.; Shimamoto, K.; Doke, N.; Yoshioka, H. Calcium-dependent protein kinases regulate the production of reactive oxygen species by potato NADPH oxidase. *Plant Cell* **2007**, *19*, 1065–1080.

77. Dubiella, U.; Seybold, H.; Durian, G.; Komander, E.; Lassig, R.; Witte, C.P.; Schulze, W.X.; Romeis, T. Calcium-dependent protein kinase/NADPH oxidase activation circuit is required for rapid defense signal propagation. *Proc. Natl. Acad. Sci. USA* **2013**, *110*, 8744–8749.
78. Asano, T.; Hayashi, N.; Kobayashi, M.; Aoki, N.; Miyao, A.; Mitsuhara, I.; Ichikawa, H.; Komatsu, S.; Hirochika, H.; Kikuchi, S.; *et al.* A rice calcium-dependent protein kinase OsCPK12 oppositely modulates salt-stress tolerance and blast disease resistance. *Plant J.* **2012**, *69*, 26–36.
79. Liese, A.; Romeis, T. Biochemical regulation of *in vivo* function of plant calcium-dependent protein kinases (CDPK). *Biochim. Biophys. Acta* **2013**, *1833*, 1582–1589.
80. Lee, J.; Rudd, J.J.; Macioszek, V.K.; Scheel, D. Dynamic changes in the localization of MAPK cascade components controlling pathogenesis-related (PR) gene expression during innate immunity in parsley. *J. Biol. Chem.* **2004**, *279*, 22440–22448.
81. Wurzinger, B.; Mair, A.; Pfister, B.; Teige, M. Cross-talk of calcium-dependent protein kinase and MAP kinase signaling. *Plant Signal. Behav.* **2011**, *6*, 8–12.
82. Gao, X.; Chen, X.; Lin, W.; Chen, S.; Lu, D.; Niu, Y.; Li, L.; Cheng, C.; McCormack, M.; Sheen, J.; *et al.* Bifurcation of *Arabidopsis* NLR immune signaling via Ca^{2+}-dependent protein kinases. *PLoS Pathog.* **2013**, *9*, e1003127.
83. Gust, A.A.; Biswas, R.; Lenz, H.D.; Rauhut, T.; Ranf, S.; Kemmerling, B.; Gotz, F.; Glawischnig, E.; Lee, J.; Felix, G.; *et al.* Bacteria-derived peptidoglycans constitute pathogen-associated molecular patterns triggering innate immunity in *Arabidopsis*. *J. Biol. Chem.* **2007**, *282*, 32338–32348.
84. Blume, B.; Nurnberger, T.; Nass, N.; Scheel, D. Receptor-mediated increase in cytoplasmic free calcium required for activation of pathogen defense in parsley. *Plant Cell* **2000**, *12*, 1425–1440.
85. Zimmermann, S.; Nurnberger, T.; Frachisse, J.M.; Wirtz, W.; Guern, J.; Hedrich, R.; Scheel, D. Receptor-mediated activation of a plant Ca^{2+}-permeable ion channel involved in pathogen defense. *Proc. Natl. Acad. Sci. USA* **1997**, *94*, 2751–2755.
86. Ma, W.; Smigel, A.; Tsai, Y.C.; Braam, J.; Berkowitz, G.A. Innate immunity signaling: Cytosolic Ca^{2+} elevation is linked to downstream nitric oxide generation through the action of calmodulin or a calmodulin-like protein. *Plant Physiol.* **2008**, *148*, 818–828.
87. Romeis, T.; Ludwig, A.A.; Martin, R.; Jones, J.D. Calcium-dependent protein kinases play an essential role in a plant defence response. *EMBO J.* **2001**, *20*, 5556–5567.
88. Freymark, G.; Diehl, T.; Miklis, M.; Romeis, T.; Panstruga, R. Antagonistic control of powdery mildew host cell entry by barley calcium-dependent protein kinases (CDPKs). *Mol. Plant Microbe Interact.* **2007**, *20*, 1213–1221.
89. Martin, M.L.; Busconi, L. Membrane localization of a rice calcium-dependent protein kinase (CDPK) is mediated by myristoylation and palmitoylation. *Plant J.* **2000**, *24*, 429–435.
90. Chehab, E.W.; Patharkar, O.R.; Hegeman, A.D.; Taybi, T.; Cushman, J.C. Autophosphorylation and subcellular localization dynamics of a salt- and water deficit-induced calcium-dependent protein kinase from ice plant. *Plant Physiol.* **2004**, *135*, 1430–1446.
91. Patharkar, O.R.; Cushman, J.C. A stress-induced calcium-dependent protein kinase from *Mesembryanthemum crystallinum* phosphorylates a two-component pseudo-response regulator. *Plant J.* **2000**, *24*, 679–691.

92. Messinese, E.; Mun, J.H.; Yeun, L.H.; Jayaraman, D.; Rouge, P.; Barre, A.; Lougnon, G.; Schornack, S.; Bono, J.J.; Cook, D.R.; *et al.* A novel nuclear protein interacts with the symbiotic DMI3 calcium- and calmodulin-dependent protein kinase of *Medicago truncatula*. *Mol. Plant Microbe Interact* **2007**, *20*, 912–921.
93. Charpentier, M.; Oldroyd, G.E. Nuclear calcium signaling in plants. *Plant Physiol.* **2013**, *163*, 496–503.
94. Curran, A.; Chang, I.F.; Chang, C.L.; Garg, S.; Miguel, R.M.; Barron, Y.D.; Li, Y.; Romanowsky, S.; Cushman, J.C.; Gribskov, M.; *et al.* Calcium-dependent protein kinases from *Arabidopsis* show substrate specificity differences in an analysis of 103 substrates. *Front Plant Sci.* **2011**, *2*, 36.
95. Shen, Q.H.; Saijo, Y.; Mauch, S.; Biskup, C.; Bieri, S.; Keller, B.; Seki, H.; Ulker, B.; Somssich, I.E.; Schulze-Lefert, P. Nuclear activity of MLA immune receptors links isolate-specific and basal disease-resistance responses. *Science* **2007**, *315*, 1098–1103.
96. Bhattacharjee, S.; Halane, M.K.; Kim, S.H.; Gassmann, W. Pathogen effectors target Arabidopsis EDS1 and alter its interactions with immune regulators. *Science* **2011**, *334*, 1405–1408.
97. Heidrich, K.; Wirthmueller, L.; Tasset, C.; Pouzet, C.; Deslandes, L.; Parker, J.E. Arabidopsis EDS1 connects pathogen effector recognition to cell compartment-specific immune responses. *Science* **2011**, *334*, 1401–1404.
98. Gao, X.; He, P. Nuclear dynamics of Arabidopsis calcium-dependent protein kinases in effector-triggered immunity. *Plant Signal. Behav.* **2013**, *8*, e23868.

12

Identification of the Abundant Hydroxyproline-Rich Glycoproteins in the Root Walls of Wild-Type Arabidopsis, an *ext3* Mutant Line, and Its Phenotypic Revertant

Yuning Chen [1,†,*], Dening Ye [1,†], Michael A. Held [1], Maura C. Cannon [2], Tui Ray [2], Prasenjit Saha [2], Alexandra N. Frye [1], Andrew J. Mort [3] and Marcia J. Kieliszewski [1]

[1] Department of Chemistry and Biochemistry, Ohio University, Athens, OH 45701, USA; E-Mails: Boxofjacky@gmail.com (D.Y.); held@ohio.edu (M.A.H.); af844006@gmail.com (A.N.F.); kielisze@helios.phy.ohiou.edu (M.J.K.)

[2] Department of Biochemistry and Molecular Biology, University of Massachusetts, Amherst, MA 01003, USA; E-Mails: mcannon@biochem.umass.edu (M.C.C.); tuiray76@gmail.com (T.R.); sahap76@gmail.com (P.S.)

[3] Biochemistry and Molecular Biology, Oklahoma State University, Stillwater, OK 74078, USA; E-Mail: Andrew.mort@okstate.edu

† These authors contributed equally to this work.

* Author to whom correspondence should be addressed; E-Mail: yuc27@psu.edu

Academic Editor: Philip J. Harris

Abstract: Extensins are members of the cell wall hydroxyproline-rich glycoprotein (HRGP) superfamily that form covalently cross-linked networks in primary cell walls. A knockout mutation in *EXT3* (*AT1G21310*), the gene coding EXTENSIN 3 (EXT3) in *Arabidopsis* Landsberg *erecta* resulted in a lethal phenotype, although about 20% of the knockout plants have an apparently normal phenotype (ANP). In this study the root cell wall HRGP components of wild-type, ANP and the *ext3* mutant seedlings were characterized by peptide fractionation of trypsin digested anhydrous hydrogen fluoride deglycosylated wall residues and by sequencing using LC-MS/MS. Several HRGPs, including EXT3, were identified in the wild-type root walls but not in walls of the ANP and lethal mutant. Indeed the ANP walls and walls of mutants displaying the lethal phenotype possessed HRGPs, but the profiles suggest that changes in the amount and perhaps type may account for the corresponding phenotypes.

Keywords: *Arabidopsis* root; cell wall; hydroxyproline-rich glycoprotein; Extensin 3; LRX; PRP4; proteomics

1. Introduction

Plants devote their extracellular space to a physically rigid cell wall comprised mainly of polysaccharides (cellulose, hemicellulose and pectin) and structural glycoproteins with a pectin-rich middle-lamella layer. Cell wall structural glycoproteins are rich in hydroxyproline (Hyp, O) and are thus named hydroxyproline-rich glycoproteins (HRGPs). HRGPs are ubiquitous components of the plant extracellular matrix and comprise up to ~10% (w/w) of the cell wall in dicots [1]. They create covalently crosslinked networks [2], generating negatively charged cushions at the membrane-wall interface [3], and thereby providing functions such as barriers against pathogen ingress [4,5], adhesives for pollen capture [6,7], primers for wall polysaccharide biosynthesis [8], and regulators of Ca^{2+} signaling [9,10].

HRGPs have characteristic repetitive peptide motifs that specify the extent and type of glycosylation on some, but not all, of the Hyp residues [11]. Three major types of HRGPs prevail in primary cell walls, namely the arabinogalactan proteins (AGPs), proline-rich proteins (PRPs), and extensins (EXTs). AGPs have short repetitive AO/SO/SOOAPAP motifs whereby most Hyp residues are heavily O-glycosylated with complex arabinogalactan polysaccharides. The AGP protein backbone accounts for less than 10% of the total molecule's mass [12]. Many, but not all, AGPs have a GPI anchor at their C-terminus [13,14] and most AGPs react with β-glucosyl Yariv reagent to form a dark red precipitant, a common method for identifying AGPs [15]. Recently an Arabidopsis AGP was found to covalently interact with wall polysaccharide components rhamnogalacturonan I and arabinoxylan, indicating a structural role for AGPs in the cell wall architecture [8].

PRPs are lightly glycosylated HRGPs compared to AGPs, as sugar only makes up ~20% of the protein on a dry weight basis and they have repetitive peptide motifs such as POVXK, where X is often Tyr or Glu [16]. PRPs are expressed in various plant tissues and are tightly regulated during plant development and in response to environmental stimuli such as wounding [17].

EXTs are a type of HRGP that contribute to cell wall integrity by creating crosslinked networks during normal development [18], in response to wounding [5] and pathogen attack [4]. They are characterized by glycosylated Ser-$(Hyp)_4$ motifs [19], whereby Ser residues are monogalactosylated and Hyp residues usually contain 1 to 4 oligoarabinosyl additions [20]. Another characteristic extensin motif is the Tyr-containing YXY motif that enables the covalent crosslinking of EXTs into the cell wall in the form of di-isodityrosine [21,22] and pulcherosine [23] presumably by wall bound peroxidases [24,25].

Earlier, Hall and Cannon identified a specific Arabidopsis EXT, designated RSH (Root Shoot Hypocotyl Defective) [26] and also known as EXT3, which is required for normal wall development. Loss of EXT3 due to a knock out mutation (the mutant designated "*rsh*") was lethal due to abnormal and incomplete cell plate development [26,27]. Indeed mutant seedlings were visibly abnormal and died by three weeks of age.

This dramatic, readily observable difference between the mutant and wild-type phenotype provided a route to test EXT3 structure in wall assembly through the creation of altered forms of EXT3 used to rescue the mutant. However, this approach was complicated by the observation that some *rsh* homozygous lines (about 20%), although true mutants, as qPCR experiments found essentially no expression of the *EXT3* gene, possessed a wild-type phenotype, and were designated apparently normal phenotype (ANP) [28]. Judging by qRT-PCR and microarray experiments, we speculated that one or more proteins in ANP lines, including some predicted to be members of the HRGP family, recovered the wild-type phenotype in the absence EXT3.

We set out to characterize the different HRGPs contributing to the networks present in wild type (WT) Arabidopsis root walls, and root walls isolated from two ANP lines and from the *rsh* mutant displaying the lethal phenotype (MT) by characterizing the major HRGPs in these walls. Our approach involved tryptic peptide mapping of HF-deglycosylated wall to isolate the major peptides present followed by peptide sequence analysis by MS/MS.

Here we identified the major HRGP contributors to the WT, ANP, and MT root cell wall protein networks and in consistent with the results of Saha *et al.* [28], we found little evidence of EXT3 in ANP and MT roots while EXT3 was a major HRGP in the WT root.

2. Results

2.1. Arabidopsis Root Cell Wall Hyp/Protein Content

The HRGP contents in root cell walls of different Arabidopsis lines were first assessed by examining the amino acid compositions of WT, ANP and MT root cell walls. Hyp was found as the most abundant amino acid component of the WT Arabidopsis root cell wall matrix and comprised about 20% (mole) of the total amino acids (Table 1), thereby showing that HRGPs are the major components of the Arabidopsis root cell wall proteome. Hyp was also a major amino acid component of the WT root walls on a dry weight basis. Proteins comprised 7.3% (w/w) of the cell wall dry weight (Table 1) in the root walls, while the weight percentage of Hyp alone was 1.5% (w/w), making Hyp a 20.5% (w/w) component of wall proteins.

ANP and MT root walls shared similar amino acid compositions to WT walls with Hyp being the most abundant amino acid accounting for more than 17% (mole) of the total amino acids (Table 1). ANP and MT walls also showed a wall protein content similar to that of WT at around 7% (w/w). All walls had the same Hyp weight percentage of around 1.5%. Thus Hyp and the corresponding HRGPs were indeed major protein components in Arabidopsis root cell walls.

2.2. HF Deglycosylation of Root Cell Walls

To allow the release of Hyp-containing peptides from root cell walls for HRGP identification we then deglycosylated the walls with HF to remove wall polysaccharides. HF cleaves all neutral and acidic sugar linkages within 1 h at 0 °C but leaves peptide bonds intact [2], thus exposing the wall protein network and making it accessible to the later protease degradation. Here HF efficiently removed ~80% (w/w) of the cell wall dry weight from WT and ANP walls and 65% (w/w) of that in the case of MT walls (Table 1). Based on the weight percentage of protein and Hyp before and after HF, HF removed up to 40% (w/w) of the wall protein but little wall-bound Hyp, judging by the high content of Hyp (>90%, w/w) recovered from the remaining HF insoluble residue (HFI) (calculated from Table 1). Hyp remained the major amino acid that comprised more than 20% (mole) of the total amino acids in the root cell wall HFI of WT and ANP and about 15% (mole) in MT HFI (Table 1). The resistance of Hyp-containing wall component to HF deglycosylation indicates that HRGPs in the cell wall form an independent network from wall polysaccharides.

Table 1. Amino Acid Composition (mole %) of the Root Cell Wall (CW), Insoluble Cell Wall Pellet Remaining after Treatment with HF (HFI), Trypsin (TI HFI) and Pronase (PI HFI).

Amino Acid	WT CW	WT HFI	WT TI HFI	WT PI HFI	ANP4 CW	ANP4 HFI	ANP4 TI HFI	ANP4 PI HFI	ANP10 CW	ANP10 HFI	ANP10 TI HFI	ANP10 PI HFI	MT CW	MT HFI	MT TI HFI
Asp	6	1	2	1	6	6	4	4	6	4	1	1	7	5	5
Glu	6	3	2	2	5	4	3	3	6	3	2	2	7	2	4
Hyp	**20**	**25**	**27**	**30**	**17**	**22**	**20**	**22**	**18**	**29**	**22**	**28**	**17**	**15**	**19**
Ser	10	14	17	18	11	9	13	18	11	8	15	19	10	2	7
Gly	11	11	12	13	14	12	17	23	12	10	12	16	12	18	20
His	2	2	3	2	2	3	1	2	2	4	2	2	2	3	4
Arg	2	2	2	2	2	3	2	1	2	3	2	2	3	5	2
Thr	3	5	4	3	3	5	3	3	3	4	5	4	3	5	4
Ala	6	6	5	5	7	9	9	6	7	9	6	3	7	12	9
Pro	7	4	4	4	7	5	4	6	6	5	5	4	6	7	5
Tyr	4	2	2	2	3	2	3	2	3	2	1	2	3	2	1
Val	7	6	4	4	7	6	3	2	7	6	7	4	7	8	7
Ile	3	3	2	1	3	2	2	0	4	2	4	2	3	4	3
Leu	5	5	4	3	4	5	5	2	4	4	6	4	5	6	5
Phe	4	3	2	2	4	2	3	0	4	2	3	2	4	1	1
Lys	4	8	8	8	5	5	8	6	5	5	7	5	4	5	4
Total	100	100	100	100	100	100	100	100	100	100	100	100	100	100	100
% Hyp [1]	1.5	6.9	5.2	4.5	1.1	3.9	2.1	1.1	1.3	6.3	4.3	4.1	1.1	2.5	3.8
% Protein [1]	7.3	25.5	17.7	13.6	6.4	16.1	7	4.4	6.8	20.5	17.4	13.1	6.2	13.7	14.6
Hyp/ Protein [2]	0.2	0.27	0.29	0.33	0.17	0.24	0.3	0.25	0.19	0.31	0.25	0.31	0.18	0.18	0.26
DW (mg) [3]	50	10	8	7	38	10	7	6	51	10	8	7	4.6	1.6	0.75
% Hyp Removed [4]		6.7	39.7	54.3		10.0	62.3	83.1		7.1	45.4	54.4		10.0	10.9
% Protein Removed [4]		30.1	44.5	62.3		34.4	69.6	83.6		41.1	32.1	55.3		20.0	60.9

[1] % Hyp or Protein: the weight percentage of Hyp and total protein; [2] Hyp/Protein: weight ratio of Hyp in total protein content; [3] DW (mg): the dry weight of cell wall used in HF deglycosylation and the remaining HFI before and after trypsin/pronase digestion; [4] % Hyp or Protein removed: the weight percentage of Hyp and protein removed after HF, trypsin or, pronase digestion calculated from Hyp/protein weight percentage of stepwise HFI, TI HFI and PI HFI.

2.3. Release of Hyp-Containing Material by Protease Degradation of Root Cell Wall HFI

Following HF deglycosylation of the root cell wall, WT, ANP, and MT HFI were sequentially treated with trypsin then pronase (WT and ANP only) to release wall peptides for protein identification. Exhaustive trypsin digestion of root wall HFI solubilized up to 60% (w/w) of the wall-bound proteins and more than 40% (w/w) of the Hyp-containing components from WT and ANP HFI, but only removed ~11% (w/w) Hyp from the MT HFI, although it removed 61% (w/w) of the MT HFI protein (Table 1). The trypsin insoluble HFI (TI HFI) of WT and ANP was further treated with pronase, which solubilized an additional ~20% (w/w) protein but a large portion of the wall bound Hyp (~50%, w/w) remained in the pellets that were resistant to proteolysis (Table 1). Hyp remained the most abundant amino acid in the protease insoluble residues, as indicated by the amino acid composition of the TI HFI and the pronase insoluble HFI (PI HFI, Table 1). Thus the HF insoluble protein network was somehow resistant to trypsin and pronase since only a portion of the Hyp-containing components in the HFI samples were released by protease degradation.

2.4. Peptide Fractionation from Trypsin Digested HFI of WT and ANP

Tryptic peptides from the WT and ANP walls were first fractionated to estimate the composition and relative abundance of specific HRGPs in these walls before applied for peptide sequencing by LC-MS/MS. Our approach combined size exclusion (SEC), strong cation exchange (SCX) and reverse phase chromatography, exploiting peptide properties to fractionate them (Figure 1).

Size fractionation of tryptic peptides liberated from WT HFI sample yielded three major fractions: a larger molecular weight major peak (fractions 7–10; Figure 2b inset), a broad shoulder peak (fractions 11–12; Figure 2c inset), and a sharp, smaller molecular weight peak (fractions 13–15; Figure 2d inset). Further separation of the SEC fractions 7–10 by SCX gave rise to five major peaks (Figure 2b) that, in turn, produced several peptide peaks after further fractionation by reverse-phase HPLC (Figure 2e–g). Four of these peaks were designated major peaks (W1, W2, W3, and W4) and several more as minor peaks (WM1-11). Similarly fractions 11–12 (Figure 2c inset) and fractions 13–15 (Figure 2d inset) from SEC were further fractionated by SCX and reverse-phase HPLC (Figure 2c,h,i,j and Figure 2d,L,m,n,o, respectively). Four major peptide peaks from the broad shoulder peak (W5, W6, W7 and W8) and two major peptide peaks from the small molecular weight SEC peak (W9 and W10) were identified.

Analysis of the Hyp content of the major WT wall peptide peaks indicated they all contained Hyp, with W3, W5, W6, and W10 being Hyp-rich, each containing more than 30% (mole) Hyp. W2 was relatively Hyp-poor and contained only 8% (mole) Hyp (Table S1).

Size fractionation of the trypsin soluble material from ANP4 and ANP10 HFI samples both resulted in profiles similar to that of WT (Figures 3a and 4a). SEC fractions of ANP samples were grouped similarly to that of WT (fractions 7–10, 11–12 and 13–15, Figure 1) and fractionated by SCX and reverse-phase HPLC. For ANP4 HFI, SEC fractions 7–10 (Figure 3b inset) produced two major peptide peaks, F1 and F2 (Figure 3f,g), both of which were Hyp-rich, containing 32% (mole) and 42% (mole) Hyp, respectively (Table S1). SEC fractions 11–12 (Figure 3c inset) produced three major peptide peaks, F3, F4, and F5 (Figure 3h,i). F3 had 28% (mole) Hyp and F4 had 14% (mole) Hyp (Table S1). The Hyp content of F5 was not determined due to lack of material. SEC fractions 13–15 (Figure 3d inset) did not yield any peptide peaks judging by reverse-phase fractionation profile.

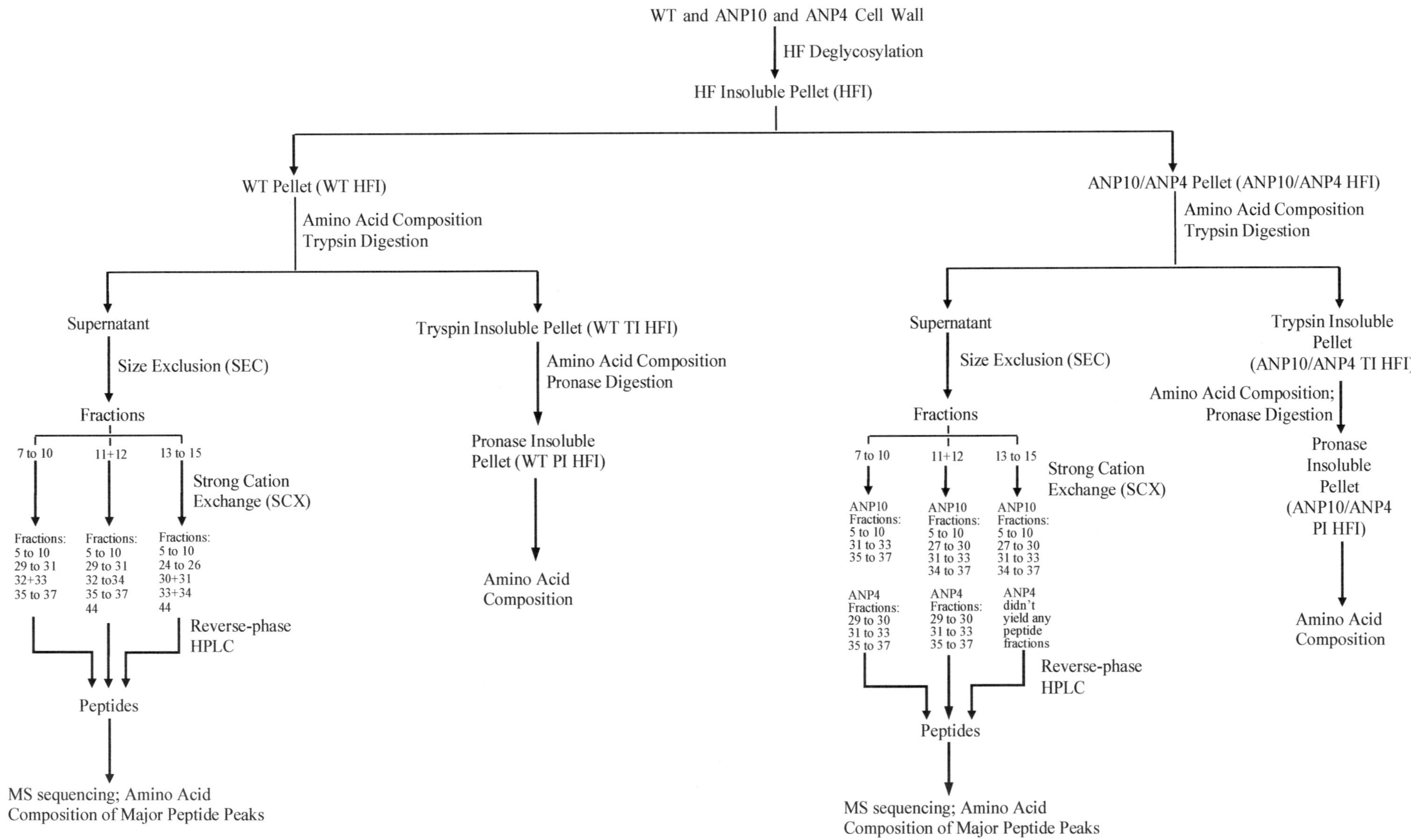

Figure 1. Flow chart outlining the isolation of WT, ANP10 and ANP4 root wall peptides. See Figures 2–4 for the peptide maps from the SEC, SCX and reverse-phase HPLC columns.

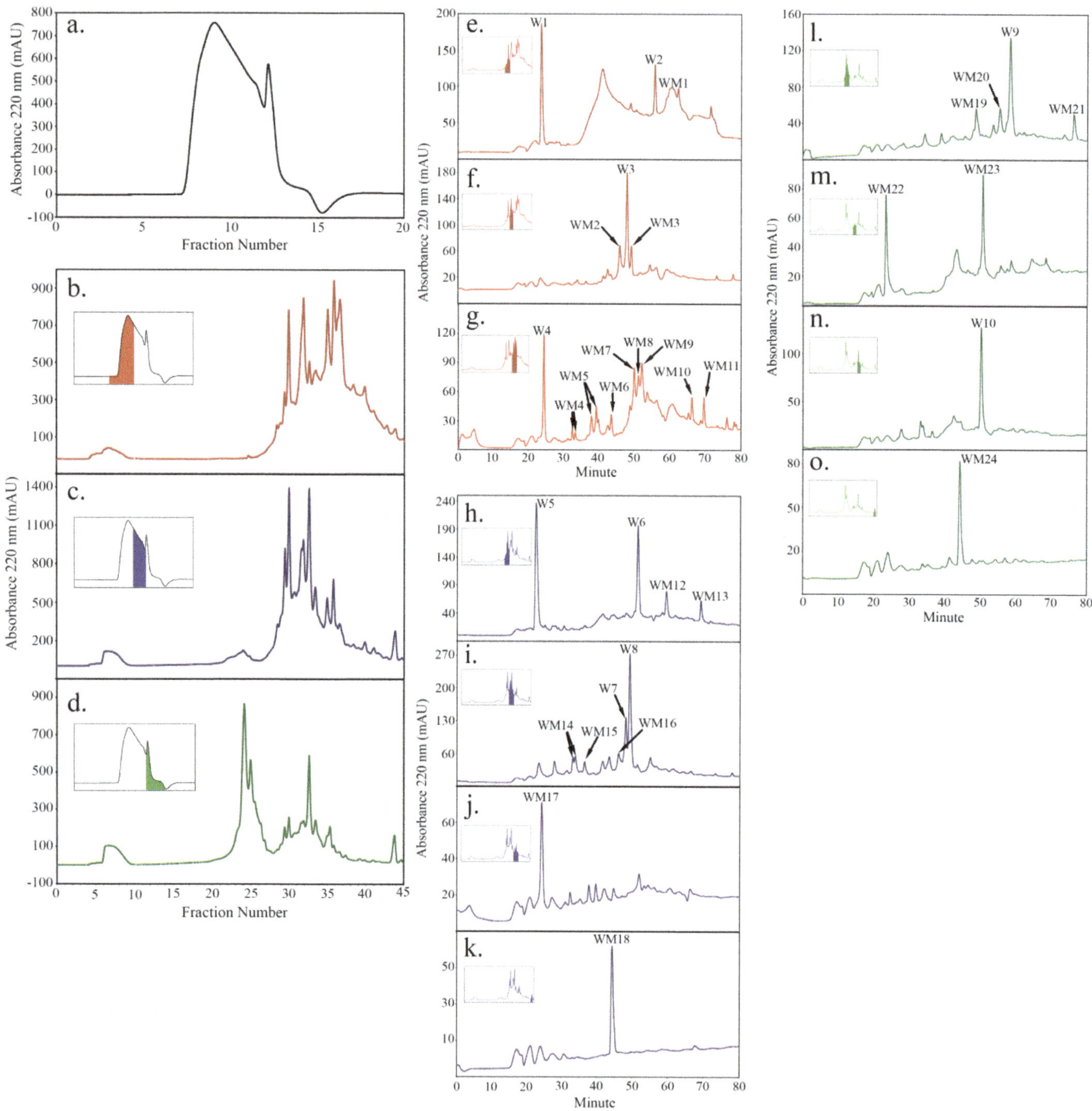

Figure 2. Peptides were isolated from trypsin digested WT HFI using a combination of size exclusion chromatography (SEC) (**a**), strong cation exchange chromatography (SCX) (**b–d**) and reverse phase chromatography (**e–o**), also summarized in Figure 1. Color-coded inserts in panels b to d identify the SEC fractions taken in turn for peptide mapping by SCX. Color-coded inserts in panels e through o identify the SCX fractions taken in turn for further peptide mapping by reverse phase chromatography. Major peptide peaks (A220 > 100) from WT HFI are labeled with "W" while minor peptides are labeled as "WM".

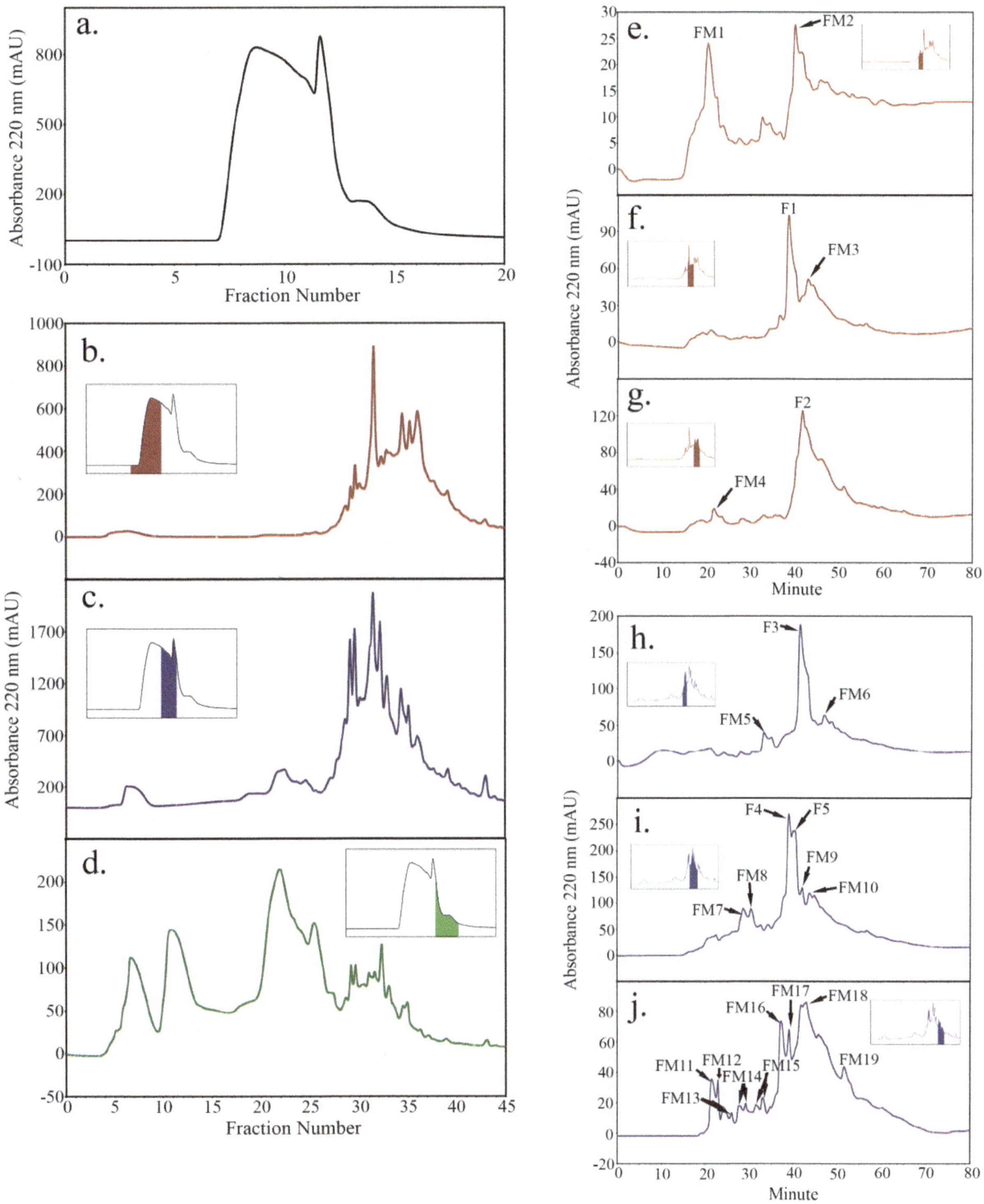

Figure 3. Peptides were isolated from trypsin digested ANP4 HFI using a combination of size exclusion chromatography (SEC) (**a**); strong cation exchange chromatography (SCX) (**b**–**d**) and reverse phase chromatography (**e**–**j**); also summarized in Figure 1. Color-coded inserts in panels (**b**) to (**d**) identify the SEC fractions taken in turn for further fractionation by SCX chromatography. Color-coded inserts in panels (**e**) to (**j**) identify the SCX fractions taken in turn for further peptide mapping by reverse phase chromatography. SCX fractions in (**d**) failed to yield any peptide peaks on further fractionation by reverse-phase HPLC. The major peptide peaks (A220 > 100) from the ANP4 HFI are labeled with "F" while minor peptides are labeled as "FM".

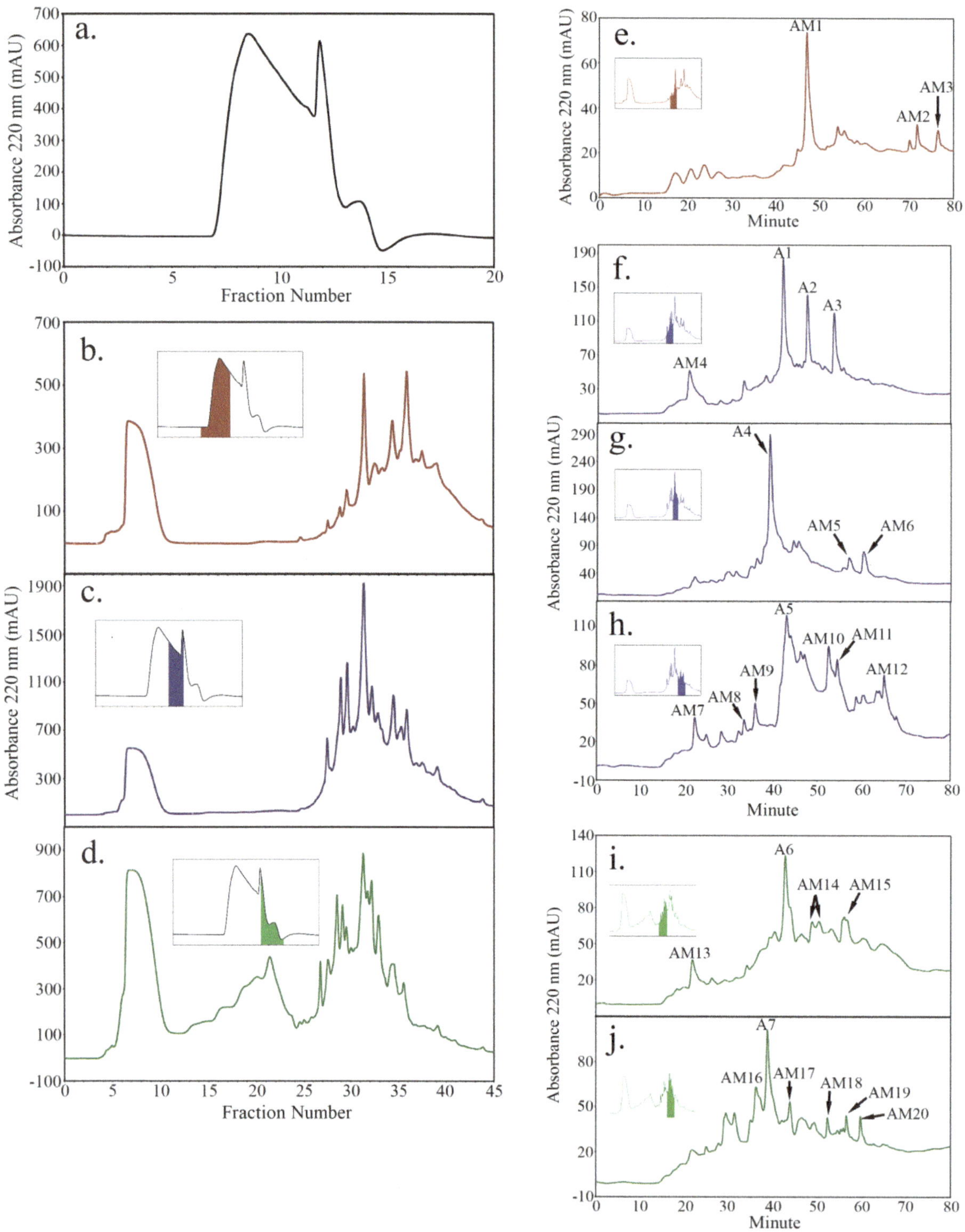

Figure 4. Peptides were isolated from trypsin digested ANP10 HFI using a combination of size exclusion chromatography (SEC) (**a**), strong cation exchange chromatography (SCX) (**b–d**) and reverse phase chromatography (**e–j**), also summarized in Figure 1. Color-coded inserts in panels b to d identify the SEC fractions taken in turn for further fractionation by SCX chromatography. Color-coded inserts in panels e to j identify the SCX fractions taken in turn for further peptide mapping by reverse phase chromatography. SCX fractions 35 to 37 in (**b**) and fractions 34 to 37 in (**d**) failed to yield any peptide peaks on further fractionation by reverse-phase HPLC. The major peptide peaks (A220 > 100) from the ANP10 HFI are labeled with "A" while minor peptides are labeled as "AM".

For ANP10 HFI, SEC fractions 7–10 (Figure 4b inset) did not yield any major peptide peaks but instead yielded three minor peptide peaks (AM1, AM2 and AM3, Figure 4e) after reverse-phase fractionation, despite the presence of several high UV-absorbing peaks evident during SCX fractionation (Figure 4b). SEC fractions 11–12 (Figure 4c inset) produced 5 major peptide peaks, A1 to A5 (Figure 4f,h) while SEC fractions 13–15 (Figure 4d inset) produced 2 major peptide peaks, A6 and A7 (Figure 4i,j) after reverse-phase fractionation. The Hyp content of the ANP10 major peptide peaks were low compared to those from WT and ANP4, with A1, A4, A5, and A6 containing ~17% (mole) Hyp (Table S1). A2 was Hyp-poor with 3% (mole) Hyp. No Hyp was detected in A3. The Hyp content of A7 was not determined due to lack of material.

The peptides released from the MT HFI were not fractionated as described above due to the low yields of MT walls. Instead, tryptic digests of the MT HFI were sequenced by LC-MS/MS without prior fractionation.

To sum up, we successfully isolated peptides from trypsin digested HFI of WT and ANP walls using a combination of chromatographic approaches and most of the major peptides were rich in Hyp, indicating the predominance of HRGPs in the root cell wall insoluble protein networks.

2.5. Peptide Sequencing and HRGP Identification

It has been suggested that the ANP phenotype may arise due to compensation by other HRGPs [27], which would likely result in changes in the protein composition of the HRGP network. Therefore, to compare the HRGP components of WT, ANP, and MT root walls, we performed LC-MS/MS peptide sequencing of peptides fractionated from WT HFI, ANP HFI, and MT HFI tryptic digests and identified HRGPs by their corresponding peptides. HRGPs that had two peptides passing the Scaffold 95% confidence filter or those having one peptide that passed the Scaffold 95% confidence filter but with a Mascot ion score larger than 40 were considered valid identifications.

Sequencing of fractionated peptides led to the identification of three EXTs (EXT1, EXT3, and EXT21), four AGPs (AGP30, AGP31, Fasciclin-like arabinogalactan proteins FLA7 and FLA15) and one PRP (PRP10) from WT HFI while a total of seven EXTs, including two classical EXTs (EXT1 and EXT17) and five chimeric EXTs (Leucine-rich Repeat/Extensin LRX1, LRX2, LRX3, LRX6 and HAE2), two AGPs (AGP31 and FLA16) and two PRPs (PRP4 and PRP10) were identified from ANP HFI (Table 2, see Figure S1 for full protein sequences of identified HRGPs).

Notably, uncertainties in the location of Hyp in some peptides were observed due to incomplete fragmentation of peptides. In those cases the total mass of a peptide suggested the existence of Pro hydroxylation on one or a few Pro residues, but the location of the Hyp residues can only be inferred from fragmented ion masses. For instance one AGP31 peptide AOVSOOAK(O/P)(O/P)VK(O/P)(O/P)VY(O/P)(O/P)TK had a mass of 2162.16, thus it contained 6 Hyp and 3 Pro residues. This peptide was identified in multiple locations on the reverse-phase HPLC fractionation peptide maps (Table 2) and showed various fragmentation patterns indicating variations in hydroxylation of the Pro residues, as reported earlier as well [29]. Similarly, this uncertainty of Hyp location was also observed in PRP4 peptides IEHPO(P/O)VOVYKOO(P/O)K and EVPO(P/O)VOVYK(P/O)(P/O)(P/O)K.

Table 2. HRGPs Identified in WT and ANP Cell Wall HFI.

Gene Accession #	Protein Identification (Plant Line of Origin)	Peptides Identified (Number of Occurrence in Protein Sequence; Peak of Peptide Identification) [1]	Peptide Mascot Ion Score [2]	Signal Peptide [3]	MM (kDa) [4]	pI [4]
EXTs						
AT1G76930	EXT1 (**WT, ANP4, ANP10**)	YYSOOOVYK (**3**; W2, W6, W8, W10, WM3, WM13; FM16; A2) HYSOOOVYK (**9**; WM6; A5), SOOOOVYK (**1**; FM8) SOOOOVK (**8**; W1, W2, W5, WM22; AM3)		1–24	32.9	10.1
AT1G21310	EXT3, RSH (**WT**)	HYSOOOVYHSOOOOK (**13**; WM6), HYTOOVK (**1**; WM4), SOOOOVK (**13**; W1, W2, W5, WM22)		1–28	49.2	10.3
AT3G54580	EXT17 (**ANP4**)	SOOOOYYSOSOK (**15**; FM8)	54.9	1–22	105.7	9.5
AT2G43150	EXT21[5] (**WT**)	SOOOOYYYHSOOOOVK (**6**; W8, W10) SOOOOVK (**10**; W1, W2, W5, WM22)	54.3	1–27	23.4	9.6
AT1G12040	LRX1 (**ANP4, ANP10**)	LQGPLPSSVGNMK (**1**; F3) LLYELDLSNNR (**1**; AM4)	42.3 44.5	1–26	81	8.2
AT1G62440	LRX2 (**ANP4, ANP10**)	SLEQLNVANNR (**1**; FM4, FM12, FM13) LSGPLPSSIGNMK (**1**; A2)	71.4 59.4		89.8	4.8
AT4G13340	LRX3 (**ANP10**)	FNEFEGTVPK (**1**; AM5, AM13, AM15, AM20)	65.9	1–20	82.2	7
AT3G22800	LRX6 (**ANP4**)	SLEQLNIAHNK (**1**; FM6)	55.7	1–28	52.3	6.2
AT3G50580	HAE2 (**ANP10**)	SOOOOTOK (**1**; AM3, AM12)	41.7	1–24	27.2	9.1
AGPs						
AT2G33790	AGP30 (**WT**)	LPAYPOAK (**1**; W2), TLVAVR (**1**; WM3), VSSLHDGGK (**1**; WM4)		1–25	25.8	10.6
AT1G28290	AGP31 (**WT, ANP4, ANP10**)	SLVAVR (**1**; W8, W10, WM3), SOVKPOVK (**1**; WM4, WM14), NITAETTTDK (**1**; FM8), AOVSOOAK(O/P)(O/P)VK(O/P)(O/P)VY(O/P)(O/P)TK[7] (**1**; W3, W4, W7, W8, W10, WM3, WM18, WM19, WM22; F1, F4, FM4, FM16; A5, AM1), LFGGDVGAELKPEK (**1**; FM7, FM8, FM18; AM10, AM11, AM18)		1–25	38.4	10.8
AT2G04780	FLA7 (**WT**)	FTDVSGTVR (**1**; W10, WM3)	61.4	1–22	26.8	6.5

Table 2. *Cont.*

Gene Accession #	Protein Identification (Plant Line of Origin)	Peptides Identified (Number of Occurrence in Protein Sequence; Peak of Peptide Identification) [1]	Peptide Mascot Ion Score [2]	Signal Peptide [3]	MM (kDa) [4]	pI [4]
AT3G52370	FLA15 (**WT**)	HHFNGEAQVK (**1**; WM24)	42.3	1–20	48.1	6.7
AT2G35860	FLA16 (**ANP4**)	EETOATEIKPAAOVVK (**1**; F1, FM4, FM13)	55.4	1–23	49.1	6.8
PRPs						
AT4G38770	PRP4 (**ANP4**)	EVPO(P/O)VOVYK(P/O)(P/O)(P/O)K[7] (**1**; F1), IEHPO(P/O)VOVYKOO(P/O)K[7] (**1**; F2, FM18)		1–29	49.1	10.3
AT5G09530	PRP10 (**WT, ANP4, ANP10**)	FPENSKPEVPK (**1**; WM8, WM9; F2, FM18; A5) VPEIPKPEETK (**1**; FM4, FM11, FM12) , MPEIQKPELPK (**1**; FM4, FM12, FM13), LPDIPK (**1**; FM16), LPEVPK (**1**; FM16), LPEFPKPELPK (**1**; AM4, AM18), MPEIPKPELPK (**1**; AM20)	64.8 (in WT)	1–35 [6]	41.6	5.9

[1] Only peptides that pass the 95% Scaffold confidence filter are shown; [2] The Mascot ion score are listed for those HRGPs that only have one peptide that pass the 95% Scaffold confidence filter but with score >40; [3] Signal peptides were predicted by SignalP (www.cbs.dtu.dk/services/SignalP); [4] Predicted protein molecular weight (MM) and isoelectric point (pI) values were obtained from TAIR (www.arabidopsis.org); [5] EXT21 shares one peptide (SOOOOVK) with EXT1 and EXT3, however the Mascot ion score of its unique peptide SOOOOYYYHSOOOOVK is larger than 40, thus validates the existence of EXT21; [6] For PRP10, SignalP prediction indicated the absence of a signal peptide while the cellular localization predicted by TargetP (http://www.cbs.dtu.dk/services/TargetP/) indicated a secretory pathway localization and thus the possession of a signal peptide. The signal peptide sequence and the putative cleavage site were determined according to the S-score (signal sequence score) and C-score (cleavage score) by SignalP; [7] The hydroxylation of some of the Pro residues in some AGP31 and PRP4 peptides could not be precisely determined, thus both possible sequences (either Pro-Hyp or Hyp-Pro) are indicated as P/O.

The peptides from the trypsin digested MT HFI were not fractionated by multiple chromatographic steps due to the low yield of the HFI material. Instead, the peptides from the trypsin digested MT HFI were directly sequenced by LC-MS/MS. This approach allowed the determination of protein relative abundance via the "total spectra count", with higher spectra count value corresponding to higher protein abundance (Table 3). This feature did not apply to the peptide mapping approach since every peptide peak was sequenced individually thus the relative abundance of proteins identified by LC-MS/MS referred to the value in that particular peak. A total of 11 HRGPs were identified in MT walls including 8 EXTs, 2 AGPs and 1 PRP (Table 3). LRX2, LRX3 and LRX4 were the most abundant HRGPs identified in MT with high total spectra counts while PAG2, PRP3 and EXT3 were minor components of the MT wall judging by their low total spectra counts (Table 3). Notably, only a trace of EXT3 was identified with a total spectra count of 1 by peptide HYSOOOVYHSOOOOK that occurs 13 times in the protein. The low peptide abundance of such a major repetitive unit in EXT3 was consistent with the result of Saha *et al.* [28] who found EXT3 expression in the mutant was 0.06 times of that in WT [28]. Thus our result confirmed the virtual absence of EXT3 in MT walls.

In conclusion, here we identified several HRGP members in the root cell walls of WT, ANP and MT and in the meantime confirmed the absence of EXT3 in ANP and MT.

2.6. Comparison of HRGP Networks

Finally, the HRGP members in the root cell wall of different Arabidopsis lines identified by this proteomics approach were compared. WT Arabidopsis and *ext3* knockout lines (ANP and MT) showed somewhat different HRGP networks but the differences did not allow a diagnosis for the ANP phenotype. For WT and ANP plants that display a normal phenotype, EXT1, AGP31 and PRP10 are in common while the two ANP lines have two chimeric LRX EXTs, LRX1 and LRX2, in common (Figure 5a). Notably, despite the absence of the LRX proteins in WT, ANP4 HRGP network also differed from that of WT with the presence of EXT17 and PRP4 (Figure 5a) that could potentially contribute to the wild-type phenotype in ANP4.

The MT and ANP plants both lack EXT3 and despite the profound phenotypic differences, they both have EXT1, LRX1, LRX2 and AGP31 (Figure 5b). Here again worth noticing is that the presence of EXT17, LRX6, PRP4, and PRP10 in ANP4 but not in MT. These HRGPs could potentially result in the rescue of MT to restore the wild-type phenotype (Figure 5b). WT and MT plants have the classical EXTs, *i.e.*, EXT1, EXT3 and EXT21, as well as AGP31 in common, even though EXT3 is a minor component in the MT walls (Figure 5c).

2.7. Potential Candidate HRGPs for the Arising of ANP

Previously Saha *et al.* [28] explored the arising of the ANP using microarray analysis and qRT-PCR. Several HRGPs were proposed to be responsible for the ANP phenotype based on their elevated gene expression levels (Table 6 in [28]). In the current study we aimed to obtain biochemical evidence to address the ANP phenotype issue and in the meantime, support the findings by Saha *et al.* [28].

Here we indeed identified several HRGPs in the ANP walls (Table 2). However, while the proteomics analysis confirmed their existence, this data alone is not sufficient to make an assessment on their involvement in rescuing of the *ext3* knockout mutant. To provide more evidence, we compared the expression of genes coding for those ANP HRGPs identified by proteomics with their expression in WT and MT roots (Table 4) to identify potential candidates for the ANP phenotype.

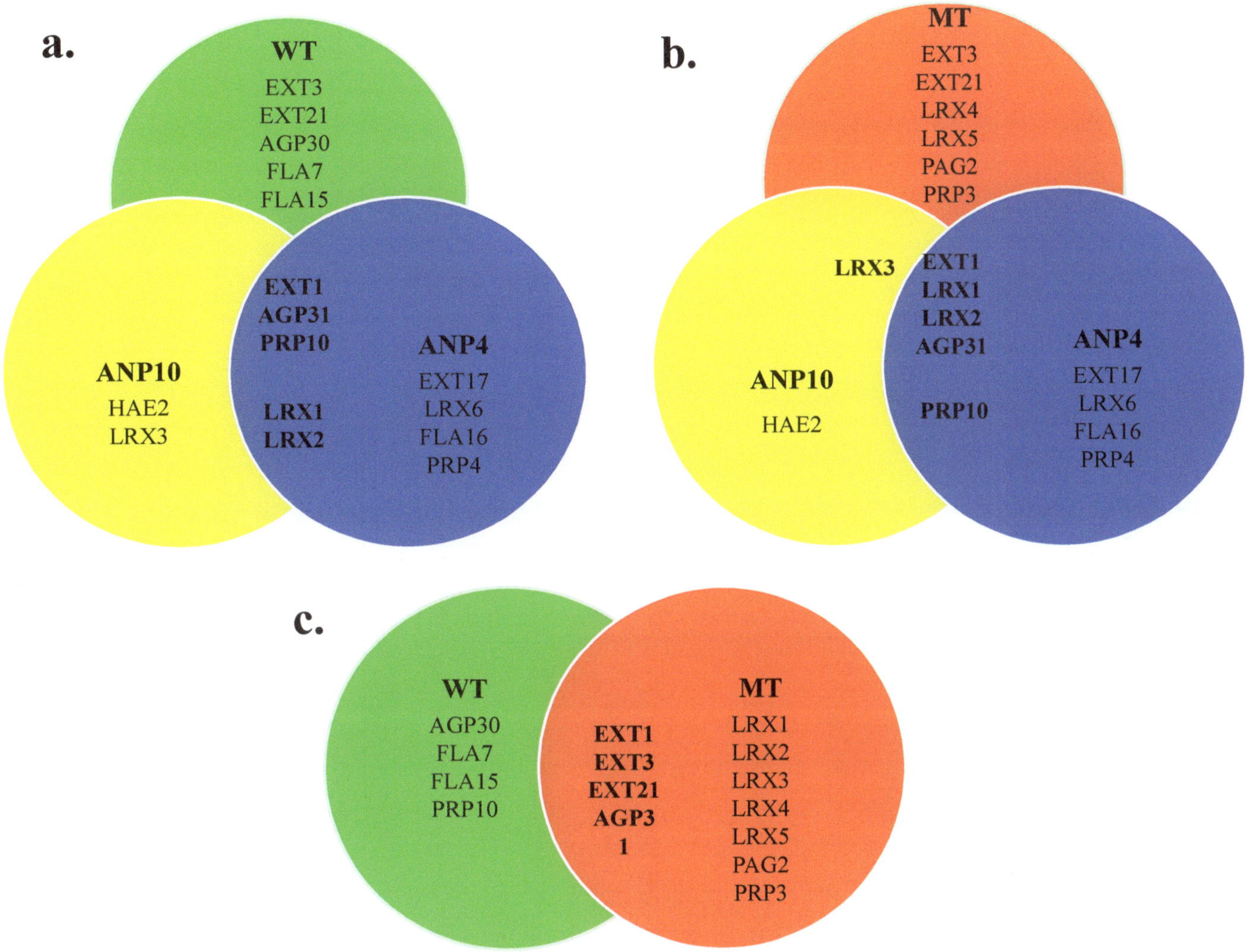

Figure 5. Comparison of members in HRGP networks of (**a**) WT and ANPs; (**b**) MT and ANPs and (**c**) WT and MT identified by proteomics approach.

In ANP4, the gene expression levels of most of the proteomics-identified HRPGs were slightly down regulated compared to WT, except for *PRP4*, which showed a more than 2-fold up-regulation. Notably, compared to MT, the expression level of three ANP4 HRGPs, *LRX1*, *PRP4* and *PRP10* showed more than 2-fold up regulation. Meanwhile, when compared to WT, the expression level of *LRX1*, *PRP4* and *PRP10* in MT showed a more than 2-fold down regulation [28]. This implies that ANP4 has increased expression of *LRX1*, *PRP4* and *PRP10* genes almost to the WT level. The peptide isolation reported here from the ANP4 wall confirmed that these proteins were indeed present and might possibly be applied by ANP4 to restore the normal phenotype. Similarly in the ANP10 line, LRX1, LRX2 and PRP10 were also candidates for the rescue (Table 4).

Table 3. HRGPs Identified in MT Cell Wall HFI.

Gene Accession #	Protein Identification	Peptides Identified [1] (Number of Occurrence in Protein Sequence)	Peptide Mascot Ion Score [2]	Total Spectra Count [3]	Signal Peptide [4]	MM (kDa) [5]	pI [5]
EXTs							
AT1G76930	EXT1	HYSOOOVYK (9), YYSOOOVYK (3)		6	1–24	32.9	10.1
AT1G21310	EXT3, RSH	HYSOOOVYHSOOOOK (13)	46.6	1	1–28	49.2	10.3
AT2G43150	EXT21	SOOOOYYYHSOOOOVK (6)	54.4	3	1–27	23.4	9.6
AT1G12040	LRX1	KVTVFDITSNR (1), VTVFDITSNR (1), VVLSLPSLK (1)		5	1–26	81	8.2
AT1G62440	LRX2	ELGLLTDLALFHLNSNR (1), FPNVVLSLPSLK (1), SLEQLNVANNR (1)		10		89.8	4.8
AT4G13340	LRX3	DLDAIFINHNR (1), FNEFEGTVPK (1), FPTVVLQLPSLK (1) FRFELPENFGDSPVSVIVLANNR (1)		11	1–20	82.2	7
AT3G24480	LRX4	FPTVVLHLPSLK (1), SAYIALQAWK (1)		11	1–25	54.7	6.9
AT4G18670	LRX5	FAGIFPTVVLQLPSLK (1), FNEFEGPVPR (1)		9	1–31 [6]	90.8	6.7
AGPs							
AT1G28290	AGP31	LFGGDVGAELKPEK (1), NGYFLLLAPK (1), TVTNFGFR (1)		5	1–25	38.4	10.8
AT2G25060	PAG2	LSLVVISPR (1)	44.7	1	1–28	19.5	7.4
PRPs							
AT3G62680	PRP3	GLTGVPLALYGYR (1), SNTEVVIYSNPTDSK (1)		2	1–22	34.4	9.7

[1] Only peptides that pass the 95% Scaffold confidence filter are shown; [2] The Mascot ion score are listed for those HRGPs that only have one peptide that pass the 95% Scaffold confidence filter but with score >40.; [3] Total spectra count indicates the relative abundance of protein in the sample; [4] Signal peptides were predicted by SignalP; [5] Predicted protein molecular weight (MM) and isoelectric point (pI) values were obtained from TAIR; [6] For LRX5, SignalP prediction indicated the absence of a signal peptide while the cellular localization predicted by TargetP indicated a secretory pathway localization and thus the possession of a signal peptide. The signal peptide sequence and the putative cleavage site were determined according to the S-score and C-score by SignalP.

Table 4. Gene Expression Patterns of Proteomics Identified ANP HRGPs using qRT-PCR: A Comparing of WT and *rsh* Mutant and ANP Revertants.

	Protein (Accession #)	Fold Change (Log_2) ANP *vs.* WT [1]	Fold Change (Log_2) ANP *vs.* MT[1]
ANP4	EXTs:		
	AtEXT1 (AT1G76930)	0.22; (0.72) [2]	−2.50; (0.00)
	AtEXT17 (AT3G54580)	−0.94; (0.13)	1.03; (0.20)
	LRX1 (AT1G12040)	−0.45; (0.29)	5.35; (0.02)
	LRX2 (AT1G62440)	−0.85; (0.05)	−0.66; (0.05)
	LRX6 (AT3G22800)	−0.50; (0.30)	−0.95; (0.05)
	AGPs:		
	AGP31 (AT1G28290)	n/d [3]	n/d
	FLA16 (AT2G35860)	n/d	n/d
	PRPs:		
	PRP4 (AT4G38770)	1.20; (0.00)	2.25; (0.00)
	PRP10 (AT5G09530)	−0.55; (0.05)	3.30; (0.01)
ANP10	EXTs:		
	AtEXT1 (AT1G76930)	0.40; (0.04)	−1.69; (0.001)
	LRX1 (AT1G12040)	−0.07; (0.86)	6.90; (0.0001)
	LRX2 (AT1G62440)	−0.07; (0.79)	0.89; (0.01)
	LRX3 (AT4G13340)	0.64; (0.02)	−2.27; (0.0003)
	HAE2 (AT3G50580)	3.33; (0.02)	0.14; (0.77)
	AGPs:		
	AGP31 (AT1G28290)	n/d	n/d
	PRPs:		
	PRP10 (AT5G09530)	−1.33; (0.00)	3.7; (7.9×10^{-6})

[1] The culture generation compared: ANP4/10-F3 *vs.* WT-F2, ANP4-F2 *vs.* MT-F1 and ANP10-F3 *vs.* MT-F1;
[2] Numbers in parenthesis: significance value, $p \leq 0.05$ are considered statistically significant; [3] n/d: not determined.

3. Discussion

3.1. Structural Protein Network in Arabidopsis Root Cell Wall

HRGPs are major protein components in the dicot plant and cell culture walls as suggested by their high Hyp contents (>10 mole% of total amino acids) [30–33]. As a model dicot, Arabidopsis is of no exception as the amino acid composition of WT Arabidopsis root cell walls showed that it was indeed a Hyp-rich matrix (Table 1).

Cell wall structural proteins have long been proposed to form a covalently linked network in the wall [34]. This network appears to be independent from the other cell wall polysaccharide networks since it is preserved after HF treatment, which cleaves all glycosidic linkages in tomato suspension culture cell walls [2]. A previous study of the legumes root cell wall EXT components further confirmed this notion where HF deglycosylation of the cell wall removed two-thirds of the wall dry weight mass, about 29% (w/w) of the wall protein, but less than 3% (w/w) of the wall-bound Hyp [35]. Here we report a similar observation with Arabidopsis root cell walls that more than 90% (w/w) of the Hyp remained after HF deglycosylation whereas more than 60% of the cell wall dry weight and up to 40% (w/w) of the wall protein was removed (Table 1), indicating the insolubilized Hyp-rich protein network was firmly held in Arabidopsis root cell wall, most likely through covalent interactions.

3.2. Protease Degradation of HFI

The wall pellets after HF deglycosylation contained the remaining wall protein components that accounted for up to 25% of the WT wall dry weight while the weight percentage of protein in ANP and MT HF insoluble pellets were close to or less than 20% (Table 1). A large portion of the pellet weight was comprised of "unknown substances" that were neither protein nor sugar considering the high efficiency of HF deglycosylation [2]. The presence of such "unknown substances" in the HF insoluble wall was reported earlier for tomato suspension culture cells [2], legume root hairs [36], legume root and root nodules [34]. The nature of these substances has yet to be determined, but secondary cell wall substances such as lignin from root xylem vessels [37] and suberin from cuticles [38,39] are possibilities. Under the conditions of HF treatment used here the phenolic linkages of lignin would not have been cleaved [40] nor would the ester linkages between the fatty acid components of cutin, which are identical to those of suberin [41,42].

Hyp remained the most abundant amino acid in the protease insoluble residues and indeed protease digestion left the WT and MT pellets enriched in Hyp judging by the increase in the Hyp/protein ratio (Table 1). This observation suggested that the Hyp containing portion of the protein network is relatively more resilient to proteases. The associations between wall proteins and the "unknown substances" in the HF insoluble wall residues could account for the incomplete removal of Hyp-containing peptides by proteases. A significant portion of the wall protein is likely embedded or crosslinked to the network of the "unknown substances" [43–45] and therefore inaccessible to the proteases.

3.3. EXT3 in Root Cell Wall

Apart from EXT3, ANP and MT cell walls showed similar protein and Hyp contents to WT (Table 1). Therefore EXT3 is either a minor but crucial component of the wall as the knock-out of *ext3* does not affect the total protein content or in the absence of EXT3 other Hyp containing proteins are present. Although the absolute amount of EXT3 in these different Arabidopsis lines has not been determined, Saha *et al*. [28] showed in previous work that many HRGP genes have altered expression (some up and others down) in the ANP and MT plants compared to WT, while there was no significant difference in the total HRGP levels between WT, ANP and MT plants, supporting the latter notion.

Also surprising, is the fact that the abundant (13 repeats in EXT3) EXT3 peptide HYSOOOVYHSOOOOK was only found in a minor peptide fraction of WT; this may be because EXT3 is a minor component of the HFI or because the peptide remained with the trypsin insoluble wall pellet due to extensive intermolecular cross-lining. The latter possibility seems likely in light of the abundance of cross-linking sites on EXT3 molecule (see below) and as shown in earlier work with Arabidopsis cell suspension cultures that identified EXT3 as the major salt-elutable extensin present in growing cells [46].

3.4. Cross-linking of HRGPs into the Cell Wall

The mechanism for the formation of the covalently cross-linked wall HRGPs network is largely unknown. Tyr residues, especially those involved in the YXY motif, have been shown to form intramolecular isodityrosine (Idt) cross-links [47,48] that lead to insolubilization by intermolecular cross-linking to produce di-isodityrosine (Di-Idt) [21,22] and pulcherosine [23]. Another putative cross-linking motif is the VYK motif in HRGPs, as the monomeric P1 type tomato extensin (contains majorly VYK motif with only small amount of Idt) [49,50] cross-linked efficiently by a wall bound peroxidase [24]. Classical EXTs usually contain multiple Idts and/or VYK motifs that are candidates for pulcherosine formation.

For the classical EXTs identified in HF insoluble Arabidopsis root cell walls, EXT1 has 1 Idt, 8 VYK motifs and 14 lone Tyr residues (Figure S1 for protein sequences). Here notably, multiple peptides containing VYK motifs were released by trypsin from WT and ANP10 HFI, in the major peptide fractions (Tables 3 and 4). This argues against the role of VYK motifs in the formation of covalent cross-links. Indeed, an EXT analog containing six SOOOOTOVYK repeats with no YXY motif was cross-linked very slowly by tomato extensin peroxidase [51] suggesting that VYK might at best be a minor participant in HRGP cross-linking. These data support the argument that EXT1 is held covalently in the protein network mainly through pulcherosine formed between EXT1 and other HRGPs.

EXT3, EXT17 and EXT21 are classical EXTs rich in Idt motifs and lone Tyr residues. EXT3 contains 16 Idt motifs and 15 lone Tyr residues that could be involved in insolubilizing EXT3 by forming Di-Idt and pulcherosine. Pulcherosine may be preferred as suggested by Cannon *et al.* [27] based on the result that it was the major *in vitro* cross-linking product of EXT3 by peroxidase. Much like EXT3, EXT17 contains 24 Idts and 126 lone Tyr residues and EXT21 has 12 Idts and 8 lone Tyr residues, all of which could contribute to their cross-linking in the wall. The high potential of extensive cross-linking of these EXTs might contribute to the observations that their corresponding peptides were only recovered as minor fractions on the peptide maps and the Hyp-rich substances remained in the pellets after protease digestion.

LRX is a family of chimeric EXTs that possess N-terminal non-HRGP Leucine-rich domains and C-terminal EXT domains. They are crucial components of the cross-linked root hair cell wall protein network and contribute to the root hair morphology [52–55]. LRX proteins were identified previously in the loosely bound cell wall proteome of Arabidopsis cell suspension cultures (LRX2, LRX3 and LRX4) [56] and etiolated hypocotyl walls (LRX3, LRX4 and LRX5) [57]. Here we identified several members of the LRX family proteins in the HFI from ANP and MT walls, confirming that LRXs are firmly bound to cell wall [55]. One common feature of the LRXs identified is that all the peptides isolated were found in the N-terminal non-HRGP Leucine-rich domain and no peptides from the C-terminal EXT domains. This is important since it could be due to the EXT domains of these LRXs being cross-linked in the protein networks. For the extensin domains of the LRX proteins identified, there are 9 Idt motifs and 30 lone Tyr residues in LRX1, 3 Idt motifs and 34 lone Tyr residues in LRX2, 2 Idt motifs and 22 lone Tyr residues in LRX3, and 3 Idt motifs and 18 lone Tyr in LRX5. LRX6 has a relatively small EXT domain, having 5 Idt motifs and 1 lone Tyr residue. One exception is LRX4, which has a very short EXT domain with 5 Tyr residues but no Idt motif. Nevertheless, LRX4 has the potential to form Tyr-related cross-links with other wall HRGPs or through other type of interactions as documented earlier [55]. Similar to LRX4 is HAE2, a short chimeric EXT with 6 lone Tyr residues but no Idt motifs, found in ANP10 HFI. Strikingly, no LRX proteins were found in WT HFI. This could indicate that (1) the LRX are minor components of WT walls, (2) they were extensively cross-linked in the wall and could not be released by trypsin, or (3) the LRX peptides were somehow lost during the peptide collecting process.

Similar to EXTs, PRP3 has 16 VYX (9X = K and 7X = T) motifs and 19 lone Tyr residues while PRP4 has 5 VYK motifs and 8 lone Tyr residues that could potentially insolubilize them in the wall protein network. The mechanism of PRP10 cross-linking in the walls remains unknown since it has no Tyr residues, thus the traditional cross-linking via Tyr does not apply to PRP10.

AGPs have been reported to interact with wall polysaccharides, especially pectin via non-covalent [58] or covalent linkages [8,59]. AGPs also self-associate with each other. The PAC domain of AGP31 was recently reported to interact non-covalently with the polygalactan chains on its Pro-rich domain [60]. Previously, AGP31 was found in etiolated Arabidopsis hypocotyl cell walls [57] and it was also one of the two abundant salt-elutable HRGPs present in Arabidopsis cell suspension cultures [46]. Here we report for the first time AGP peptides isolated from HF deglycosylated walls. For AGP31, due to the removal of sugars on the glycan chain of the Pro-rich domain by HF, the interaction between the PRP-AGP containing Cys (PAC) domain and the Pro-rich domain was disrupted thus a self-association mechanism for the presence of AGP31 in the HFI is unlikely. One possible mechanism for the occurrence of AGP31 could be due to hydrophobic interactions with other wall protein components during the removal of HF. AGP30 has a structure similar to AGP31 with a PAC domain that shares 61% sequence identity with that of AGP31, thus it is likely that these two AGPs were associated with the walls via identical mechanisms.

The FLA proteins are characterized by their fasciclin-like domains that are involved in protein-protein recognition via non-covalent forces, such as hydrophobic interactions [61,62], an interaction that tended to be dissociated by HF treatment [63]. However, upon the removal of HF, these interactions might direct the re-association of FLAs and hold them in the HFI. Other interactions between AGPs or AGP and wall proteins also exist. One example of AGPs covalent cross-linking was reported in sugar beet leaves and this covalent coupling of two AGP molecules was H_2O_2 mediated and potentially follows a mechanism similar to the Tyr-involved cross-linking of extensins [64]. This could be one potential mechanism for the cross-linking of the AGPs identified in the wall HFI—through the formation of Tyr derivatives judging by their high Tyr content (Figure S1).

3.5. Potential Candidate HRGPs for the WT Phenotype Reversion

EXT3 has unique structural features and functions in Arabidopsis root cell wall by self-assembling into a scaffold that is further rigidified by peroxidase cross-linking [27]. This scaffold formed by EXT3 was proposed [26,27] to serve as a template for the correct deposition of wall polysaccharides that contribute to the cell wall integrity. Structural features of EXT3 include (1) an extended poly-proline II confirmation that ensures maximum contact area between EXT3 molecules during self-assembly and it is contributed by the Ser-$(Hyp)_4$ motif along with the arabinosylation on the Hyp residues [65]; (2) the Idt motifs and free Tyr residues ensures the correct alignment of EXT3 molecules in self-assembly and later cross-linking process [27]; (3) a rich content of His and Lys residues enables the ionic interaction between EXT3 and cell wall pectic polysaccharides under physiological conditions [66,67]. Thus to compensate the loss of EXT3 and revise the *EXT3* knockout to wild-type phenotype, the ANPs require an HRGP or a group of HRGPs that possess structural characteristics resemble that of EXT3.

Among the candidate ANP HRGPs for the auto-rescuing of *ext3* knockout phenotype (see Section 2.7), the two LRX proteins (LRX1 and LRX2) both have a C-terminal extensin domain that is similar to EXT3 in protein size and in content of the Ser-$(Hyp)_4$ glycomotifs (Figure S1). Both extensin domains also have abundant Tyr-based cross-linking components (elaborated in the previous section). Thus LRX1 and LRX2 could potentially serve as EXT3 substitutes to form a well cross-linked network despite the relative low Lys and His contents in their extensin domains.

PRP4, on the other hand, has a protein size similar to that of EXT3 and its high Lys and His content will enable the ionic interaction with wall polysaccharides (Figure S1). PRP4 also has abundant cross-linking sites to form a covalent protein network (Figure S1). However, PRP4 lacks the Ser-$(Hyp)_4$ motifs that reinforce the protein shape (Figure S1). Instead, it contains many repeats of XPPP motif (X = K, H, V, I, Figure S1). The XPPP motifs are sufficient to form a poly-Proine II confirmation with contiguous *trans* polyproline [68] and thus stabilize the protein structure of PRP4. Combining all the structural features, PRP4 also has the potential to replace EXT3 in the cell wall to reverse the WT phenotype.

3.6. Gene Expression vs. Protein Identification by Proteomics

Our results serve as a general survey of the HRGPs from WT and *ext3* knockout lines. However, our findings of the wall HRGP members in ANP and MT walls via proteomics approach showed poor correspondence with the members proposed in the self-rescuing model of ANPs by microarray and qRT-PCR [28], with only one HRGP, PRP4, common to both studies. Both approaches should be considered in determining the molecular composition of the cell walls as each has its limitations: HRGPs are structural proteins that stably remain bound in the network after their production, however, not all of them are equally accessible during protein extraction procedures, and the most abundant of them are likely to be the best represented, while the less abundant may be providing a vital wall function, but not detectable. In the case of taking gene expression as an indication of HRGP presence in the wall, it must be considered that the technology is measuring a snap shot of the messages at a particular time, and not necessarily all of that message will be translated or modified to become stable networked HRGPs. Meanwhile, the uncoupling of gene expression and RNA translation, for instance regulated by endogenous double-stranded RNA interference [69] or microRNA [70], could also be responsible for the difference in gene expression and protein abundance.

4. Experimental Section

4.1. Arabidopsis Seedling Culture

Seeds of wild-type *Arabidopsis* Landsberg *erecta* (WT, purchased from Lehle Seeds, Brushy Creek, TX, USA), ANP lines #4 and #10 (ANP4 and ANP10), and heterozygote *RSH*/*rsh* were surface sterilized before being planted on solid ½ MS medium plates (pH 5.8) containing 8 g/L agar. Seeds of WT and ANP lines were planted along a line 3 cm from the top of the plates and the plates were positioned vertically and cultured for a total of 14 days. *RSH*/rsh heterozygote seeds were scattered onto plates and placed horizontally for 21 days. To synchronize seed germination, the plated seeds were first incubated at 4 °C for two days before maintained at room temperature. All seeded plates were kept under constant fluorescent light.

4.2. Root Cell Wall Isolation

Root cell walls were prepared using a protocol by York *et al.* [71]. The roots were homogenized by grinding in liquid N_2 to a fine powder followed by sonication in 35 mL 0.5 M K-P (K_2HPO_4: KH_2PO_4 = 2:1

molar ratio) buffer. Cell breakage was evaluated by optical microscopy. The wall fragments were pelleted by centrifugation at 11,000 *g* for 15 min, washed three times with 45 mL 0.5 M K-P buffer followed by four washes with 0.1 M K-P buffer and four washes with cold water to remove cellular contents. The wall pellets were then extracted four times with an extraction solution (chloroform: methanol = 1:1, v/v) to remove membrane components and collected by centrifugation at 20,000 *g* for 15 min after each extraction. The wall pellets were washed with 95% ethanol and water to remove extraction mixture and then lyophilized.

4.3. Wall Deglycosylation with HF

Cell wall from WT, ANP, and MT were deglycosylated with anhydrous hydrogen fluoride (HF) using methods described previously [2]. The HF insoluble pellets (HFI) were dialyzed against distilled water to remove excessive HF then collected by centrifugation. The HFIs were re-suspended in water and washed four times before lyophilization.

4.4. Trypsin Digestion of HFI

WT and ANP HFI were resuspended in 500 μL of water and incubated at 110 °C for 10 min. The suspensions were cooled and the following reagents were added: 1280 μL water, 100 μLof 1 M NH_4HCO_3 (final [NH_4HCO_3] = 0.05 M, pH 8.0), 100 μL of 1 mg/mL trypsin (pellet: trypsin = 100:1, w/w) and 20 μL of 1 M $CaCl_2$ (final [Ca^{2+}] = 10 mM) to make a 2 mL reaction mixture with a HFI concentration at 5 mg/mL. The reactions were incubated at room temperature with constant stirring (200 rpm) for 24 h and another 100 μL of trypsin solution was added for another 24-h incubation. The trypsin insoluble HFI (TI HFI) was collected by centrifugation, washed four times with water and then lyophilized. The supernatant from each digestion and the water washes were pooled, lyophilized and used for peptide isolation.

MT HFI was re-suspended in 100 μL water with 2 μL Tris(2-carboxyethyl) phosphine hydrochloride and incubated at 95 °C for 10 min. The suspension was cooled and pre-treated with 2 μL 0.75 M Iodoacetamide in dark for 1 h and 2 μL 0.25 M DL-Dithiothreitol for 10 min. After pretreatment, the solution was mixed with 47 μL H_2O, 20 μL 0.5M NH_4HCO_3, 20 μL acetonitrile and 7 μL of 1 mg/mL trypsin (Trypsin Gold, Promega) to make a total volume of 200 μL with a HFI concentration at 5 mg/mL and kept at 37 °C with shaking (200 rpm) for 24 h. Seven microliters of trypsin was added after the first 24-h incubation period for another 24-h incubation. The TI HFI was washed and collected as mentioned earlier and the supernatant was lyophilized and sent for proteomic analysis.

4.5. Pronase Degradation

TI HFI from WT and ANP were digested with pronase using the trypsin digestions protocol described above. After digestion, the pronase insoluble HFI (PI HFI) were collected by centrifugation, washed four times with water and lyophilized.

4.6. Tryptic Peptide Isolation

Typtic peptides were isolated from HFI following the workflow shown in Figure 1. Lyophilized supernatants from trypsin digested WT and ANP HFI were dissolved at 2 mg/mL in 50 mM formic acid and fractionated by a PolyHYDROXYETHYL A column in size exclusion (SEC) mode. The column was eluted isocratically at a flow rate of 0.2 mL/min with the eluate monitored at 220 nm. One-minute fractions were collected and combined into three groups: (1) fractions 7 to 10; (2) fractions 11 and 12; and (3) fractions 13 to 15. Fraction groups were lyophilized, dissolved in loading buffer (10 mM KH_2PO_4 with 20% acetonitrile, pH 2.8), and further fractionated on a strong cation exchange (SCX) PolySULFOETHYL A column. The column was eluted at 0.2 mL/min using a 60 min gradient of 0 to 100% wash buffer (10 mM KH_2PO_4 with 0.8 M KCl, pH 2.8) followed by 15 min elution with 100% wash buffer. The eluate was monitored at 220 nm and two-minute fractions were collected.

Fractions from SCX were combined in groups (Figure 1), lyophilized, and dissolved in Buffer A (0.1% TFA, v/v) for further fractionation on a reverse-phase PRP-1 column using a 100 min gradient of 0 to 50% Buffer B (0.1% TFA, 80% acetonitrile, v/v) at a flow rate of 0.5 mL/min for WT and ANP10 peptides and a 100-min gradient of 0 to 70% Buffer B for ANP4 peptides. The eluate absorbance was monitored at 220 nm and the eluted peaks were collected and lyophilized before sequencing by MS/MS.

4.7. Peptide Sequencing by Mass Spectrometry

Peptide sequencing of isolated peptide peaks (WT and ANP) and the unfractionated MT tryptic digest were performed at the Proteomic Facility at Michigan State University (E. Lansing, MI, USA). Detailed protocols were listed under supplementary material.

5. Conclusions

Here we reveal unique HRGP networks in wild-type Arabidopsis and *ext3* knockout lines using peptide isolation and sequencing approaches. Our methodology released wall HRGP components for identification. EXT3 is a major HRGP component in the WT cell wall and our approach did not find EXT3 in any of the ANP walls, confirming that the auto phenotypic auto-revertant ANPs can assemble functional walls by an alternative mechanism to that involving EXT3. Furthermore we found indications of a difference between the walls of WT and ANPs, thereby providing an opportunity to compare the molecular structure and related functional mechanisms of two different cell wall types of apparently similar function.

Acknowledgments

The authors would like to thank Joellen Kemper and Nick Cummings for their assistance in plant tissue culture. This work was supported by the National Science Foundation (Grant # IOS955569 to Marcia J. Kieliszewski and Grant # IOS0955805 to Maura C. Cannon) and was conducted in a facility constructed with support from Research Facilities Improvement Program Grant Number C06 RR-014575-01 from the National Center for Research Resources, National Institutes of Health.

Author Contributions

Yuning Chen and Marcia J. Kieliszewski designed the research. Yuning Chen, Dening Ye, Alexandra N. Frye and Andrew J. Mort carried out the experiments. Tui Ray, Prasenjit Saha and Maura C. Cannon prepared and provided the ANP and MT seeds. Yuning Chen, Michael A. Held,

Maura C. Cannon and Marcia J. Kieliszewski analyzed the data. Yuning Chen, Michael A. Held, Maura C. Cannon and Marcia J. Kieliszewski wrote the paper. All authors read and approved the final manuscript.

Conflicts of Interest

The authors declare no conflict of interest.

References

1. Cassab, G.I.; Varner, J.E. Cell wall proteins. *Ann. Rev. Plant. Physiol. Plant. Mol. Biol.* **1998**, *39*, 321–353.
2. Mort, A.J.; Lamport, D.T.A. Anhydrous hydrogen fluoride deglycosylates glycoproteins. *Anal. Biochem.* **1977**, *82*, 289–309.
3. Komalavilas, P.; Zhu, J.K.; Nothnagel, E.A. Arabinogalactan-proteins from the suspension culture medium and plasma membrane of rose cell. *J. Biol. Chem.* **1991**, *268*, 15956–15965.
4. Showalter, A.M.; Bell, J.N.; Cramer, C.L.; Bailey, J.A.; Varner, J.E.; Lamb, C.J. Accumulation of Hydroxyproline-rich glycoprotein mRNAs in response to fungal elicitor and infection. *Proc. Natl. Acad. Sci. USA* **1985**, *82*, 6551–6555.
5. Bradley, D.J.; Kjellbom, P.; Lamb, C.J. Elicitor- and wound-induced oxidative cross-linking of a proline-rich plant cell wall protein: A novel, rapid defense response. *Cell* **1992**, *70*, 21–30.
6. Clarke, A.; Gleeson, P.; Harrison, S.; Knox, R.B. Pollen-stigma interactions: identification and characterization of surface components with recognition potential (cell surface receptors/plant glycoproteins/adhesion/concanavalin A/tridacnin). *Proc. Natl. Acad. Sci. USA* **1979**, *76*, 3358–3362.
7. Levitin, B.; Richter, D.; Markovich, I.; Zik, M. Arabinogalactan proteins 6 and 11 are required for stamen and pollen function in Arabidopsis. *Plant J.* **2008**, *56*, 351–363.
8. Tan, L.; Eberhard, S.; Pattathil, S.; Warder, C.; Glushka, J.; Yuan, C.; Hao, Z.; Zhu, X.; Avci, U.; Miller, J.S.; *et al.* An Arabidopsis cell wall proteoglycan consists of pectin and Arabinoxylan covalently linked to an Arabinogalactan protein. *Plant Cell* **2013**, *25*, 270–287.
9. Lamport, D.T.A.; Varnai, P. Periplasmic Arabinogalactan glycoproteins act as a calcium capacitor that regulates plant growth and development. *New Phytol.* **2013**, *197*, 58–64.
10. Lamport, D.T.A.; Varnai, P.; Seal, C.E. Back to the future with the AGP-Ca^{2+} flux capacitor. *Ann. Bot.* **2014**, doi:10.1093/aob/mcu161.
11. Shpak, E.; Leykam, J.F.; Kieliszewsk, M.J. Synthetic genes for glycoprotein design and the elucidation of hydroxyproline-O-glycosylation codes. *Proc. Natl. Acad. Sci. USA* **1999**, *96*, 14736–14741.
12. Du, H.; Simpson, R.J.; Moritz, R.L.; Clarke, A.E.; Bacic, A. Isolation of the protein backbone of an Arabinogalactan-protein from the styles of *Nicotiana alata* and characterization of a corresponding cDNA. *Plant Cell* **1994**, *6*, 1643–1653.
13. Youl, J.J.; Bacic, A.; Oxley, D. Arabinogalactan-proteins from *Nicotiana alata* and *Pyrus communis* contain glycosylphosphatidylinositol membrane anchors. *Proc. Natl. Acad. Sci. USA* **1998**, *95*, 7921–7926.

14. Oxley, D.; Bacic, A. Structure of the glycosylphosphatidylinositol anchor of an arabinogalactan protein from *Pyrus communis* suspension-cultured cells. *Proc. Natl. Acad. Sci. USA* **1999**, *96*, 14246–14251.
15. Schultz, C.J.; Johnson, K.L.; Currie, G.; Bacic, A. The classical arabinogalactan protein gene family of arabidopsis. *Plant Cell* **2000**, *12*, 1751–1767.
16. Fowler, T.J.; Bernhardt, C.; Tierney, M.L. Characterization and expression of four proline-rich cell wall protein genes in arabidopsis encoding two distinct subsets of multiple domain proteins. *Plant Physiol.* **1999**, *121*, 1081–1091.
17. Wyatt, R.E.; Nagao, R.; Key, J.L. Patterns of soybean proline-rich protein gene expression. *Plant Cell* **1992**, *4*, 99–110.
18. Lamport, D.T.A.; Kieliszewski, M.J.; Chen, Y.; Cannon, M.C. Role of the extensin superfamily in primary cell wall architecture. *Plant Physiol.* **2011**, *156*, 11–19.
19. Sadava, D.; Chrispeels, M.J. Hydroxyproline-rich cell wall protein (extensin): Role in the cessation of elongation in excised pea epicotyls. *Dev. Biol.* **1973**, *30*, 49–55.
20. Lamport, D.T.A.; Katona, L.; Roerig, S. Galactosylserine in extensin. *Biochem. J.* **1973**, *133*, 125–131.
21. Brady, J.D.; Sadler, I.H.; Fry, S.C. Di-isodityrosine, a novel tetrameric derivative of tyrosine in plant cell wall proteins: A new potential cross-link. *Biochem. J.* **1996**, *315*, 323–327.
22. Held, M.A.; Tan, L.; Kamyab, A.; Hare, M.; Shpak, E.; Kieliszewski, M.J. Di-isodityrosine is the intermolecular cross-link of Isodityrosine-rich extensin analogs cross-linked *in vitro*. *J. Biol. Chem.* **2004**, *279*, 55474–55482.
23. Brady, J.D.; Sadler, I.H.; Fry, S.C. Pulcherosine, an oxidatively coupled trimer of tyrosine in plant cell wall: Its role in cross-link formation. *Phytochemistry* **1998**, *47*, 349–353.
24. Schnabelrauch, L.S.; Kieliszewski, M.J.; Upham, B.L.; Alizedeh, H.; Lamport, D.T.A. Isolation of pI 4.6 extensin peroxidase from tomato cell suspension cultures and identification of Val-Tyr-Lys as putative intermolecular cross-link site. *Plant J.* **1996**, *9*, 477–489.
25. Brady, J.D.; Fry, S.C. Formation of di-lsodityrosine and loss of lsodityrosine in the cell walls of tomato cell-suspension cultures treated with fungal elicitors or H_2O_2. *Plant Physiol.* **1997**, *11*, 87–92.
26. Hall, Q.; Cannon, M.C. The cell wall hydroxyproline-rich glycoprotein RSH is essential for normal embryo development in *Arabidopsis*. *Plant Cell* **2002**, *14*, 1161–1172.
27. Cannon, M.C.; Terneus, K.; Hall, Q.; Tan, L.; Wang, Y.; Wegenhart, B.L.; Chen, L.; Lamport, D.T.A.; Chen, Y.; Kieliszewski, M.J. Self-assembly of the plant cell wall requires an extensin scaffold. *Proc. Natl. Acad. Sci. USA* **2008**, *105*, 2226–2231.
28. Saha, P.; Ray, T.; Tang, Y.; Duttal, I.; Evangelous, N.R.; Kieliszewski, M.J.; Chen, Y.; Cannon, M.C. Self-rescue of an EXTENSIN mutant reveals alternative gene expression programs and candidate proteins for new cell wall assembly in Arabidopsis. *Plant J.* **2013**, *75*, 104–116.
29. Hijazi, M.; Durand, J.; Pichereaux, C.; Pont, F.; Jamet, E.; Albenne, C. Characterization of the arabinogalactan protein 31 (AGP31) of *Arabidopsis thaliana*: New advances on the Hyp-O-glycosylation of the pro-rich domain. *J. Biol. Chem.* **2012**, *287*, 9623–9632.
30. Sadava, D.; Walker, F.; Chrispeels, M.J. Hydroxyproline-rich cell wall protein (extensin): Biosynthesis and accumulation in growing pea epicotyls. *Dev. Biol.* **1973**, *30*, 42–48.

31. O'Neill, M.A.; Selvendran, R.R. Glycoproteins from the cell wall of *Phaseolus coccineus*. *Biochem. J.* **1980**, *187*, 53–63.
32. Lamport, D.T.A. The isolation and partial characterization of hydroxyproline-rich glycopeptides obtained by enzymatic degradation of primary cell wall. *Biochemistry* **1969**, *8*, 1115–1163.
33. Heath, M.F.; Northcote, D.H. Glycoprotein of the wall of sycamore tissue-culture cells. *Biochem. J.* **1971**, *125*, 953–961.
34. Keegstra, K.; Talmadge, K.W.; Bauer, W.D.; Albersheim, P. The structure of plant cell walls III. A model of the walls of suspension-cultured sycamore cells based on the interconnections of the macromolecular components. *Plant Physiol.* **1973**, *51*, 188–196.
35. Frueauf, J.B.; Dolata, M.; Leykam, J.F.; Lloyd, E.A.; Gonzales, M.; VandenBosch, K.; Kieliszewski, M.J. Peptides isolated from cell walls of *Medicago. truncatula* nodules and uninfected root. *Phytochemistry* **2000**, *55*, 429–438.
36. Mort, A.J.; Grover, P.B.J. Characterization of root hair cell walls as potential barriers to the infection of plants by rhizobia: the carbohydrate component. *Plant Physiol.* **1988**, *86*, 638–641.
37. Dolan, L.; Janmaat, K.; Willemsen, V.; Linstead, P.; Poethig, S.; Roberts, K.; Scheres, B. Cellular organization of the *Arabidopsis thaliana* root. *Development* **1993**, *119*, 71–84.
38. Baxter, I.; Hosmani, P.S.; Rus, A.; Lahner, B.; Borevitz, J.O.; Muthukumar, B.; Mickelbart, M.V.; Schreiber, L.; Franke, R.B.; Salt, D.E. Root suberin forms an extracellular barrier that affects water relations and mineral nutrition in Arabidopsis. *PLOS Genet.* **2009**, *5*, e1000492.
39. Ranathunge, K.; Schreiber, L. Water and solute permeabilities of Arabidopsis roots in relation to the amount and composition of aliphatic suberin. *J. Exp. Bot.* **2011**, *62*, 1961–1974.
40. Hardt, H.; Lamport, D.T.A. Hydrogen fluoride saccharification of wood: Lignin fluoride content, isolation of alpha-D-glucopyranosyl fluoride and posthydrolysis of reversion products. *Biotechnol. Bioeng.* **1982**, *24*, 903–918.
41. Fang, X.; Qiu, F.; Yan, B.; Wang, H.; Mort, A.J.; Stark, R.E. NMR studies of molecular structure in fruit cuticle polyesters. *Phytochemistry* **2001**, *57*, 1035–1042.
42. Tian, S.; Fang, X.; Wang, W.; Yu, B.; Cheng, X.; Qiu, F.; Mort, A.J.; Stark, R.E. Isolation and identification of oligomers from partial degradation of lime fruit cutin. *J. Agric. Food Chem.* **2008**, *56*, 10318–10325.
43. Mort, A.J. Partial Characterization of Extensin by Selective Destruction of Cell Walls. Ph.D. Thesis, Michigan State University, East Lancing, MI, USA, 1978.
44. Whitmore, F.W. Lignin-carbohydrate complex formed in isolated cell walls of callus. *Phytochemistry* **1978**, *17*, 421–425.
45. McDougall, G.J.; Stewart, D.; Morrison, I.M. Tyrosine residues enhance cross-linking of synthetic proteins into lignin-like dehydrogenation products. *Phytochemistry* **1996**, *41*, 43–47.
46. Terneus, K.A. The Isolation and Characterization of a Tobacco Extensin Precursor and Two *Arabidopsis* Hydroxyproline-rich Glycoproteins. Ph.D. Thesis, Ohio University, Athens, OH, USA, 2006.
47. Fry, S.C. Isodityrosine, A new cross-linking amino acid from plant cell-wall glycoprotein. *Biochem. J.* **1982**, *204*, 449–455.
48. Epstein, L.; Lamport, D.T.A. An intramolecular linkage involving isodityrosine in extensin. *Phytochemistry* **1984**, *23*, 1241–1246.

49. Smith, J.J.; Muldoon, E.P.; Lamport, D.T.A. Isolation of extensin precursors by direct elution of intact tomato cell suspension cultures. *Phytochemistry* **1984**, *23*, 1233–1239.
50. Smith, J.J.; Muldoon, E.P.; Willard, J.J.; Lamport, D.T.A. Tomato extensin precursors P1 and P2 are highly periodic structures. *Phytochemistry* **1986**, *25*, 1021–1030.
51. Lloyd, E.A.; Diezman, J.M.; Leykam, L.F.; Kieliszewski, M.J. Synthetic Genes for Elucidating Cross-linking Amino Acids in P1 Type Extensins. In Proceedings of the Annual Meeting of the American Society of Plant Biologists, Honolulu, HA, USA, 25–30 July 2003.
52. Baumberger, N.; Ringli, C.; Keller, B. The chimeric leucine-rich repeat/extensin cell wall protein LRX1 is required for root hair morphogenesis in *Arabidopsis thaliana. Gene Dev.* **2001**, *15*, 1128–1139.
53. Baumberger, N.; Doesseger, B.; Guyot, R.; Diet, A.; Parsons, R.L.; Clark, M.A.; Simmons, M.P.; Bedinger, P.; Goff, S.A.; Ringli, C.; *et al.* Whole-genome comparison of leucine-rich repeat extensins in Arabidopsis and rice. A conserved family of cell wall proteins form a vegetative and a reproductive clade. *Plant Physiol.* **2003**, *131*, 1313–1326.
54. Baumberger, N.; Steiner, M.; Ryser, U.; Keller, B.; Ringli, C. Synergistic interaction of the two paralogous Arabidopsis genes LRX1 and LRX2 in cell wall formation during root hair development. *Plant J.* **2003**, *35*, 71–81.
55. Ringli, C. The hydroxyproline-rich glycoprotein domain of the *Arabidopsis* LRX1 requires Tyr for function but not for insolubilization in the cell wall. *Plant J.* **2010**, *63*, 662–669.
56. Bayer, E.; Bottrill, A.; Walshaw, J.; Vigouroux, M.; Naldrett, M.; Thomas, C.; Maule, A. *Arabidopsis* cell wall proteome defined using multidimensional protein identification technology. *Proteomics* **2006**, *6*, 301–311.
57. Irshad, M.; Canut, H.; Borderies, G.; Pont-Lezica, R.; Jamet, E. A new picture of cell wall protein dynamics in elongating cells of *Arabidopsis thaliana*: Confirmed actors and newcomers. *BMC Plant Biol.* **2008**, *8*, doi:10.1186/1471-2229-8-94.
58. Baldwin, T.C.; McCann, M.C.; Roberts, K. A Novel Hydroxyproline-deficient arabinogalactan protein secreted by suspension-cultured cells of *Daucus carota. Plant Physiol.* **1993**, *103*, 115–123.
59. Immerzeela, P.; Eppink, M.M.; de Vries, S.C.; Schols, H.A.; Voragen, A.G.J. Carrot arabinogalactan proteins are interlinked with pectins. *Physiol. Plantarum.* **2006**, *128*, 18–28.
60. Hijazi, M.; Roujol, D.; Nguyen-Kim, H.; Del Rocio Cisneros Castillo, L.; Saland, E.; Jamet, E.; Albenne, C. Arabinogalactan protein 31 (AGP31), a putative network-forming protein in *Arabidopsis thaliana* cell walls? *Ann. Bot.* **2014**, doi:10.1093/aob/mcu038.
61. Johnson, K.L.; Jones, B.J.; Bacic, A.; Schultz, C.J. The fasciclin-like arabinogalactan proteins of Arabidopsis. A multigene family of putative cell adhesion molecules. *Plant Physiol.* **2003**, *133*, 1911–1925.
62. Johnson, K.L.; Kibble, N.A.J.; Bacic, A.; Schultz, C.J. A fasciclin-like arabinogalactan-protein (FLA) mutant of *Arabidopsis thaliana*, FLA1, shows defects in shoot regeneration. *PLOS ONE* **2011**, *6*, e25154.
63. Roubal, W.T.; Tappel, A.L. Polymerization of proteins induced by free-radical lipid peroxidation. *Arch. Biochem. Biophys.* **1966**, *113*, 150–155.
64. Kjellbom, P.; Snogerup, L.; Stöhr, C.; Reuzeau, C.; McCabe, P.F.; Pennell, R.I. Oxidative cross-linking of plasma membrane arabinogalactan proteins. *Plant J.* **1997**, *12*, 1189–1196.

65. Stafstrom, J.P.; Staehelin, L.A. The role of carbohydrate in maintaining extensin in an extended conformation. *Plant Physiol.* **1986**, *81*, 242–246.
66. Cooper, J.B.; Varner, J.E. Cross-linking of soluble extensin in isolated cell walls. *Plant Physiol.* **1984**, *76*, 414–417.
67. Valentin, R.; Cerclier, C.; Geneix, N.; Aguié-Béghin, V.; Gaillard, C.; Ralet, M.C.; Cathala, B. Elaboration of extensin-pectin thin film model of primary plant cell wall. *Langmuir* **2010**, *26*, 9891–9898.
68. Park, S.H.; Shalongo, W.; Stallwagen, E. The role of PII conformations in the calculation of peptide fractional helix content. *Protein Sci.* **1997**, *6*, 1694–1700.
69. Chuang, C.F.; Meyerowitz, E.M. Specific and heritable genetic interference by double-stranded RNA in *Arabidopsis thaliana*. *Proc. Natl. Acad. Sci. USA* **2000**, *97*, 4985–4990.
70. Li, J.; Reichel, M.; Li, Y.; Millar, A.A. The functional scope of plant microRNA-mediated silencing. *Trends Plant Sci.* **2014**, *19*, 785–756.
71. York, W.S.; Darvill, A.G.; McNeil, M.; Stevenson, T.T.; Albersheim, P. Isolation and characterization of plant cell walls and cell wall components. *Methods Enzymol.* **1986**, *118*, 3–40.

13

Quantification of (1→4)-β-D-Galactans in Compression Wood Using an Immuno-Dot Assay

Ramesh R. Chavan, Leona M. Fahey and Philip J. Harris *

School of Biological Sciences, The University of Auckland, Private Bag 92019, Auckland 1142, New Zealand; E-Mails: r.chavan@auckland.ac.nz (R.R.C.); lfah338@aucklanduni.ac.nz (L.M.F.)

* Author to whom correspondence should be addressed; E-Mail: p.harris@auckland.ac.nz

Academic Editor: Dilantha Fernando

Abstract: Compression wood is a type of reaction wood formed on the underside of softwood stems when they are tilted from the vertical and on the underside of branches. Its quantification is still a matter of some scientific debate. We developed a new technique that has the potential to do this based on the higher proportions of (1→4)-β-D-galactans that occur in tracheid cell walls of compression wood. Wood was milled, partially delignified, and the non-cellulosic polysaccharides, including the (1→4)-β-D-galactans, extracted with 6 M sodium hydroxide. After neutralizing, the solution was serially diluted, and the (1→4)-β-D-galactans determined by an immuno-dot assay using the monoclonal antibody LM5, which specifically recognizes this polysaccharide. Spots were quantified using a dilution series of a commercially available (1→4)-β-D-galactan from lupin seeds. Using this method, compression and opposite woods from radiata pine (*Pinus radiata*) were easily distinguished based on the amounts of (1→4)-β-D-galactans extracted. The non-cellulosic polysaccharides in the milled wood samples were also hydrolysed using 2 M trifluoroacetic acid followed by the separation and quantification of the released neutral monosaccharides by high performance anion exchange chromatography. This confirmed that the compression woods contained higher proportions of galactose-containing polysaccharides than the opposite woods.

Keywords: *Pinus radiata*; coniferous gymnosperms; compression wood; (1→4)-β-D-galactans; immuno-dot assay; LM5 monoclonal antibody; monosaccharide composition; delignification

---

1. Introduction

Compression wood is a type of reaction wood formed by coniferous gymnosperms (softwoods) on the underside of stems that have been tilted from the vertical [1–6]. This type of wood is important to the tree because it restores normal, vertical growth to the stem. It can also be formed on the underside of branches where it has a role in maintaining the angle of the branch to the stem. However, it is regarded as a defect in the timber industry as, on drying, it exhibits higher longitudinal shrinkage than normal wood or wood on the opposite side of the stem (opposite wood) and contributes to warping. It is also more difficult to work with and pulp than normal or opposite wood [5,6]. Compression wood is often darker in colour than normal and opposite wood, but the chemical basis of this coloration is unknown. Furthermore, dark coloured compression wood can be confused with late wood [6]. There is thus a need to devise reliable methods for detecting compression wood.

Compression wood differs from normal and opposite woods anatomically and chemically. Anatomically, the tracheids, the predominant cell type in softwoods, are shorter and more rounded in cross section in compression wood than in normal or opposite woods. The tracheid walls in compression wood are thicker, but lack the third secondary wall layer (S3) that in normal and opposite wood tracheids occurs adjacent to the lumen [5,6]. Chemically, the tracheid walls of normal and opposite woods are composed of microfibrils of cellulose set in a matrix of lignin composed predominantly of guaiacyl units (G-units), and the non-cellulosic polysaccharides (hemicelluloses) heteromannans (*O*-acetyl-galactoglucomannans) and smaller proportions of heteroxylans [arabino(4-*O*-methylglucurono)xylans] [7–11]. The tracheid walls of compression wood contain less cellulose, heteromannans and heteroxylans, but more lignin, which in addition to G-units contains *p*-hydroxyphenyl units (H-units) that are almost absent from normal and opposite woods. These walls also contain much higher proportions of (1→4)-β-D-galactans (up to ~10%) than normal or opposite woods [5,10,12]. All of these features of compression wood apply to what is frequently termed severe compression wood, although mild compression woods of different severities with anatomical, and possibly chemical, features intermediate between severe compression wood and normal and opposite woods are also known to occur, but have been subjected to much less research [6,13].

Pinus radiata (radiata pine or Monterey pine) is a fast-growing softwood widely grown in plantations in many temperate countries, including New Zealand. The compression wood of this species can be distinguished from normal and opposite woods by detailed chemical analyses of the cell-wall polymers [12,13]. The aim of the present study was to develop a method of detecting compression wood based on the occurrence of higher proportions of (1→4)-β-galactans in the tracheid walls of compression wood than normal or opposite woods. The method involves extracting non-cellulosic polysaccharides from milled wood and determining the content of (1→4)-β-galactans using an immuno-dot assay, with lupin seed (1→4)-β-galactans as the quantification standard. The extracted polysaccharides are bound to nitrocellulose membranes as dots and then treated with the monoclonal antibody LM5, which binds specifically to (1→4)-β-galactans [14], as the primary antibody. An enzyme-labelled secondary antibody is then used followed by a substrate that gives an insoluble coloured product [15]. The intensity of the colour is measured by image analysis. Essentially, we developed a simplified manual version of the technique known as comprehensive microarray polymer profiling [16,17], but with only a single extraction solvent and primary antibody.

2. Results

2.1. Assays Using Reference Lupin Seed (1→4)-β-Galactans

A plot of spot integral intensities against the amounts of lupin seed (1→4)-β-galactans for a dilution series from 0.001 to 1.0 μg/dot was curvilinear (Figure 1), with the linear portion in the lower dilution range (up to 0.006 μg/dot). It is thus important to use (1→4)-β-galactan amounts in this dilution range for quantification purposes.

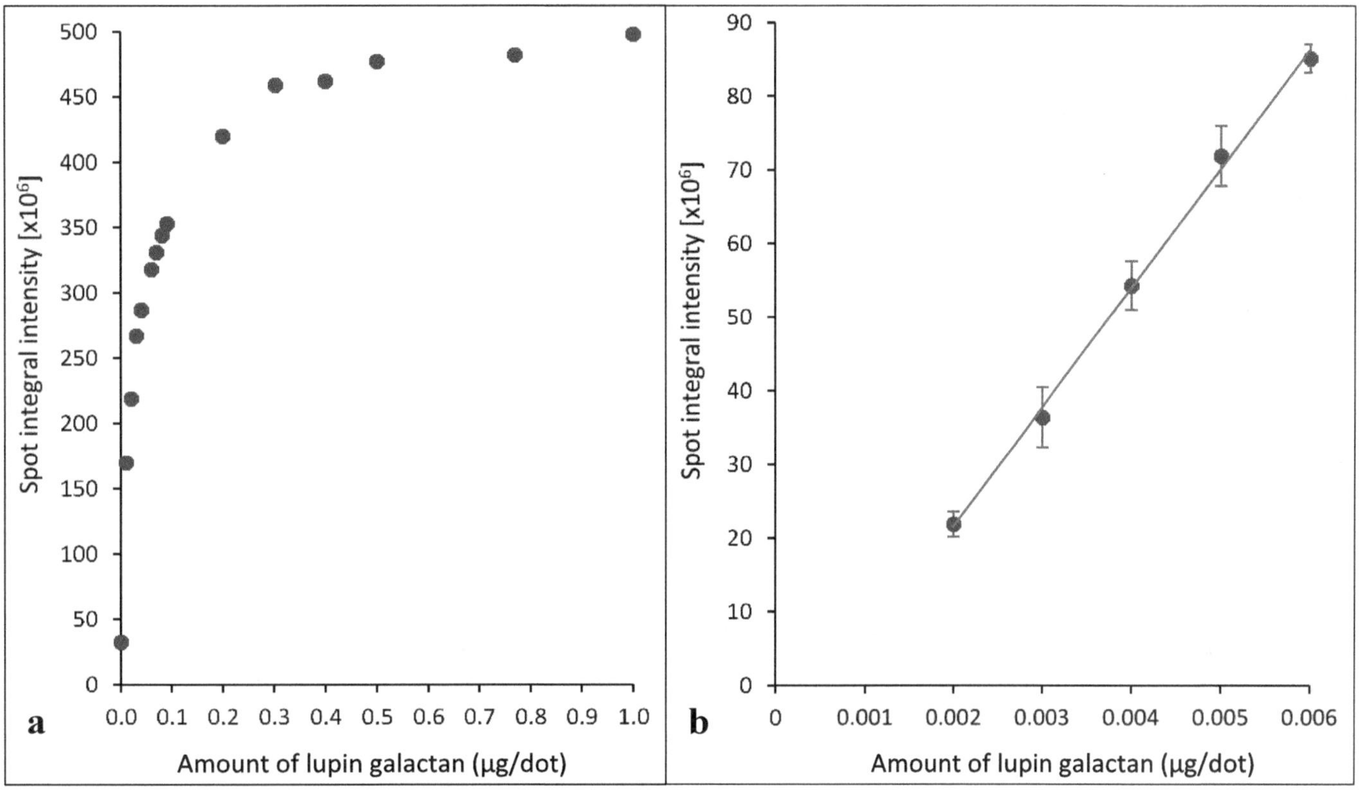

Figure 1. (**a**) Curvilinear relationship of spot integral intensity to amount of lupin seed (1→4)-β-galactans in the dilution series generated with Xplore software; (**b**) linear relationship of spot integral intensity to amount of lupin seed (1→4)-β-galactans at lower dilutions showing a strong positive correlation of amounts of (1→4)-β-galactans to the spot integral intensity ($R^2 = 0.99$; $y = (2(10^{10})x-1(10^7))$; ($R^2$ = Coefficient of determination; y = spot integral intensity; x = amount of galactan (μg/dot)). Error bars indicate the standard errors.

Blue dot images obtained from the dilution series of the standard lupin seed (1→4)-β-galactans are shown in Figure 2.

All lupin seed (1→4)-β-galactan amounts, from 0.001 μg/dot upwards, were readily detected by eye (Figure 2). A similar control dilution series without primary antibody showed no coloration, as did a no-(1→4)-β-galactan control dot.

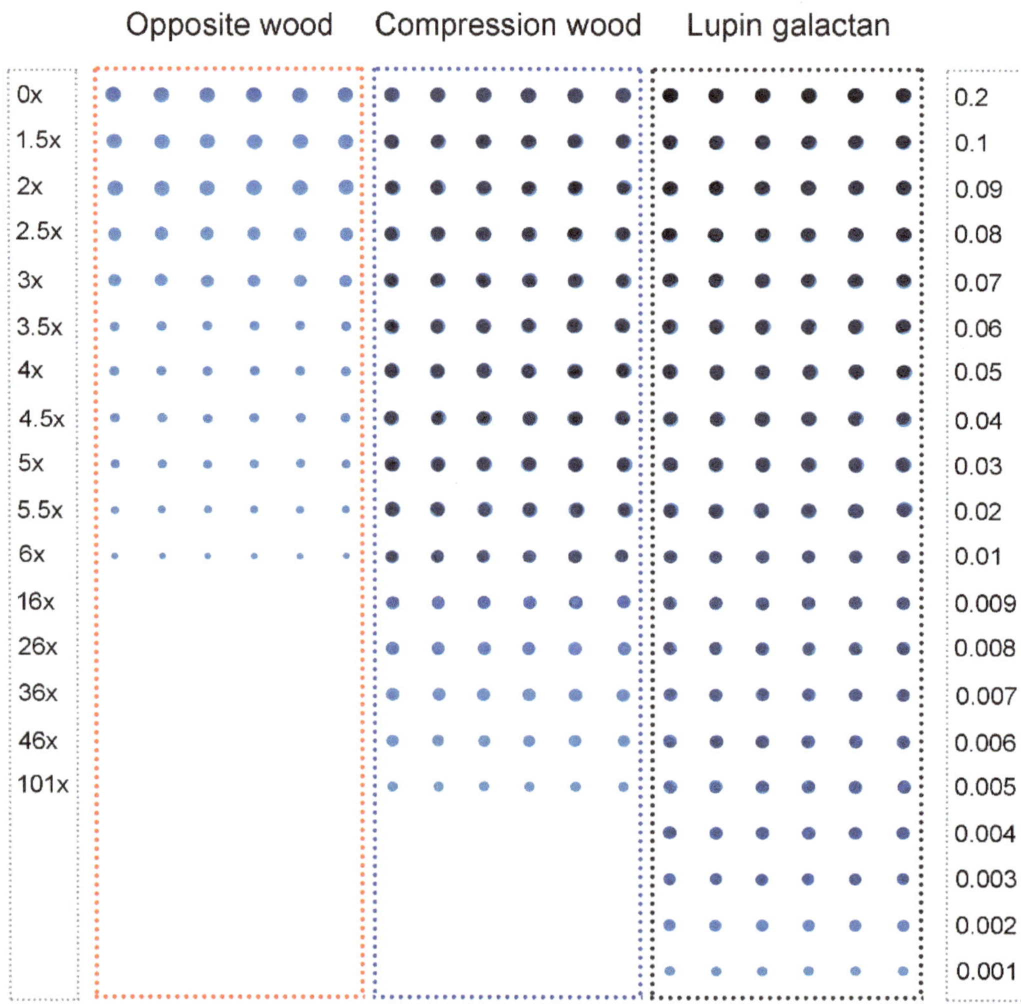

Figure 2. Immuno-dot assays of (1→4)-β-galactans extracted from non-delignified wood from Tree 1. Opposite wood (10 dilutions 1.5× to 6×), compression wood (15 dilutions 1.5× to 101×) and lupin seed (1→4)-β-galactan (twenty dilutions giving 0.001 to 0.2 μg/μL or dot) plotted in increasing order of dilution; each dilution has six replicates. This shows a scanned image (1200 dpi) of the nitrocellulose membrane with blue dots.

2.2. Assays Using Extracts of Radiata Pine Wood

The dilution series of the extracts from non-delignified compression wood and opposite wood from Tree 1 (Figure 2) showed that (1→4)-β-galactans could be detected in the opposite wood extract down to only a 6× dilution, but in the compression wood extract down to a 101× dilution, indicating a higher concentration of (1→4)-β-galactans in the compression wood extract than the opposite wood extract. Quantification, using dilutions on the linear part of the curve, showed 0.03% and 0.35% for the extracted (1→4)-β-galactan content of the opposite wood and compression wood, respectively (Table 1). (The 0.03% and 0.35% values mean 0.003 and 0.035 mg (1→4)-β-galactan were extracted into the 6 M NaOH from each 10 mg of dry wood, respectively). However, when the three-year-old trees (Trees 2–7) were compared, no differences in extracted (1→4)-β-galactan contents were found between the compression wood and opposite wood of each tree (Table 1). The extracted (1→4)-β-galactan content of the opposite and compression wood was similar to that found in the opposite wood from Tree 1.

Table 1. Immuno-dot assay of (1→4)-β-galactans extracted from opposite and compression woods of *Pinus radiata* with and without delignification.

Tree	Amounts of (1→4)-β-Galactans Extracted Without Delignification			Amounts of (1→4)-β-Galactans Extracted after Delignification		
	OW	CW	CW/OW	OW	CW	CW/OW
	Mean ± SE	Mean ± SE		Mean ± SE	Mean ± SE	
Tree 1	0.03 ± 0.00	0.35 ± 0.03	11.7	0.38 ± 0.09	9.95 ± 1.39	26.2
Tree 2	0.02 ± 0.00	0.01 ± 0.00	0.5	0.60 ± 0.11	8.34 ± 0.60	14.0
Tree 3	0.01 ± 0.00	0.01 ± 0.00	1.0	0.42 ± 0.08	8.25 ± 0.02	19.5
Tree 4	0.03 ± 0.00	0.02 ± 0.00	0.7	0.60 ± 0.03	6.63 ± 0.95	11.1
Tree 5	0.01 ± 0.00	0.02 ± 0.00	2.0	0.39 ± 0.02	5.92 ± 0.76	15.3
Tree 6	0.03 ± 0.00	0.04 ± 0.00	1.3	0.22 ± 0.04	4.27 ± 0.50	19.5
Tree 7	0.03 ± 0.00	0.03 ± 0.00	1.0	0.74 ± 0.08	3.14 ± 0.10	4.2

OW = Opposite wood; CW = Compression wood. CW/OW = ratio of amounts of (1→4)-β-galactans extracted from compression and opposite woods. Amounts of galactans extracted are expressed as weight percentages extracted from dry wood (means of three aliquots of milled wood).

We therefore examined the effect of partially delignifying the milled wood on the amounts of (1→4)-β-galactans extracted by the 6 M NaOH solution. After partial delignification, much greater amounts of (1→4)-β-galactans were extracted, particularly from the compression wood. For example, with the wood from Tree 7, (1→4)-β-galactans were detected in the opposite wood extract down to only a 50× dilution, but in the compression wood extract, (1→4)-β-galactan was still detected at a 500× dilution (Figure 3). The particular dilutions were chosen because 20× to 50× for opposite wood and 200× to 500× for compression wood were on the linear part of the curve. The calculated extracted concentrations ranged from 0.22% to 0.74% for opposite wood and 3.14% to 9.95% for compression wood, with the highest concentrations of extracted (1→4)-β-galactans being in the compression wood of Tree 1 (Table 1).

The upper value for extracted (1→4)-β-galactans in opposite wood was 0.74% (Tree 7) and could represent the upper value for wood not containing any compression wood. The lowest concentration for compression wood was 3.14% (Tree 7) and could represent one of the lowest that may be obtained from severe compression wood. However, although the Tree 7 compression wood sample was dark coloured, it was not examined microscopically and may not be pure severe compression wood. It may have also contained some tracheids better described as mild compression wood tracheids, which may have extracted (1→4)-β-galactan concentrations intermediate between opposite (or normal) wood and severe compression wood [13]. We thus tested the immuno-dot assay method on mixtures of compression wood and opposite wood to simulate different mild compression woods. Such mixtures would also simulate samples containing adjacent compression and normal (or opposite) woods. As expected, the relationship between the content of extracted (1→4)-β-galactans and percentage of compression wood in the mixture was linear (Figure 4).

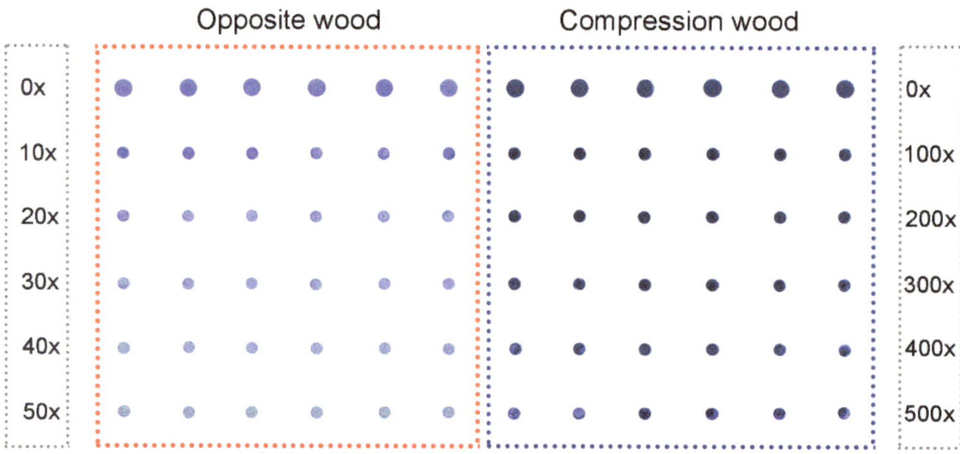

Figure 3. Immuno-dot assays of (1→4)-β-galactans in extracts of delignified wood of Tree 7. Extracts of opposite wood (five dilutions 0× to 50×) and compression wood (five dilutions 0× to 500×; each dilution has six replicates. This shows a scanned image (1200 dpi) of the nitrocellulose membrane with blue dots.

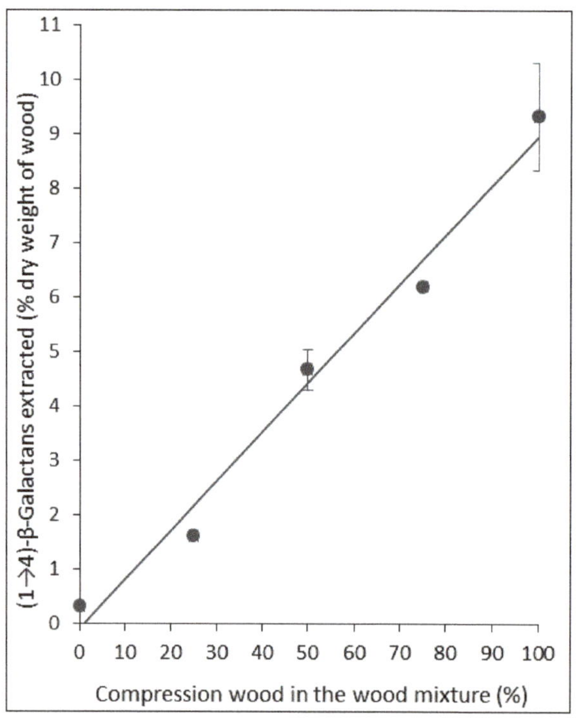

Figure 4. The amount of (1→4)-β-galactans extracted from delignified mixtures of compression wood and opposite wood from Tree 1 showing a linear relationship with amount of compression wood in the mixture, with a strong positive correlation ($R^2 = 0.98$; $y = 0.0905x - 0.0866$). (R^2 = Coefficient of determination; y = spot integral intensity; x = amount of galactan (μg/dot)). Error bars indicate standard errors (some values have negligible standard errors, so the error bars are not visible).

2.3. Monosaccharide Compositions of Radiata Pine Wood

To show that the compression and opposite woods examined in the present study contained different proportions of galactosyl-containing polysaccharides, the woods were acid hydrolysed and analysed.

The proportions of the neutral monosaccharides in the 2 M TFA hydrolysates of the opposite and compression woods are shown in Table 2. The highest proportions of particular monosaccharides were for galactose in compression wood hydrolysates, probably mostly from (1→4)-β-galactans. The proportions ranged from 38.1% (Tree 7) to 58.5% (Tree 3). Galactose also showed the greatest differences in proportions between compression and opposite woods, with the proportions in the opposite woods ranging from 7.7% (Tree 3) to 10.5% (Trees 2 and 5). This gave ratios of the proportion of galactose in the compression wood to opposite wood ranging from 4.3 (Tree 7) to 7.6 (Tree 3). The proportions of xylose (mostly from heteroxylans) and mannose (mostly from heteromannans) were higher in hydrolysates of opposite than compression woods, giving ratio (opposite/compression) ranges of 1.8–3.4 and 1.5–2.5, respectively. A similar ratio range, 1.3–2.3, was also found for arabinose (probably mostly from heteroxylans), but the proportions were lower. In contrast, the proportions of glucose in hydrolysates of opposite and compression woods were similar, giving a ratio (opposite/compression) range of 0.8–1.2.

Table 2. Neutral-monosaccharide compositions of opposite and compression woods (% of all neutral monosaccharides).

		Monosaccharides					
		Arabinose	**Xylose**	**Galactose**	**Glucose**	**Mannose**	**CW/OW Galactose**
Tree	**Wood Type**	**Mean ± SE**	**Mean ± SE**	**Mean ± SE**	**Mean ± SE**	**Mean ± SE**	
Tree 1	OW	10.2 ± 0.53	32.2 ± 0.12	8.4 ± 0.07	13.6 ± 0.26	35.6 ± 0.24	6.3
	CW	4.7 ± 0.36	15.3 ± 0.29	52.7 ± 0.25	11.4 ± 0.11	16.0 ± 0.28	
Tree 2	OW	13.2 ± 0.23	35.8 ± 0.47	10.5 ± 0.20	9.4 ± 0.30	31.2 ± 0.18	5.5
	CW	5.8 ± 0.06	13.1 ± 0.13	56.9 ± 0.49	9.8 ± 0.23	14.3 ± 0.25	
Tree 3	OW	12.0 ± 0.33	38.7 ± 0.11	7.7 ± 0.10	10.1 ± 0.37	31.8 ± 0.46	7.6
	CW	5.3 ± 0.26	11.4 ± 0.17	58.5 ± 0.28	12.2 ± 0.27	12.6 ± 0.29	
Tree 4	OW	13.3 ± 0.35	31.4 ± 0.71	10.2 ± 0.17	12.4 ± 0.25	32.7 ± 1.03	4.9
	CW	6.5 ± 0.24	15.9 ± 0.18	49.3 ± 0.12	10.4 ± 0.34	17.8 ± 0.78	
Tree 5	OW	10.2 ± 1.44	30.4 ± 0.61	10.5 ± 0.05	13.1 ± 0.31	35.8 ± 0.89	4.6
	CW	7.6 ± 0.29	15.7 ± 0.08	47.7 ± 0.89	11.9 ± 0.16	17.3 ± 0.39	
Tree 6	OW	12.3 ± 0.21	29.5 ± 0.31	7.9 ± 0.08	12.5 ± 0.23	37.8 ± 0.19	6.8
	CW	6.8 ± 0.02	11.9 ± 0.07	53.7 ± 0.41	11.2 ± 0.29	16.5 ± 0.11	
Tree 7	OW	10.2 ± 0.90	34.4 ± 0.35	8.9 ± 0.09	11.1 ± 0.23	35.4 ± 0.29	4.3
	CW	7.4 ± 0.13	18.9 ± 0.39	38.1 ± 1.46	11.3 ± 0.29	24.4 ± 0.85	

Mean of determinations on three hydrolysates. CW/OW Galactose is the ratio of the percentage of galactose in compression wood and opposite wood.

The rankings of the compression woods of the three-year-old trees (Trees 2–7) for percentages of galactose in hydrolysates and percentages of (1→4)-β-galactans in extracts were similar; the only major difference was for Tree 6, which ranked higher in terms of percentage galactose than percentage of (1→4)-β-galactan extracted.

As with the extracted (1→4)-β-galactan concentrations, upper limits of ranges can be set for opposite wood and compression wood. The highest galactose percentage for opposite wood of 10.5% could be used as the upper limit for wood not containing compression wood. The lowest galactose percentage for compression wood, 38.1%, could be used as the lowest percentage for severe compression wood, although the compression wood for this tree may not be pure severe compression wood.

For the mixtures of compression wood and opposite wood, the relationship between percentage galactose and percentage of compression wood in the mixture was again linear [18].

3. Discussion

Comprehensive microarray polymer profiling and its methods for polysaccharide extractions was developed and has subsequently been used most frequently on plant tissues containing principally non-lignified primary cell walls [16,17]. However, in wood science, the non-cellulosic polysaccharides are usually extracted from wood cell walls after they have first been delignified, yielding a fraction known as holocellulose. This is because non-cellulosic polysaccharides are difficult to extract from lignified walls, particularly those of softwoods [19,20]. This is consistent with our finding that partial delignification of the milled radiata pine wood with acid chlorite [21] resulted in the subsequent extraction of more (1→4)-β-galactans. Similar (1→4)-β-galactans were also more easily extracted from the compression wood of Norway spruce (*Picea abies*) after acid chlorite delignification [22]. The original acid chlorite delignification process [21] aimed to achieve complete delignification and involves prolonged treatment with fresh charges of sodium chlorite and acetic acid being added at intervals. However, in the present study, we simplified the delignification treatment by omitting fresh charges. This was done to simplify the process, and to reduce the likelihood of the (1→4)-β-galactans being degraded or extracted during the delignification. Although the delignification is unlikely to have been complete, the process can easily be reproduced and is sufficient. Using the immuno-dot assay, the percentage of (1→4)-β-galactans extracted from the compression wood of Tree 1, 9.95%, compared well with the 10.3% for the galactose content of compression wood from a tree of the same genotype (an analogous ramet), but a different age, using a two-stage sulphuric acid hydrolysis [12]. Using the immuno-dot assay, we found only trace amounts of (1→4)-β-galactans in the acid chlorite solution after treatment of compression wood. In a study of the delignification of black spruce (*Picea mariana*) by the chlorite process [21], it was found that during the first 60% of delignification only lignin was dissolved, and beyond that point small quantities of galactoglucomannans were also dissolved [23].

The increased amounts of (1→4)-β-galactans extracted from both compression and opposite woods of radiata pine after partial delignification suggests these polysaccharides are held in the tracheid walls by being covalently linked to lignin. Lignin-carbohydrate complexes, containing such covalent bonds, have been isolated from a variety of woods [10], including a complex containing lignin and (1→4)-β-galactans from the compression wood of Japanese red pine (*Pinus densiflora*) [24]. These lignin-polysaccharide cross links are formed by reactions involving quinone methide intermediates, produced during lignin synthesis, and polysaccharides [25]. Consistent with covalent links between lignin and (1→4)-β-galactans has been the finding, using immunomicroscopy with LM5, that lignins and (1→4)-β-galactans are colocated in the tracheid walls of radiata pine. In normal and opposite woods, the small proportion of (1→4)-β-galactans found in these woods have been localized to the tracheid primary wall [26,27]. This is consistent with the (1→4)-β-galactans occurring as side chains of the pectic polysaccharide rhamnogalacturonan I (RG-1), which is known to occur in the primary cell walls of coniferous gymnosperms [12,28]. The highest concentrations of lignin in these wood types also occur in the compound middle lamella comprising the middle lamella and primary wall. In the mature tracheids of compression woods, the (1→4)-β-galactans have been localized to the outer S2 layer, a region in which the highest concentration of lignin also occurs and is hence referred to as the S2L layer. Using the same antibody, the (1→4)-β-galactans were also localized to the same layer in mature compression wood tracheids of Japanese cedar (*Cryptomeria japonica*) [29]. Interestingly, in that study, the intensity of the labelling over the S2L layer was higher after acid chlorite delignification of the wood.

The immuno-dot method could also be extended to other species of coniferous gymnosperms. It is likely to be particularly useful for species where other galactose-containing polysaccharides are known to occur in the wood because the method is specific to (1→4)-β-galactans. For example, the wood of larches (*Larix* spp.), especially the heartwood, contains high proportions of water-soluble arabino-3,6-galactans, which are non-structural polysaccharides. However, compression wood of larches also does contain (1→4)-β-galactans [30]. Lower proportions of arabino-3,6-galactans occur in the heartwoods of other coniferous gymnosperms, including Norway spruce (*Picea abies*) and Scots pine (*Pinus sylvestris*) [10,11,31–33]. Previous studies using 2D NMR [12] on compression and opposite wood samples from a tree of the same genotype (an analogous ramet), as Tree 1, but a different age, showed that the higher proportions of galactose in the compression wood hydrolysates were due to higher proportions of (1→4)-β-galactans; there was no evidence for the presence of arabino-3,6-galactans.

Although a TFA hydrolysis method was used in the present study to show differences in proportions of galactosyl-containing polymers between the opposite and compression woods, this method of acid hydrolysis could itself be used as a method to detect compression wood, although it is, of course, not specific for (1→4)-β-galactans. In wood science, the usual method of hydrolysing wood cell walls is the two-stage sulphuric acid method [12,34], which hydrolyses all the cell-wall polysaccharides, including cellulose. However, in studies of primary cell walls, the TFA hydrolysis method [35], which does not hydrolyse crystalline cellulose [36], is often used to study non-cellulosic polysaccharides. The TFA can be easily removed, so analyses can be carried out on very small samples, if required. We found this method of hydrolysis excellent for comparing the percentages of neutral monosaccharides from opposite and compression woods, with galactose being a high proportion of the neutral monosaccharides released from compression woods, but not from opposite woods. By simply using relative percentages of neutral monosaccharides rather than absolute yields of each monosaccharide on a wood dry weight basis, accurate weighing of the wood samples and addition of an internal quantification standard are unnecessary. We separated and quantified the neutral monosaccharides in the TFA hydrolysate using HPAEC, but these monosaccharides could also be converted to alditol acetates and separated and quantified by gas chromatography [37,38].

In contrast to HPAEC and gas chromatography, the immuno-dot method of detecting compression wood uses relatively simple equipment, but it is time consuming and labour intensive, particularly grinding of the wood. However, the comprehensive microarray polymer profiling technique [16,17], on which it is based, uses robotics. A high-throughput platform, which also includes a grinding step, has been developed for screening milligram quantities of plant biomass to determine cell-wall digestibility, and uses the iWALL robot [39]. Similar robotic techniques could be adapted for quantifying (1→4)-β-galactans in wood using the immuno-dot method. This method could then be used in wood research and to screen the wood of saplings in breeding programmes.

4. Experimental

4.1. Wood Samples

Samples of compression and opposite woods were obtained from seven saplings of *Pinus radiata* (D. Don) grown tilted at ~45° from the vertical; we refer to these as Trees 1–7. Tree 1 (clone A; 96047/98015, Arborgen Australasia, TeTeko, Whakatane, New Zealand) was grown in an unheated glasshouse at the University of Canterbury, Christchurch, New Zealand, for nine months [40] and then outside for 11 months. Trees 2–7, from six different control-pollinated families (one tree per family), were grown for three years in the field at Amberley, Canterbury, New Zealand. The families were as follows: Tree 2 (family 05–854), Tree 3 (family 99–405), Tree 4 (family 05–829), Tree 5 (family 05–862), Tree 6 (06–524) and Tree 7 (05–844) [41].

A basal segment (about ~15 cm long) was cut from each tree, debarked, cut longitudinally into compression wood and opposite wood (based on the dark coloration of the compression wood) with a band saw and chisel, and dried at 35 °C to ~4% moisture content. The wood was then milled to pass a 60-mesh screen (251 μm pore size) using a Wiley® mini-mill (Thomas Scientific, Swedesboro, NJ, USA), dried at 105 °C and stored over silica gel. Mixtures of milled compression wood and opposite wood (75, 50 and 25% w/w compression wood) were also prepared from Tree 1 to simulate samples containing adjacent compression and normal (or opposite) woods or mild compression woods of different severities [42].

4.2. Extraction of Non-Cellulosic Polysaccharides from the Wood Samples

Milled wood was extracted either directly or after partial delignification that was done as follows. Milled wood (10 mg) was treated with a freshly-prepared aqueous mixture of sodium chlorite (1% w/v) (1 mL) and 17.4 M acetic acid (20 μL) for 3 h at 70 °C using a Thermomixer Comfort (Eppendorf AG, Hamburg, Germany) operated at 1400 rpm [21]. After cooling to room temperature, the mixture was centrifuged (15,700× g, 10 min), the supernatant removed, and the pellet washed (3×) with water (1 mL) followed by centrifugation (5 min). The remaining holocellulose preparation or dry milled wood (10 mg) was extracted on an orbital shaker at 180 rpm for 16 h at 37 °C with 6 M NaOH containing 1% $NaBH_4$ (500 μL) to prevent alkaline peeling. After centrifuging (10 min), the supernatant was removed and neutralized with 17.4 M acetic acid, using a 2:1 (v/v) ratio of supernatant to acetic acid to achieve a pH >5 <6. A dilution series was prepared from each of these extracts using water. These dilutions are referred to as × dilutions; for example a 1.5× dilution was prepared by adding 0.5 volumes of water to a 1.0 volume of extract.

4.3. Immuno-Dot Quantification of (1→4)-β-Galactans

Six replicates (1 μL) of each successive extract dilution were applied as dots to a Protran® nitrocellulose membrane (0.45 μm pore size; Schleicher & Schuell BioScience, Dassel, Germany) and dried at 30 °C for 1 h [16,17]. The membrane was incubated in phosphate-buffered saline (PBS) (0.01 M sodium phosphate buffer, pH 7.4; 0.14 M NaCl) containing 5% (w/v) milk powder (0.1% fat; Alpine, Dairyworks Ltd, Christchurch, New Zealand) (MP-PBS) for 2 h at 30 °C to block all nonspecific binding sites. It was then incubated with the monoclonal antibody LM5 (PlantProbes, Leeds, UK) diluted 1:10 in MP-PBS, followed by goat anti-rat IgG (H + L chains) secondary antibody conjugated with alkaline phosphatase (Invitrogen, Auckland, New Zealand) diluted 1:100 in MP-PBS. Both antibodies were incubated for 2 h at 30 °C, and after each incubation, the membrane was washed (5×) with PBS on an orbital shaker (600 rpm). Finally, the membrane was treated for 15 min with substrate solution containing 5-bromo-4-chloro-3-indolyl phosphate (BCIP), nitro-blue tetrazolium (NBT) and water (1:1:8 v/v) (BCIP/NBT substrate kit, Invitrogen, New Zealand). Alkaline phosphatase reacts with the substrate to produce an insoluble blue product. The membrane was then washed in running water for 20 min and dried at 30 °C. Two types of negative control were used: a no primary antibody control and a no extract control in which 6 M NaOH (containing 1% $NaBH_4$) after the addition of acetic acid (see above) was used.

The dried nitrocellulose membranes were scanned to produce TIFF images (1200 dpi) using a desktop scanner (Model DCP-130C; Brother International, Bridgewater, NJ, USA) that were then converted into inverted 16-bit grey-scale images using Adobe Photoshop (Creative Suite 6; Adobe Systems, San Jose, CA, USA). Xplore image processing software (LabNEXT Inc, West New York, NJ, USA), which includes grids, was then used to generate spot integral intensities (heat map values) of the spots, which were exported as comma separated values (csv) and opened in Excel format. The spot integral intensity is a specific numerical value generated by the Xplore program and is based on the numbers of spots per unit area defined by a grid. Mean values for each set of six replicate spot integral intensities were computed. The final spot integral intensities were calculated by subtracting the values of blank (no sample) spots. These were converted into mg of (1→4)-β-galactans extracted per 100 mg of dry wood (*i.e.*, weight percentages) by comparing the intensities with those given by the lower range (0–0.006 μg/dot) (Figure 1) of a dilution series of lupin seed (1→4)-β-galactans. In the final calculation, it was assumed that the lupin seed (1→4)-β-galactans preparation contained 91% (1→4)-β-galactan as indicated in the Megazyme literature.

4.4. Neutral Monosaccharide Compositions of the Wood Samples

Triplicate samples (5 mg) of dry milled opposite and compression woods were hydrolysed using 2 M trifluoroacetic acid (TFA) (0.5 mL, 121 °C, 1 h) in a sealed tube under argon [35,38]. After cooling, D-ribose (500 μg) was added as a retention time standard, and the hydrolysate evaporated to dryness in a stream of air. The residue was dissolved in water (5 mL), filtered using a polytetrafluoroethylene disposable filter (diameter 13 mm, pore size 0.22 μm; Shandong Hapool Medical Technology, Heze, China), and the filtrate made up to 50 mL.

Neutral monosaccharides were separated and quantified using high-performance anion-exchange chromatography with pulsed amperometric detection (HPAEC-PAD) on a Dionex BioLC system (Dionex, Sunnyvale, CA, USA) fitted with an ED50 electrochemical detector and GP50 gradient pump. A CarboPac PA20 guard column (3 × 30 mm) and a CarboPac PA20 analytical column (3 × 150 mm) were used. Column temperature was kept at 25 °C by a TCC-100 thermostatted column compartment. Separation was achieved using isocratic elution (1 mM NaOH for 30 min) and the column was then washed for 5 min with 200 mM NaOH before equilibration for 5 min with 1 mM NaOH. The injection volume was 20 μL, and the flow rate 0.4 mL min^{-1}. The order of elution of monosaccharides was confirmed by running solutions of individual reference monosaccharide. Before the hydrolysate runs, a water blank was run followed by a reference solution containing 10 μg mL^{-1} each of L-arabinose, D-galactose, D-glucose, D-xylose, D-mannose, D-xylose and D-ribose, which was used to determine the relative responses of equal weights of each monosaccharide.

5. Conclusions

The present study shows that the immuno-dot assay method using the LM5 antibody can be applied, after partial delignification, to milled (ground) wood of radiata pine for the quantification of (1→4)-β-galactans. The percentages of extracted (1→4)-β-galactans from opposite wood and compression wood were quite different and the method could thus be used to identify the presence of compression wood in wood samples. The present study is a proof of principle, but more extensive studies need to be carried out to determine the range in percentages of extracted (1→4)-β-galactans in microscopically characterized compression woods, both severe and mild, as well as normal and opposite woods, in trees of different species, genotypes and ages.

Acknowledgments

We thank Isabel Moller for her valuable suggestions; Martin Schomber and Guangwu Jin for their technical help; and the New Zealand Foundation for Research, Science and Technology (now Ministry of Business, Innovation, and Employment) for funding (project PROJ-12401-PPS-UOC, "Compromised Wood Quality"); and our colleagues at the University of Canterbury, John Walker and Luis Apiolaza, for providing the wood samples.

Author Contributions

Philip Harris conceived and coordinated the study, Ramesh Chavan developed the methods and carried out the immunological experimental work, and Leona Fahey carried out the monosaccharide analyses. Ramesh Chavan and Philip Harris wrote most of the manuscript, but all authors edited and approved the final version.

Conflicts of Interest

The authors declare no conflict of interest.

References

1. Westing, A.H. Formation and function of compression wood in gymnosperms. *Bot. Rev.* **1965**, *31*, 381–480.
2. Westing, A.H. Formation and function of compression wood in gymnosperms, II. *Bot. Rev.* **1968**, *34*, 51–78.
3. Robards, A.W. The effect of gravity on the formation of wood. *Sci. Prog. (Oxf.)* **1969**, *57*, 513–532.
4. Scurfield, G. Reaction wood: Its structure and function. *Science* **1973**, *179*, 647–655.
5. Timell, T.E. *Compression Wood in Gymnosperms*; Springer-Verlag: Berlin, Germany, 1986.

6. Donaldson, L.A.; Singh, A.P. Formation and structure of compression wood. In *Cellular Aspects of Wood Formation*; Fromm, J., Ed.; Springer-Verlag: Berlin, Germany, 2013; pp. 225–256.
7. Harwood, V.D. Studies on the cell wall polysaccharides of *Pinus radiata* 1. Isolation and structure of a xylan. *Sven. Papp.* **1972**, *75*, 207–212.
8. Harwood, V.D. Studies on the cell wall polysaccharides of *Pinus radiata* II. Structure of a glucomannan. *Sven. Papp.* **1973**, *76*, 377–379.
9. Whistler, R.L.; Chen, C.-C. Hemicelluloses. In *Wood Structure and Composition*; Lewin, M., Goldstein, I.S., Eds.; Marcel Dekker: New York, NY, USA, 1991; pp. 287–319.
10. Harris, P.J.; Stone, B.A. Chemistry and molecular organization of plant cell walls. In *Biomass Recalcitrance: Deconstructing the Plant Cell Wall for Bioenergy*; Himmel, M.E., Ed.; Blackwell Publishing: Oxford, UK, 2008; pp. 61–93.
11. Teleman, A. Hemicelluloses and pectin. In *Pulp and Paper Chemistry and Technology. Vol. 1. Wood Chemistry and Biotechnology*; Ek, M., Gellerstedt, G., Henriksson, G., Eds.; de Gruyter: Berlin, Germany, 2009; pp. 101–120.
12. Brennan, M.; McLean, J.P.; Altaner, C.; Ralph, J.; Harris, P.J. Cellulose microfibril angles and cell-wall polymers in different wood types of *Pinus radiata*. *Cellulose* **2012**, *19*, 1385–1404.
13. Nanayakkara, B.; Manley-Harris, M.; Suckling, I.D.; Donaldson, L.D. Quantitative chemical indicators to assess the gradation of compression wood. *Holzforschung* **2009**, *63*, 431–439.
14. Jones, L.; Seymour, G.B.; Knox, J.P. Localization of pectic galactan in tomato cell walls using a monoclonal antibody specific to (1→4)-β-D-galactan. *Plant Physiol.* **1997**, *113*, 1405–1412.
15. Willats, W.G.T.; Orfila, C.; Limberg, G.; Buchholt, H.C.; van Alebeek, G.-J.W.M.; Voragen, A.G.J.; Marcus, S.E.; Christenen, T.M.I.E.; Mikkelsen, J.D.; Murray, B.S.; *et al.* Modulation of the degree and pattern of methyl-esterification of pectic homogalacturonan in plant cell walls. *J. Biol. Chem.* **2001**, *276*, 19404–19413.
16. Moller, I.; Sørensen, I.; Bernal, A.J.; Blaukopf, C.; Lee, K.J.D.; Øbro, J.; Pettolino, F.; Roberts, A.; Mikkelsen, J.D.; Knox, J.P.; *et al.* High-throughput mapping of cell-wall polymers within and between plants using novel microarrays. *Plant J.* **2007**, *50*, 1118–1128.
17. Moller, I.E.; Pettolino, F.A.; Hart, C.; Lampugnani, E.R.; Willats, W.G.T.; Bacic, A. Glycan profiling of plant cell wall polymers using microarrays. *JoVE* **2012**, *70*, e4238.
18. Jin, G. Analytical approaches to assessing poor wood quality in *Pinus radiata*. M.Sc. Thesis, The University of Auckland, Auckland, New Zealand, 2011.
19. Wise, L.E. Quantitative isolation of hemicelluloses from coniferous woods. Preliminary communication. *Ind. Eng. Chem. Anal. Ed.* **1945**, *17*, 63–64.
20. Lindberg, B. Recent advances in methods of isolating and purifying hemicelluloses. *Pure Appl. Chem.* **1962**, *5*, 67–76.
21. Wise, L.E.; Murphy, M.; D'Addieco, A.A. Chlorite holocellulose, its fractionation and bearing on summative wood analysis and on studies on hemicelluloses. *Paper Trade J.* **1946**, *122*, 35–43.
22. Bouveng, H.O.; Meier, H. Studies on a galactan from Norwegian spruce compression wood (*Picea abies* Karst.). *Acta Chem. Scand.* **1959**, *13*, 1884–1889.
23. Ahlgren, P.A.; Goring, D.A.I. Removal of wood components during chlorite delignification of black spruce. *Can. J. Chem.* **1971**, *49*, 1272–1275.

24. Watanabe, T.; Ohnishi, J.; Yamasaki, Y.; Kaizu, S.; Koshijima, T. Binding-site analysis of the ether linkages between lignin and hemicelluloses in lignin-carbohydrate complexes by DDQ-oxidation. *Agric. Biol. Chem.* **1989**, *53*, 2233–2252.
25. Ralph, J.; Brunow, G.; Boerjan, W. Lignins. In *Encyclopedia of Life Sciences*; John Wiley & Sons: Hoboken, NJ, USA, 2007; pp. 1–10.
26. Altaner, C.M.; Tokareva, E.N.; Jarvis, M.C.; Harris, P.J. Distribution of (1→4)-β-galactans, arabinogalactan proteins, xylans and (1→3)-β-glucans in tracheid cell walls of softwoods. *Tree Physiol.* **2010**, *30*, 782–793.
27. Donaldson, L.A.; Knox, J.P. Localization of cell wall polysaccharides in normal and compression wood of radiata pine: Relationships with lignification and microfibril orientation. *Plant Physiol.* **2012**, *158*, 642–653.
28. Harris, P.J. Diversity in plant cell walls. In *Plant Diversity and Evolution: Genotypic and Phenotypic Variation in Higher Plants*; Henry, R.J., Ed.; CAB International Publishing: Wallingford, Oxon, UK, 2005; pp. 201–227.
29. Kim, J.S.; Awano, T.; Yoshinaga, A.; Takabe, K. Immunolocalization of β-1-4-galactan and its relationship with lignin distribution in developing compression wood of *Cryptomeria japonica*. *Planta* **2010**, *232*, 109–119.
30. Jiang, K.S.; Timell, T.E. Polysaccharides in compression wood of tamerack (*Larix laricina*) 4. Constitution of an acidic galactan. *Sven. Papp.* **1972**, *75*, 522–594.
31. Willför, S.; Holmbom, B. Isolation and characterisation of water soluble polysaccharides from Norway spruce and Scots pine. *Wood Sci. Technol.* **2004**, *38*, 173–179.
32. Willför, S.; Sundberg, A.; Hemming, J.; Holmbom, B. Polysaccharides in some industrially important softwood species. *Wood Sci. Technol.* **2005**, *39*, 245–258.
33. Goellner, E.M.; Utermoehlen, J.; Kramer, R.; Classen, B. Structure of arabinogalactan from *Larix laricina* and its reactivity with antibodies directed against type-II-arabinogalactans. *Carbohydr. Polym.* **2011**, *86*, 1739–1744.
34. Saeman, J.F.; Moore, W.E.; Millett, M.A. Sugar units present. Hydrolysis and quantitative paper chromatography. In *Methods in Carbohydrate Chemistry*; Whistler, R.L., Ed.; Academic Press: New York, NY, USA, 1963; Volume 3, pp. 54–69.
35. Albersheim, P.; Nevins, D.G.; English, P.D.; Karr, A. A method for analysis of sugars in plant cell-wall polysaccharides by gas-liquid chromatography. *Carbohydr. Res.* **1967**, *5*, 340–345.
36. Mankarios, A.T.; Jones, C.F.G.; Jarvis, M.C.; Thelfall, D.R.; Friend, J. Hydrolysis of plant polysaccharides and GLC analysis of their constituent neutral sugars. *Phytochemistry* **1979**, *18*, 419–422.
37. Blakeney, A.B.; Harris, P.J.; Henry, R.J.; Stone, B.A. A simple and rapid preparation of alditol acetates for monosaccharide analysis. *Carbohydr. Res.* **1983**, *113*, 291–299.
38. Smith, B.G.; Harris, P.J. Polysaccharide composition of unlignified cell walls of pineapple [*Ananas comosus* (L.) Merr.] fruit. *Plant Physiol.* **1995**, *107*, 1399–1409.
39. Santoro, N.; Cantu, S.L.; Tornqvist, C.-E.; Falbel, T.G.; Bolivar, J.L.; Patterson, S.E.; Pauly, M.; Walton, J.D. A high-throughput platform for screening milligram quantities of plant biomass for lignocellulose digestibility. *Bioenergy Res.* **2010**, *3*, 93–102.

40. Apiolaza, L.A.; Butterfield, B.; Chauhan, S.S.; Walker, J.C.F. Characterization of mechanically perturbed young stems: Can it be used for wood quality screening? *Ann. For. Sci.* **2011**, *68*, 407–414.
41. Apiolaza, L.A.; Chauhan, S.S.; Walker, J.C.F. Genetic control of very early compression and opposite wood in *Pinus radiata* and its implications for selection. *Tree Genet. Genomes* **2011**, *7*, 563–571.
42. McLean, J.P.; Jin, G.; Brennan, M.; Nieuwoudt, M.K.; Harris, P.J. Using NIR and ATR-FTIR spectroscopy to rapidly detect compression wood in *Pinus radiata*. *Can. J. For. Res.* **2014**, *44*, 820–830.

14

Redox Modulation Matters: Emerging Functions for Glutaredoxins in Plant Development and Stress Responses

Shutian Li

Department of Botany, Osnabrück University, 49076 Osnabrück, Germany;
E-Mail: shutian.li@biologie.uni-osnabrueck.de

Abstract: Glutaredoxins (GRXs) are small ubiquitous glutathione (GSH)-dependent oxidoreductases that catalyze the reversible reduction of protein disulfide bridges or protein-GSH mixed disulfide bonds via a dithiol or monothiol mechanism, respectively. Three major classes of GRXs, with the CPYC-type, the CGFS-type or the CC-type active site, have been identified in many plant species. In spite of the well-characterized roles for GRXs in *Escherichia coli*, yeast and humans, the biological functions of plant GRXs have been largely enigmatic. The CPYC-type and CGFS-type GRXs exist in all organisms, from prokaryotes to eukaryotes, whereas the CC-type class has thus far been solely identified in land plants. Only the number of the CC-type GRXs has enlarged dramatically during the evolution of land plants, suggesting their participation in the formation of more complex plants adapted to life on land. A growing body of evidence indicates that plant GRXs are involved in numerous cellular pathways. In this review, emphasis is placed on the recently emerging functions for GRXs in floral organ development and disease resistance. Notably, CC-type GRXs have been recruited to participate in these two seemingly unrelated processes. Besides, the current knowledge of plant GRXs in the assembly and delivery of iron-sulfur clusters, oxidative stress responses and arsenic resistance is also presented. As GRXs require GSH as an electron donor to reduce their target proteins, GSH-related developmental processes, including the control of flowering time and the development of postembryonic roots and shoots, are further discussed. Profiling the thiol redox proteome using high-throughput proteomic approaches and measuring cellular redox changes with fluorescent redox biosensors will help to further unravel the redox-regulated physiological processes in plants.

Keywords: glutaredoxins (GRXs); glutathione (GSH); thioredoxins (TRXs); redox regulation; development; biotic and abiotic stresses; iron-sulfur clusters

1. Introduction

Glutaredoxins (GRXs) are small ubiquitous glutathione (GSH)-dependent oxidoreductases that are widely known to play a crucial role in the response to oxidative stress in *E. coli*, yeast and humans [1,2]. Together with thioredoxins (TRXs), protein disulfide isomerases (PDIs), glutathione-S-transferases (GSTs) and glutathione peroxidases, GRXs are classified into the TRX superfamily sharing a conserved β1-α1-β2-α2-β3-β4-α3 TRX fold [3]. Both TRXs and GRXs belong to a multigenic family of proteins, which are represented by numerous isoforms localized in multiple subcellular compartments [4]. GRXs are able to reversibly reduce disulfide bridges of substrate proteins through two different mechanisms (Figure 1) and thereby posttranslationally influence protein activities, as well as DNA binding and/or transcriptional activities in the case of transcription factors [1,2,5]. In the dithiol mechanism (Figure 1A), GRXs use GSH as electron donors to reduce target proteins through a dithiol-disulfide exchange reaction in a manner similar to that of TRXs. Alternatively, GRXs can also efficiently and specifically catalyze the reversible reduction of protein GSH-mixed disulfides via the monothiol mechanism (Figure 1B), a process known as deglutathionylation [1,6]. In accordance with the predicted amino acid sequences and the composition of the active site motif, GRXs are divided into three major classes, including the CPYC-type, the CGFS-type and the CC-type [6]. Not all of the dithiol GRXs of the CPYC-type contain the exact CPYC active site, as P is replaced by G and S in GRXC1 and GRXC5, respectively [6]. Moreover, the monothiol GRXS12 with the CSYS active motif is phylogenetically more related to GRXC5 than to the CGFS-type GRXs and consequently falls into the CPYC-type [6]. The first structural insight into poplar GRXC1 with a CGYC active site enables us to perform homology modeling for a given GRX using it as a template (Figure 2) [7,8]. Whereas the dithiol CPYC-type and the monothiol CGFS-type are common to all pro- and eukaryotes, the CC-type class is only specific for land plants [4,6,9,10]. Despite the well-recognized roles of plant TRXs in chloroplastic and mitochondrial processes, seed development and germination, as well as self-incompatibility [2], little has been reported to date with respect to the physiological roles of plant GRXs.

The knowledge of plant GRXs has been significantly advanced by the genomic and EST sequence datasets of several plant model species, which uncovers an unexpected number of genes encoding GRXs and, hence, allows for the comparative and evolutionary analysis of GRX classes in aerobic photosynthetic organisms representing different evolutionary stages of plants (Table 1) [4,6,10]. The green alga, *Chlamydomonas reinhardtii*, and the cyanobacterium, *Synechocystis* sp. PCC6803, are unicellular, and both possess the CPYC-type and CGFS-type GRXs, but no CC-type GRXs exist in these primitive photosynthetic organisms (Table 1). The bryophyte *Physcomitrella patens* represents the basal nonvascular land plant and contains only two CC-type GRXs (Table 1). The number of CC-type GRXs expands from the bryophyte to the gymnosperm, *Pinus taeda*, and then to the angiosperms, including the monocot, *Oryza sativa* (rice), as well as the dicots, *Populus trichocarpa* (poplar) and *Arabidopsis thaliana* (Table 1). By contrast, the size of the other two classes remains relatively constant during the land plant evolution, varying from four to six for the CPYC-type and four to eight for the CGFS-type (Table 1). Compositional analysis of three GRX classes in the evolutionarily representative plant species indicates that the exclusive expansion of the land plant-specific CC-type GRXs contributes to plant terrestrial adaptations concomitant with the development of sophisticated and differentiated architectures.

Figure 1. GRXs reduce disulfide bridges via a dithiol (**A**) or monothiol (**B**) mechanism. (**A**) Dithiol mechanism: electrons are transferred from NADPH to glutathione reductase (GR), then to glutathione (GSH) and further to glutaredoxin (GRX), eventually leading to the GRX-mediated reduction of disulfides in target proteins. (**B**) Monothiol mechanism: reduction of protein-GSH mixed disulfides is termed deglutathionylation (black arrows). The reverse reaction is called glutathionylation (gray arrows), resulting in the production of glutathionylated proteins. Adapted from [10].

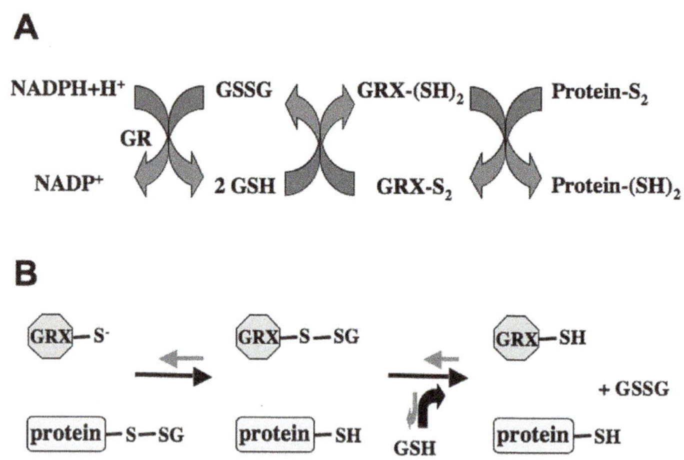

Figure 2. Structural characterization of plant GRXs. (**A**) Ribbon representation of poplar GRXC1. Four central β-sheets are surrounded by five α-helices. The position of the conserved glycine in the GSH binding site is indicated in red, and the CGYC active site is blue. (**B**) Ribbon representation of *Arabidopsis* ROXY1 obtained by homology modeling using poplar GRXC1 as a template. The conserved glycine (G110, red) and the CCMC active site (blue) are respectively located on the same β-sheet and α-helix as in poplar GRXC1. Adapted from [7].

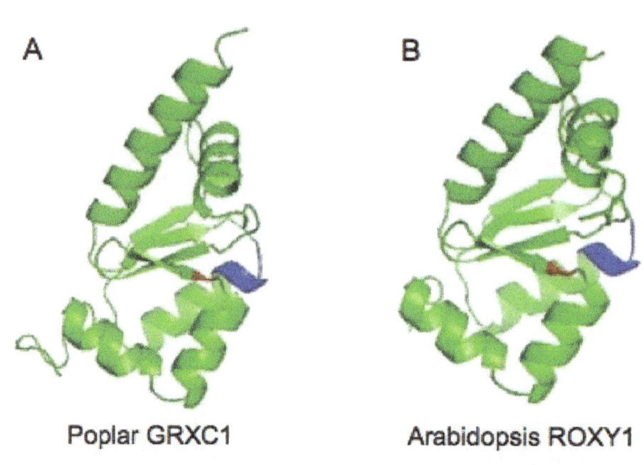

Table 1. The respective number of three GRX types in the evolutionarily representative plant species.

Species	CGFS-Type	CPYC-Type	CC-Type
Chlamydomonas reinhardtii (green alga)	4	2	0
Synechocystis sp. *PCC6803* (cyanobacteria)	1	2	0
Physcomitrella patens (moss)	6	4	2
Pinus taeda (loblolly pine)	8	4	5
Oryza sativa (rice)	5	5	17
Populus trichocarpa (black cottonwood)	6	5	22
Arabidopsis thaliana (*Arabidopsis*)	4	6	21

CC-type GRXs from *Arabidopsis*, rice and *Zea mays* (maize) have been implicated in floral organ development, defense responses to pathogens and protection against oxidative stress [11–16]. This land plant-specific GRX class might have been recruited to participate in these different pathways during the evolution of land plants [12,13,16]. Emerging evidence supports additional roles for plant GRXs in the assembly and transfer of iron-sulfur clusters, oxidative stress responses, as well as arsenic resistance [17–20]. In addition, GRXs may directly participate in GSH-associated developmental processes, as they require GSH to reduce substrate proteins. Partially inactivating the first enzyme of GSH biosynthesis coupled with knocking out NADPH-dependent TRX reductase genes (*NTRA* and *NTRB*) interferes with auxin homeostasis and leads to the growth inhibition of postembryonic roots and shoots [21–26], demonstrating that one of the GRX and TRX systems is required for the postembryonic activities of apical meristems and that the genetic cross-talk between these two redundant systems provides a safety backup to ensure normal postembryonic development. Here, we summarize established roles for plant GRXs in known biological processes with an emphasis on recent advances in developmental processes and defense responses. Furthermore, high-throughput proteomic approaches and fluorescent redox biosensors are also discussed regarding their respective application in profiling the thiol redox proteome and measuring cellular redox changes.

2. GRXs with a CGYC or CGFS Active Site Act in the Assembly of Iron-Sulfur Clusters

Iron-sulfur (Fe-S) clusters are ubiquitous in all living organisms and exist in a variety of Fe-S proteins involved in various essential biological processes, such as photosynthesis, metabolism and respiration [27–30]. In plants, the biogenesis of Fe-S clusters mainly occurs in mitochondria and chloroplasts, utilizing the components of the Fe-S cluster assembly (ISC) machinery and the sulfur mobilization (SUF) machinery as potential scaffold proteins, respectively [31–32]. These two plant organelles contain many Fe-S proteins, such as [2Fe-2S] ferredoxin and [4Fe-4S] ferredoxin-thioredoxin reductase, and are known to participate in the biosynthesis of organellar and cytosolic Fe-S proteins [33]. However, less is currently known about the detailed mechanism controlling the assembly and transfer of Fe-S clusters in plants.

Biochemical and spectroscopic studies have indicated specific roles for plant GRXs with a CGFS or CGYC active site motif in the biosynthesis and delivery of Fe-S clusters [8,17,19,34]. The subunit-bridging [2Fe-2S] cluster present in holo GRXs is ligated by the catalytic cysteines of two GRXs, as well as the cysteines of two external GSH molecules [8,19,35]. GRXC1 and its orthologs are cytosolic CPYC-type GRXs with a CGYC active site and occur exclusively in dicotyledonous plants [8,19,34]. GRXC1 exists predominantly as a holodimeric form, suggestive of the presence of a Fe-S cluster *in vivo* [34]. The reductive activity of *Arabidopsis* GRXC1 is modulated by the redox-dependent stability of the subunit-bridging [2Fe-2S] cluster, suggesting that GRXC1 acts as a redox sensor to reduce downstream signaling steps under oxidative conditions [34]. Cytosolic GRXC2 is the closest paralog of GRXC1, possesses a CPYC active site, occurs in all seed plants and is unable to accommodate a Fe-S cluster [34]. Mutagenesis analysis of poplar GRXC1 indicates that the incorporation of a [2Fe-2S] cluster is likely characteristic of plant GRXs possessing a glycine adjacent to the catalytic cysteine [19]. In support of this notion, the monothiol CGFS-type GRXs of *Arabidopsis* and poplar have the potential to function as scaffold proteins for the assembly and delivery of [2Fe-2S] clusters or as the storage and delivery proteins of [2Fe-2S] clusters to mediate the cluster transfer from the ISC or SUF scaffold proteins to the physiologically relevant acceptor proteins [17,35]. Yeast GRX5 is a mitochondrial CGFS-type GRX, and the *grx5* null mutation leads to the enrichment of free iron, oxidative damage, compromised respiratory growth, as well as the impaired assembly and transfer of Fe-S clusters [33,36]. Among the four CGFS-type GRXs encoded by plant genomes, cytosolic GRXS17 (PICOT) and chloroplastic GRXS14 and GRXS16 can fully rescue the defects of the yeast *grx5* mutant, whereas mitochondrial GRXS15 cannot [17]. This complementation experiment demonstrates the capacity of plant monothiol GRXs for the proper biogenesis of Fe-S proteins. Although the dicotyledon-specific GRXC1 is capable of incorporating a [2Fe-2S] cluster to constitute a holoprotein, it fails to further deliver the [2Fe-2S] cluster to an acceptor protein and, thus, lacks the ability to complement the yeast *grx5* mutant [17,34]. The capacity to transfer a [2Fe-2S] cluster to an acceptor protein seems to be specific and exclusive for monothiol GRXs. Comparative studies of GRXS14 and GRXC1 show that this is likely due to the increased accessibility and lability for Fe-S cofactors in monothiol, but not dithiol GRXs [17]. To further address the specificity of plant monothiol GRXs to the biogenesis of Fe-S clusters, it will be of great value to explore their ability to assemble a variety of [2Fe-2S]-containing holoproteins, as well as to evaluate the possibility of their involvement in the maturation of [3Fe-4S] and [4Fe-4S] clusters.

Clearly, the CPYC-type GRXs with the CGYC active site can encompass a stable [2Fe-2S] cluster of unknown function in a homodimeric fashion, and members of the CGFS-type GRXs can also occur as a haloform that binds a labile Fe-S cofactor [17,34]. However, whether the natural CC-type GRXs possess the capacity to assemble a Fe-S cluster still remains an open question. The lack of biochemical properties for this land plant-specific class is due to the inability to produce soluble recombinant proteins [37]. Engineering CC-type GRXs with two soluble poplar CPYC-type GRXs, including GRXC1 and GRXC4, successfully circumvents this obstacle [37]. Divergent from GRXC1, GRXC4 possesses the CPYC active site and fails to bind a [2Fe-2S] cluster [19,34]. Surprisingly, as observed for wild-type GRXC1, each of the mutagenized GRXs with the artificial CCMC/S active site enables the incorporation of a Fe-S cofactor [37], suggesting the capacity of the natural CC-type GRXs to accommodate a Fe-S cluster.

3. Land Plant-Specific CC-Type GRXs in Flower Development and Defense Responses

3.1. CC-Type GRXs Are Required for Floral Organ Development and Microspore Formation

Genetic dissection of floral homeotic mutants unveils that the combinatorial action of floral homeotic genes specifies floral organ identity in flowering plants [38]. Nevertheless, floral organ primordia are initiated prior to the activation of floral homeotic genes. Up to now, the knowledge about the genes required for the initiation of floral organ primordia has been scarce. Unexpectedly, *GRXC7/ROXY1*, which encodes a CC-type GRX, is required for petal initiation and organogenesis in *Arabidopsis* [11]. The *roxy1* mutant initiates a reduced number of petal primordia and forms 2.5 petals on average instead of four petals in wild-type flowers. During further petal morphogenesis, developmental defects, such as the lack of blade expansion and abnormal blade bending, are also observed in the *roxy1* mutant [11]. Additional mutants of CC-type GRXs from different plant species have been functionally characterized, shedding light on a novel unexpected role for these GRXs in anther development and microspore formation [7,15,16]. Although the *Arabidopsis roxy2* mutant does not exhibit any floral abnormalities, cytological characterization of the *roxy1 roxy2* double mutant flowers reveals a disruption of anther lobe differentiation and microspore formation, indicative of a redundant role of these two closely related CC-type GRXs in the control of *Arabidopsis* male gametogenesis [7]. Similar to the case in dicots, CC-type GRXs are also required for microspore formation in monocots, including rice and maize [15,16]. *MICROSPORELESS1* (*MIL1*) codes for a rice CC-type GRX that is localized on meiocyte centromeres [16]. The *mil1* mutant disables the meiotic entry of sporogenous cell progenies and is therefore male sterile [16]. Apparently, rice MIL1 participates in an anther-specific meiosis initiation pathway to commence meiotic division. Furthermore, *GRXC7/ROXY1* homologs from rice are isolated and characterized [14]. Although rice flowers have a characteristic morphology distinct from that of *Arabidopsis* flowers, *OsROXY1* and *OsROXY2* exhibit highly dynamic floral expression patterns and are transiently expressed in young floral organs, similar to those of their *Arabidopsis* counterparts [14]. Microscopic reconstruction of maize anther development in the *male sterile converted anther1* (*msca1*) mutant offers evidence underlining a crucial role for a CC-type GRX to function as a determinant of male germline fate in maize [15]. In *msca1* anthers, centrally positioned presumptive archesporial cells fail to proceed, but instead divide longitudinally to differentiate into vascular bundles eventually [15]. However, reductive, but not oxidative, treatment rescues archesporial cell specification and triggers meiotic fate acquisition in *msca1* anthers, demonstrating that a hypoxic and reducing environment suffices to modify unknown targets of the maize CC-type GRX MSCA1 and, thus, causes archesporial cell formation.

Floral CC-type GRXs have been recently revealed as interacting partners of TGA transcription factors [13,16]. The *Arabidopsis* genome encodes 10 TGA proteins, including TGA1–7, PERIANTHIA (PAN), TGA9 and TGA10 [39]. TGA2, TGA5 and TGA6 operate redundantly to regulate systemic acquired resistance (SAR), whereas TGA1, TGA3 and TGA4 participate in basal resistance [40–43]. By contrast, PAN is involved in the determination of flower organ number in *Arabidopsis*. The *pan* mutant converts tetramerous flowers to pentamerous ones, a feature of ancestral flowering plants [44,45]. All TGA proteins, including PAN, are identified by yeast two-hybrid assays and bimolecular fluorescence complementation (BiFC) experiments as GRXC7/ROXY1-interacting partners [13,46]. Deletion analysis of GRXC7/ROXY1 unravels the importance of the GRXC7/ROXY1 C-terminal L**LL motif for the interactions with TGA proteins and for the GRXC7/ROXY1 function in petal initiation and morphogenesis [13,47,48]. In agreement with their nuclear interactions, intracellular localization studies of GRXC7/ROXY1 combined with complementation experiments of the *roxy1* mutant reveal that a novel nuclear activity of GRXC7/ROXY1 is required for petal development [13]. Mutagenesis studies of all of the PAN cysteines followed by complementation experiments unveil the indispensability of only the cysteine 340 (Cys-340) for the function of PAN in flower development [13]. Notably, Cys-340 is conserved in several other TGA proteins, such as TGA1 and TGA4 [48]. However, direct evidence in support of this conserved cysteine acting as a redox switch has been lacking to date. Genetic analysis of the *roxy1* and *pan* single mutants shows opposing effects of GRXC7/ROXY1 and PAN on petal primordia initiation [13]. In the *roxy1 pan* double mutant, pentamerous flowers develop, indicating that PAN is epistatic to GRXC7/ROXY1 in the regulation of petal primordia initiation [13]. The chimeric repressor gene-silencing technology (CRES-T) successfully uncouples a dual role of TGA factors, namely the negative effect of PAN on petal initiation and the positive effect of additional TGA factors on later petal morphogenesis [13]. Given the opposite effects of GRXC7/ROXY1 and PAN on petal primordial initiation, GRXC7/ROXY1 seems to inhibit PAN activity. Conversely, GRXC7/ROXY1 likely exerts a positive effect on other TGA factors during later petal development, enabling proper petal morphogenesis [13].

Arabidopsis plants mutant for both TGA9 and TGA10 are male sterile and display defects in male gametogenesis [46], which is similar to what is observed in the *roxy1 roxy2* double mutant [7]. The male-sterile phenotypes seen in the *tga9 tga10* double mutant demonstrate a redundant role for *TGA9* and *TGA10* in anther development and microspore formation [46]. More specifically, the development of the adaxial and abaxial anther lobes is differentially affected in both the *tga9 tga10* and *roxy1 roxy2* double mutants, with early steps blocked in adaxial lobes and later steps impaired in abaxial lobes [7,46]. In conjunction with the nuclear interactions of TGA9 and TGA10 with GRXC7/ROXY1 and GRXC8/ROXY2 [46], these genetic data indicate that TGA9 and TGA10 act in concert with GRXC7/ROXY1 and GRXC8/ROXY2 to promote anther development. Like *Arabidopsis* CC-type GRXs, the centromere-localized maize MIL1 also interacts with rice TGA proteins [16], implying that MIL1 might also influence the male meiosis initiation through associating with TGA proteins.

Collectively, reduced petal number in the *roxy1* mutant and the male-sterile phenotype of the aforementioned CC-type GRX mutants suggest that redox regulation plays a significant role in floral organ development and in the production of male germline cells. Nuclear interactions between floral CC-type GRXs and floral TGA proteins further advance our knowledge as to how some TGA proteins contribute to flower development. Flowering plants seem to have evolved an anther-specific mechanism, whereby CC-type GRXs dictate meiotic entry and male gametogenesis via modulating redox homeostasis.

3.2. CC-Type GRXs Participate in Disease Resistance

Plant pathogens generally fall into two categories: biotrophs and necrotrophs. To fend off these attackers that differ in lifestyles and infection strategies, plants have evolved complex defense mechanisms. The R protein-mediated hypersensitive response (HR) and salicylic acid (SA)-mediated basal resistance play a crucial role in defense against biotrophs. whereas jasmonic acid (JA) activates a set of *pathogenesis-related* (*PR*) genes distinct from those induced by SA and exerts basal resistance against necrotrophs [49,50]. Nevertheless, the SA- or JA-dependent signaling pathways are not always activated exclusively in response to biotrophs or necrotrophs [51]. When pathogen infection simultaneously induces the synthesis of both SA and JA, plants are able to take advantage of the cross-talk between SA- and JA-mediated signaling pathways to optimize the defense against a single attacker [51]. Furthermore, the cross-communication between SA- and JA-dependent defense pathways also occurs in the defense against multiple pathogens with distinct lifestyles [52]. Several key regulators, such as NON-EXPRESSOR OF PR GENES1 (NPR1), MAP kinase 4 (MPK4), WRKY transcription factors and TGA proteins, are demonstrated to be involved in the cross-talk between SA and JA signaling pathways [53,54].

In recent years, the *Arabidopsis* CC-type GRXs, GRXC9/ROXY19 and *GRXS13/ROXY18*, have been shown to act as novel players implicated in the cross-talk between SA and JA signaling pathways [12,55]. Transcriptional activation of *GRXC9/ROXY19* expression by SA requires TGA proteins, but is partially dependent on NPR1 [12,56]. Moreover, constitutive expression of *GRXC9/ROXY19* impairs JA/ethylene (ET)-induced *PDF1.2* transcription, but fails to affect the expression of *LOX2* and *VSP*, two marker genes of the JA response pathway, implying that this CC-type GRX is involved in the SA-induced suppression of the JA/ET signaling pathways [12]. It is well known that the APETALA2/ETHYLENE RESPONSE FACTOR transcription factor ORA59 acts as the global regulator of JA/ET-induced defense responses [57]. *ORA59* transcription is stimulated by JA or ET, but is contrarily repressed by SA [58,59]. Activation of the *ORA59* promoter by the transcription factor ETHYLENE INSENSITIVE3 (EIN3) is suppressed by coexpressing *GRXC9/ROXY19*. Besides, ET-induced *ORA59* expression is also compromised in transgenic plants ectopically expressing *GRXC9/ROXY19* [60]. Together, these observations highlight the SA-JA/ET antagonism mediated by SA-induced expression of *GRXC9/ROXY19*.

GRXS13/ROXY18 encodes a second disease resistance-related CC-type GRX, which is phylogenetically closely related to *GRXC9/ROXY19* and participates in defense signaling via functioning as a disease susceptibility gene [55,56]. *GRXS13/ROXY18* transcription is stimulated by the virulent strain of the plant necrotrophic pathogen, *Botrytis cinerea.* Moreover, expression of *GRXS13/ROXY18* is SA-inducible, but is conversely suppressed by JA [55]. After *B. cinerea* infection, mRNA levels of *GRXS13/ROXY18* are significantly increased in JA-related mutants, but are strongly reduced in SA-related mutants [55], demonstrating that SA and JA signaling pathways regulate *GRXS13/ROXY18* expression in opposing ways. Whereas the loss-of-function mutation in *GRXS13/ROXY18* causes enhanced resistance to *B. cinerea*, overexpression of this GRX gene in wild-type plants results in augmented susceptibility to this fungus, lending support to the notion that *GRXS13/ROXY18* is required for the successful colonization of *Arabidopsis* plants by *B. cinerea*.

Interestingly, ectopic expression of *GRXC7/ROXY1*, as well as its rice homologs, *OsROXY1* and *OsROXY2*, in *Arabidopsis* leads to increased accumulation of hydrogen peroxide (H_2O_2) and renders transgenic plants highly susceptible to infection by *B. cinerea* [14]. In addition, OsWRKY13 operates not only as an activator of the SA-dependent pathway, but also as a suppressor of the JA-dependent pathway [61]. Overexpression of *OsWRKY13* induces the expression of two rice CC-type GRX genes and enhances rice resistance to *Xanthomonas oryzae* pv. *oryzae* (*Xoo*) and *Magnaporthe grisea*, hinting at the possible involvement of these two CC-type GRXs in the SA-JA antagonism [61,62].

SA-induced GRXC9/ROXY19 interacts with various TGA proteins and negatively regulates JA-responsive *PDF1.2* transcription [12]. The suppressive effect of GRXC9/ROXY19 on *PDF1.2* transcription is abolished in the *tga2 tga5 tga6* triple mutants, indicating that the interaction between GRXC9/ROXY19 and TGA proteins is indispensable for GRXC9/ROXY19-dependent cross-talk between SA and JA signaling pathways [12]. Furthermore, ectopically expressed *GRXC9/ROXY19* suppresses ET-induced expression of *ORA59*, and this repression depends on the C-terminal L ** LL motif of GRXC9/ROXY19 [60]. This short conserved motif was previously found to be essential for the interaction of floral GRXC7/ROXY1 with TGA proteins [47]. Interestingly, 10 of the 21 CC-type GRXs in *Arabidopsis* interact with TGA proteins in yeast two-hybrid assays and repress the *ORA59* promoter activity [60].

Functional analysis of *GRX13/ROXY18*, which encodes the closest paralog of GRXC9/ROXY19, indicates a role for this CC-type GRX gene in facilitating pathogen infection [55]. In wild-type plants, expression of *GRXS13/ROXY18* is undetectable, but is strongly inducible by SA. Nonetheless, the *tga2 tga5 tga6* triple mutation results in slightly elevated expression of *GRXS13/ROXY18* and a severe reduction of SA-induced transcription of *GRXS13/ROXY18* [55,56], suggesting that SA-induced expression of *GRXS13/ROXY18* requires the subclass II of TGA transcription factors (TGA2, TGA5 and TGA6). These TGA proteins are identified as redundant modulators of defense responses in *Arabidopsis* plants challenged with *B. cinerea* [63]. Yeast two-hybrid assays further reveals the capacity of GRXS13/ROXY18 to associate with TGA proteins [55], implying that GRXS13/ROXY18 act together with these TGA factors to trigger a cascade of events leading to disease susceptibility.

Taken together, land plant-specific CC-type GRXs participate in defense responses by cooperating with stress-related TGA proteins. The specific recruitment of CC-type GRXs in plant defense signaling suggests that these small proteins act to maintain cellular redox homeostasis after pathogen infection, which is essential for plant adaptations to life on land under adverse environmental conditions during the evolution of land plants.

3.3. Functional Redundancy of a Subset of CC-Type GRXs with a Conserved C-Terminal Motif

Yeast two-hybrid assays and BiFC experiments reveal that both stress-related and floral TGA proteins interact with CC-type GRXs [12,13,16,46,55]. Complementation experiments of the *roxy1* mutant unveil the functional conservation of complementing CC-type GRXs and the functional importance of their C-terminal A [L/I]WL motif in flower organ development [13]. One of these rescuing CC-type GRXs is encoded by *GRXC11/ROXY4* (At3g62950), which is upregulated by DELLA proteins and participates in the gibberellin (GA) signaling pathway and floral organ development [64]. Unexpectedly, GRXC9/ROXY19 is a disease resistance-related CC-type GRX, possesses this C-terminal short motif and is able to replace GRXC7/ROXY1 if properly expressed [12,13]. Besides, OsROXY1 and OsROXY2 share the C-terminal ALWL motif and can fully rescue the *roxy1* mutant [14]. The exchangeability between CC-type GRXs is also corroborated by a study focused on the inhibitory effects of both GRXC9/ROXY19 and its homologs on *ORA59* transcription [60]. These suppressive CC-type GRXs have a common C-terminal ALWL motif and are capable of interacting with TGA proteins [60]. Thus, nucleotide changes in the regulatory, but not coding, regions of these CC-type GRX genes contribute to their sub- and neo-functionalization [65]. In sum, these observations demonstrate a conserved function for a subgroup of

CC-type GRXs in flower development and disease resistance, *i.e.*, effectively recognizing TGA proteins involved in either of the two signaling pathways. Since CC-type GRXs are required for microspore production and defense response, they can potentially be utilized to engineer disease-resistant crops and to develop male sterile lines for hybrid seed production.

4. GRXs Participate in Abiotic Stress Responses

In a continuously changing environment, plants are constantly challenged by a plethora of abiotic stresses, such as heavy metals, extreme temperatures, high light and toxic substances. A common characteristic of plant responses to such externally environmental stimuli is an oxidative burst of reactive oxygen species (ROS), which alters cellular redox homeostasis and produces oxidative stress. To overcome such oxidative stress, plants have orchestrated a sophisticated antioxidant mechanism to counteract the deleterious effects of ROS and avoid cellular oxidation [66]. Redox signaling involves posttranslational modifications of protein thiols that can be oxidized into different reversible states, such as *S*-glutathionylation, *S*-nitrosylation and intra-/intermolecular disulfide bridges (Figure 3). These oxidized forms of cysteine residues can be effectively reduced by GRXs, indicating a potential key role for GRXs in oxidative stress signaling (Figure 3) [66,67].

Figure 3. Oxidation and reduction of protein thiols. Under oxidative conditions, free and accessible protein thiols undergo several different posttranslational modifications, which can be either reversible or not. Protein cysteines can be oxidized by ROS into sulfenic acid (SOH), which can be reduced by XRX (GRX or TRX). Further oxidation of sulfenic acid by ROS can result in the formation of sulfinic acid (SO_2H), which can be reversed by sulfiredoxins (SRX) or irreversibly oxidized to sulfonic acid (SO_3H). The presence of oxidants and/or GSH allows the glutathionylation of protein cysteines to occur through different mechanisms. Deglutathionylation can be catalyzed by GRX. Reversible formation of intra-/inter-molecular disulfide bridges is mediated by XRX. Direct H_2O_2-dependent oxidation of cysteines to intra-/intermolecular disulfides and peroxidase (POX)-catalyzed H_2O_2 sensing can be reversed by XRX. Besides, protein cysteines also undergo nitrosylation in the presence of reactive nitrogen species (RNS), a reversible process that could be catalyzed by TRX. Adapted from [48].

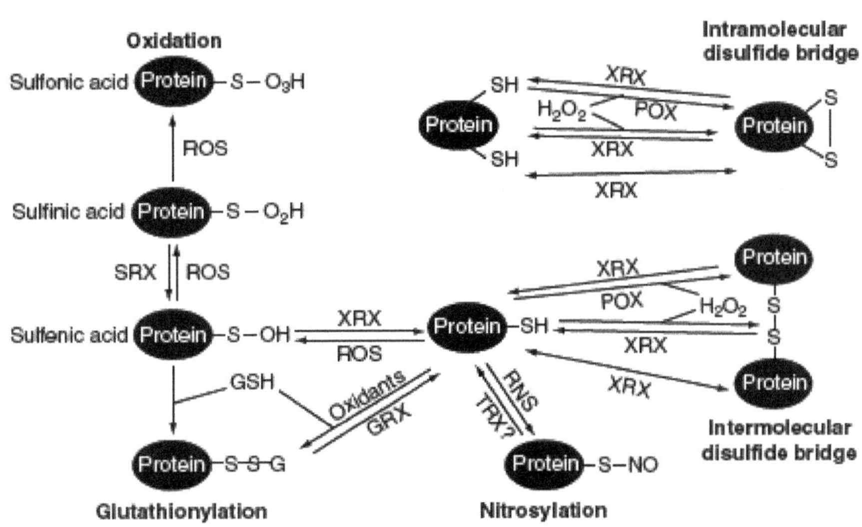

Plants adapt themselves to oxidative stress by coordinately regulating a battery of antioxidant genes. In *Arabidopsis*, at least 152 genes are possibly involved in managing the level of ROS, and 27 of them encode GRXs [68]. Reverse genetic analysis combined with transgenic approaches has identified some *Arabidopsis* GRXs as critical players in protection against oxidative stress [18,34,69–71]. Except for GRXS16, all of the other three monothiol GRXs, namely GRXS14 (also known as GRXcp), GRXS15 and GRXS17 (PICOT), seem to exert such antioxidant functions [18,69,70]. Analysis of the *grxs14* mutant demonstrates developmental defects in early seedling growth under oxidative stress [18]. Besides, augmented protein carbonylation is detected within *grxs14* chloroplasts [18]. Despite a ubiquitous expression pattern, transcription levels of *GRXS15* are affected by various abiotic stresses. Genetic experiments reveal that *grxs15* are sensitive to oxidants, implying important functions for GRXS15 in plant growth and development under extreme conditions [69]. The *grxs17* mutant is hypersensitive to high temperature, and the expression of *GRXS17* is induced by elevated temperature [70]. Unexpectedly, auxin sensitivity and polar auxin transport are perturbed in this mutant. Under high temperature, *grxs17* plants accumulate high levels of ROS and exhibit phenotypic characteristics reflecting defects in the cell cycle control [70]. Ectopic expression of *GRXS17* in tomato minimizes chlorophyll photooxidation and lessens the oxidative damage of the cell membrane system under heat stress [72]. These observations firmly establish a mechanistic link between redox homeostasis, auxin signaling and temperature-dependent postembryonic development. Knocking down the CGFS-type GRX gene *SlGRX1* using virus-induced gene silencing increases the sensitivity of tomato plants to the oxidative and salt stresses and reduces drought tolerance, whereas overexpression of *SlGRX1* in *Arabidopsis* significantly improves plant tolerance to the oxidative, drought and salt stresses [73]. Arsenate-activated *Pteris vittata GRX5* encodes a CGFS-type GRX and represents an ortholog of *Arabidopsis* GRXS14 [20]. Compared with the arsenic-sensitive fern *P. ensiformis*, *P. vittata* is able to hyperaccumulate arsenic in fronds and displays significantly elevated tolerance to oxidative stress [74]. Expression of *PvGRX5* in *E. coli* mutants suggests that PvGRX5 depends on aquaglyceroporins to play a role in cellular arsenic resistance [20]. In most organisms, aquaglyceroporins act as arsenite channels to move arsenite into and out of cells [75]. Therefore, it is most likely that PvGRX5 regulates intracellular arsenite levels via modulating aquaglyceroporins. Ectopic expression of *PvGRX5* in *Arabidopsis* decreases arsenic accumulation, reduces oxidative damage to intracellular proteins and improves tolerance to arsenic and high temperature [76,77]. Apart from facilitating *B. cinerea* infection of *Arabidopsis* plants as described above [55], GRXS13/ROXY18 with the CCLG active site motif also protects plants from photooxidative stresses [71]. Treatment of *Arabidopsis* seedlings with Mev and high light (HL) enhances *GRXS13/ROXY18* transcription and engenders photooxidative stress responses characterized by heightened superoxide levels [71]. Knocking down *GRXS13/ROXY18* leads to the elevated production of superoxide radicals, inhibits plant growth and lowers plant tolerance to Mev and HL. In agreement, overexpression of this GRX gene enhances protection against the Mev- and HL-induced damage [71], suggesting that GRXS13/ROXY18 limits ROS accumulation and protects plants from damage caused by photooxidative stresses. Divergent from GRXC2 found in all seed plants, the dicotyledon-specific GRXC1 is able to assemble a Fe-S cluster [34]. Under various environmental stresses, the *Arabidopsis grxc1* or *grxc2* single mutant shows a wild-type phenotype, yet the *grxc1 grxc2* double mutant displays a lethal phenotype at early steps after pollination, implying a redundant and vital function shared by GRXC1 and GRXC2 [34].

In response to oxidative stress, GRXs limit ROS production, participate in redox signaling and play an antioxidant role, thereby protecting plants from the cellular oxidative damage. This conserved function might be performed by either direct ROS scavenging or redox regulation of target proteins via their disulfide oxidoreductase or deglutathionylation activities. Alternatively, GRXs might indirectly mediate this function through protein interactions or transcriptional control of gene expression. Despite the demonstrated involvement of CC-type GRXs in the control of ROS levels, it should be noted that the oxidoreductase and deglutathionylation activities of this land plant-specific class have not been characterized so far. Therefore, further biochemical experiments are needed to ascertain the enzymatic properties of CC-type GRXs.

5. GRXs Cross-Talk with TRXs in Plant Development and Stress Responses

Proteomic approaches have identified numerous common targets of GRXs and TRXs, supporting the cross-talk between these two plant antioxidant systems [78]. GRXs mainly differ from TRXs by the fact that they use GSH as an electron donor instead of TRX reductases [6]. Plants have an unusually complex complement of TRXs, which consists of six well-defined types (f, m, x, y, h, o and z) and resides in different cellular compartments, including the cytosol, the nucleus, mitochondria and chloroplasts [66,79,80]. Typically, chloroplastic TRXs are reduced in light by ferredoxin-dependent heterodimeric TRX reductases (FTR) [81]. By contrast, the reduction of plant cytosolic and mitochondrial TRXs is catalyzed by NADPH-dependent TRX reductases (NTR) [6]. In spite of the divergent reduction modes exerted by these two redox systems, accumulating evidence suggests that GRXs interplay with TRXs to constitute a complex network of redox signaling [25,26,82]. Besides, as oxidized GRXs are recycled by GSH for both the monothiol and dithiol mechanisms, it is likely that GRXs are directly involved in GSH-associated biological pathways.

In plants, GSH is enzymatically synthesized from its constituent amino acids via two ATP-dependent steps, which are catalyzed by γ-glutamylcysteine synthase (γ-GCS) and GSH synthetase (GS), respectively. In *Arabidopsis*, both γ-GCS and GS are encoded by a unique gene [21]. Complete inactivation of *Arabidopsis GSH1*, which encodes γ-GCS, causes the formation of white, embryo-lethal seeds [21,83]. Artificially changing endogenous GSH levels coupled with phenotypic analysis of *GSH1* mutants, including *root meristemless1* (*rml1*), *cadmium sensitive2* (*cad2*) and *phytoalexin-deficient2-1* (*pad2-1*), unravels that GSH participates in developmental processes and plant responses to biotic and abiotic stresses, such as heavy metals, ultraviolet (UV)-induced ROS, drought, chilling and pathogens [22–26,84–89]. The strong mutant allele *rml1* encodes a very inefficient γ-GCS and reduces the GSH synthesis by 95%, resulting in sterile plants with an extremely short mature root that comprises the same number of cells as the embryonic root [22,88]. This arrest with respect to root growth demonstrates that the *rml1* mutation disables postembryonic cell division in the root. By contrast, cell division still occurs in the apical shoot meristem of the *rml1* mutant, albeit generating a small shoot with leaves, flowers and seeds containing abnormal embryos [88]. Comparatively, the two weak mutant alleles, *cad2* and *pad2-1*, lead to a decline of 80% and 84% in GSH levels, respectively. Except that *cad2* displays a late flowering phenotype [23,24], both the *cad2* and *pad2-1* single mutants are developmentally indistinguishable from wild-type plants [21,89]. In addition, *pad2-1* is sensitive to several pathogens, whereas *cad2* is hypersensitive to cadmium [84,89].

The key reductants of cytosolic and mitochondrial TRXs are NTRs encoded by *NTRA* and *NTRB*, both of which code for cytosolic and mitochondrial isoforms in *Arabidopsis* [90]. Knocking out *NTRA* or *NTRB* fails to reveal any discernible mutant phenotype, indicative of a functional redundancy between them. Although the inactivation of both *NTRs* retards plant growth, changes seed shape, compromises pollen fitness and hyperaccumulates anthocyanins, the *ntra ntrb* double mutant is fully fertile and does not display a significantly altered sensitivity to biotic and abiotic oxidative stresses [25,82]. Moreover, cytosolic TRXh3 is only partially oxidized in the double mutant, implying an alternative mechanism for reducing TRXs [25]. Crossing *ntra ntrb* to the wild-type-like cytosolic GSH reductase mutant *gr1* generates the *ntra ntrb grl* plants with lethal pollen [82]. Associating *ntra ntrb* with *rml1* completely suppresses both the shoot and root growth [25]. Consistently, both root growth and shoot development are inhibited by the highly specific γ-ECS inhibitor, buthionine sulfoximine (BSO), in the *ntra ntrb* mutant, and the cytosolic TRXh3 is totally oxidized in the double knockout under this treatment, indicating that a GSH-dependent pathway takes charge of the alternative reduction of cytosolic TRXh3 [25]. A subgroup of the poplar h-type TRX isoforms is found to be reduced by GRXs [91]. Biochemical experiments demonstrate that GRXs, but not GSH, are able to reduce *Arabidopsis* TRXh3, suggesting that GRXs function as alternative reductants of TRXs in the *ntra ntrb* mutant [25]. Together, these results support an overlapping role for NTRs and GRXs in establishing the postembryonic activity in the apical meristems [25]. Genetic and biochemical characterization of the *ntra ntrb cad2* mutant further unravels that these two thiol reduction pathways interfere with plant developmental processes via modulating auxin signaling [26]. This triple mutant develops normally at the rosette stage, but later forms naked stems devoid of flower development. Besides, the *ntra ntrb cad2* plants show the lack of apical dominance, defects in vascular formation and reduced secondary root production [26]. All of these anomalous phenotypes are reminiscent of those transgenic and mutant plants impaired in auxin transport, signaling or biosynthesis [92–98]. Obviously, auxin levels and auxin transport capacities are significantly reduced in the *ntra ntrb cad2* mutant [26], highlighting a mechanistic link between thiol-based redox regulation and auxin signaling. In contrast with the *ntra ntrb rml1* mutant in which both the root and shoot apical meristematic activity is completely abrogated [25], the two apical meristems are not profoundly impaired in the *ntra ntrb cad2* mutant [26]. This phenotypic disparity conveys a threshold GSH concentration that is necessary for the development of postembryonic roots and shoots. Furthermore, the *ntra ntrb cad2* mutant defines a second threshold required for proper flower development, implying that flower development is more sensitive to redox perturbation than leaf development [26].

Mediator is an evolutionarily conserved transcriptional coregulator complex in all eukaryotes ranging from yeast to human. The plant Mediator is composed of 21 conserved and six plant-specific subunits [99]. Very recently, functional characterization of the *Arabidopsis* Mediator subunit 18 (MED18) has suggested a potentially synergistic role for GRXs and TRXs in plant immunity [100]. MED18 interacts with the putative repression domain of the zinc finger transcription factor YIN YANG1 (YY1), and YY1 is able to associate with the promoter regions of *GRXS13/ROXY18*, *GRXC9/ROXY19* and *TRXh5* [100]. Moreover, the *yy1* and *med18* mutants exhibit deregulated expression of *GRXS13/ROXY18*, *GRXC9/ROXY19* and *TRXh5*, which might contribute to their elevated susceptibility to necrotrophic fungal pathogens [100]. Accordingly, ectopic expression of *GRXS13/ROXY18* and *GRXC9/ROXY19* results in transgenic plants with a severe susceptibility to fungal infection [55,100]. Conversely, the *grxs13/roxy18* mutant shows increased resistance to *B. cinerea* [55]. In addition, the *trxh5* mutant is insensitive to the victorin toxin produced by the necrotrophic fungus, *Cochliobolus victoriae* [101]. These data together point to the direct transcriptional regulation of GRX and TRX genes by YY1, thereby preventing the expression of these disease susceptibility genes [100]. TRXs and GRXs are central players of the redox regulatory network, and virulence factors of necrotrophic fungi may interfere with the cellular redox equilibrium to promote disease susceptibility. Alternatively, pathogen-stimulated GRXs and TRXs may posttranslationally modify downstream signaling proteins, thus inactivating a specific resistance mechanism.

GRXs and TRXs function as major disulfide reduction enzymes that regulate the redox state of protein thiol groups. Biochemical and genetic studies demonstrate that these two systems share some common target proteins and operate in parallel in both developmental processes and defense responses. Elucidation of the functional redundancy between GRXs and TRXs, as well as dissection of the cross-communication between GRXs, TRXs and GSH are the key aspects of future studies on cellular redox signaling.

6. Conclusions and Perspectives

Over the last decade, our knowledge about the biological functions of GRXs in plants has drastically expanded (Table 2). Genetic dissection of available GRX mutants across the plant kingdom has enabled us to gain deep insight into their contribution to a plethora of plant processes [7,11,15,16,18,34,46,55,69–71]. An accumulating body of evidence indicates that GRX-mediated regulation of cellular redox homeostasis plays a potentially crucial role in posttranslational modifications of target proteins involved in organ development and defense responses against biotic and abiotic stresses in plants [25,46,47]. Given the ubiquity of the CPYC-type and CGFS-type GRXs to prokaryotes and eukaryotes, as well as the specificity of the CC-type GRXs to land plants, the involvement of GRXs in the assembly and delivery of iron-sulfur clusters seems to represent an ancestral GRX function. All three types of GRXs participate non-exclusively in oxidative stress responses caused by abiotic factors, revealing a conserved role for GRXs in scavenging ROS and maintaining the cellular redox balance. Strikingly, CC-type GRXs interact directly with TGA transcription factors to modulate two seemingly distinct biological processes, *i.e.*, floral organ development and plant immunity, suggesting a possible role for CC-type GRXs in posttranslationally modifying transcription factors and thereby affecting their DNA-binding and/or transcriptional activity. Only the number of CC-type GRXs increases during the evolution of land plants, underlining the notion that CC-type GRXs might have been recruited to develop complex plant architectures and adapt plants for constantly changing environments. Evidently, GRXs are involved in a wide variety of redox-controlled biological pathways, elucidating a key role for GRXs in modulating protein functions in a redox-dependent manner.

Monitoring physiological oxidants and measuring intracellular redox changes have recently been feasible due to the development of fluorescent redox biosensors, such as the redox-sensitive green fluorescent protein (roGFP), the redox-sensitive yellow fluorescent protein (rxYFP) and the GFP-based bioindicator, *HyPer*, specific for H_2O_2 [102–104]. A fascinating advantage of roGFP and HyPer over rxYFP is their ratiometric fluorogenic property. As genetically encoded redox probes, these biosensors can be targeted to specific subcellular compartments, enabling the global analysis of dynamic redox changes *in planta*. Oxidized roGFP can be reduced by the *Arabidopsis* CPYC-type GRXC1 [103], manifesting the potential of roGFP to monitor the intracellular redox homeostasis. Measuring the redox potential using roGFP reveals that the root cap and the meristematic zone have a more reduced redox status than the elongation zone in *Arabidopsis* roots [102], circumstantiating the key role for TRXs and GRXs in establishing the postembryonic root meristem [25,26]. The redox-sensing fluorescent probes can be engineered into model plants, offering powerful tools to dynamically image cellular redox processes.

Table 2. A list of functionally characterized GRXs from different plant species.

GRX	Type	Species	Function *
GRXC7/ROXY1 [a]	CC	*Arabidopsis*	petal initiation and organogenesis, anther development and microspore formation [7,11,46]
GRXC8/ROXY2 [a]	CC	*Arabidopsis*	anther development and microspore formation [7,46]
OsMIL1 [a]	CC	rice	anther development and microspore formation [16]
MSCA1 [a]	CC	maize	anther development and microspore formation [15]
GRXC11/ROXY4 [b]	CC	*Arabidopsis*	GA-signaling and floral organ development [64]
GRXS13/ROXY18 [a]	CC	*Arabidopsis*	disease susceptibility, photooxidative stresses [55,71,100]
GRXC9/ROXY19 [b]	CC	*Arabidopsis*	Crosstalk between the SA and JA/ET defense pathways, disease susceptibility [12,60,100]
PvGRX5 [c]	CGFS	fern	arsenic resistance, oxidative abiotic stresses [20,76,77]
SlGRX1 [a]	CGFS	tomato	abiotic oxidative stresses [73]
GRXS17 [a]	CGFS	*Arabidopsis*	temperature-dependent postembryonic growth and development, thermotolerance [70,72]
GRXS14 [a]	CGFS	*Arabidopsis*	protection against protein oxidation [18]
GRXS14, 16 [d]	CGFS	poplar	assembly and transfer of Fe-S clusters [17]
GRXS15 [d]	CGFS	*Arabidopsis*	abiotic oxidative stresses [69]
GRXC1 [a,d]	CPYC	*Arabidopsis*	assembly of Fe-S clusters, early steps after pollination [34]
GRXC2 [a]	CPYC	*Arabidopsis*	early steps after pollination [34]
GRXC1 [d]	CPYC	poplar	assembly of Fe-S clusters [8,19]

* The numbers within braces represent references; [a–d] The function of the GRX is determined by the phenotype of the corresponding null mutant [a], overexpression studies [b], ectopic expression in *E. coli* and *Arabidopsis* [c] or experiments performed *in vitro* and in yeast [d].

In spite of recent advances in deciphering plant GRXs, many gaps still remain in our understanding of their cellular functions, specific targets and biochemical activities. Photosynthetic organisms, particularly flowering plants, encompass a large number of GRXs [9]. However, only a few plant GRXs have been functionally characterized to date by means of genetic and biochemical approaches [7,11,15,16,18,34,46,55,69–71]. Genome-wide expression analysis suggests diverse roles of plant GRXs during various developmental stages, as well as in response to phytohormones, biotic and abiotic stresses [105,106]. Further functional analysis of plant GRXs and their potential targets will elucidate their contribution to different cellular processes. Apart from the three classical types of GRXs, the existence of GRX-like proteins in plants, such as the 4CxxC family, the CPFC/A class of the GRX/GST proteins, as well as the GRL-type GRXs, adds another layer of complexity to dissecting the GRX-mediated redox modulation of cellular processes [105,107]. Many attempts have been made to explore plant GRX targets via profiling the disulfide proteome and the S-glutathionylation proteome [78,108–112]. High-throughput proteomic techniques, including alkylation of free thiols followed by reduction of disulfide bonds and thiol affinity chromatography [110], incorporation of biotinylated GSH combined with two-dimensional polyacrylamide gel electrophoresis (2D-PAGE) and matrix-assisted laser desorption/ionization time of flight mass spectrometry (MALDI-TOF MS) [111], liquid chromatography with a mutated poplar GRX coupled to tandem MS [78], as well as cysteine labeling with a cysteine-reactive tandem mass tag (cysTMT) combined with tandem MS [112], allow for the large-scale identification of GRX targets implicated in all aspects of plant life processes. Up till now, combining proteomic approaches with protein interaction assays has uncovered a myriad of plant GRX targets involved in diverse cellular events [78,108–112]. Most of these target proteins are encoded by ubiquitously expressed housekeeping genes, and signaling proteins or transcription factors, which are generally encoded by spatiotemporally expressed genes, may escape these detection approaches. Then, *in planta* interactions of GRXs with their respective targets need further experimental confirmation. In particular, genetic experiments are required to explore whether GRXs act with their potential targets in the same cellular pathway. Despite a large array of putative GRX targets unraveled, it still remains a challenge to delineate the complete thiol-disulfide redox proteome modified by GRXs. Besides, the enzymatic properties of the CC-type GRXs have not yet been elucidated biochemically, though the CGFS-type GRXs possess the deglutathionylation activity and the CPYC-type GRXs display both the disulfide oxidoreductase and deglutathionylation activities [113,114]. As genetic experiments reveal that both the catalytic cysteine and the conserved glycine involved in GSH binding are indispensable for GRXC7/ROXY1 to exert its proper function, it has been postulated that the biochemical role of this type of GRXs is related to their oxidoreductase and deglutathionylation activities [11]. A second piece of experimental evidence in support of the CC-type GRXs having a catalytic activity arises from the artificially engineered CC-type GRXs, where the CP/GYC active site of the CPYC-type GRXs is replaced by the CCMC/S active site [37]. In contrast to the CPYC-type GRXs, all of the variants with the artificial CC-type active motif exhibit an impaired oxidoreductase activity, raising the possibility that the CCMC/S active sites of the genuine CC-type GRXs are permissive for the oxidoreductase activity, albeit being far less efficient than the CPYC-type GRXs. To substantiate that CC-type GRXs operate as oxidoreductases, it will be essential to produce soluble proteins corresponding to natural CC-type GRXs and further investigate their enzymatic activities.

Acknowledgments

I acknowledge all scientists that contributed their valuable published data to this review.

Conflicts of Interest

The author declares no conflict of interest.

References

1. Fernandes, A.P.; Holmgren, A. Glutaredoxins: Glutathione-dependent redox enzymes with functions far beyond a simple thioredoxin backup system. *Antioxid. Redox Signal.* **2004**, *6*, 63–74.
2. Buchanan, B.B.; Balmer, Y. Redox regulation: A broadening horizon. *Annu. Rev. Plant Biol.* **2005**, *56*, 187–220.
3. Martin, J.L. Thioredoxin—A fold for all reasons. *Structure* **1995**, *3*, 245–250.
4. Lemaire, S.D. The glutaredoxin family in oxygenic photosynthetic organisms. *Photosyn. Res.* **2004**, *79*, 305–318.
5. Heine, G.F.; Hernandez, J.M.; Grotewold, E. Two cysteines in plant R2R3 MYB domains participate in redox-sensitive DNA binding. *J. Biol. Chem.* **2004**, *279*, 37878–37885.
6. Rouhier, N.; Gelhaye, E.; Jacquot, J.P. Plant glutaredoxins: Still mysterious reducing systems. *Cell. Mol. Life Sci.* **2004**, *61*, 1266–1277.
7. Xing, S.; Zachgo, S. *ROXY1* and *ROXY2*, two *Arabidopsis* glutaredoxin genes, are required for anther development. *Plant J.* **2008**, *53*, 790–801.
8. Feng, Y.; Zhong, N.; Rouhier, N.; Hase, T.; Kusunoki, M.; Jacquot, J.P.; Jin, C.; Xia, B. Structural insight into poplar glutaredoxin C1 with a bridging iron-sulfur cluster at the active site. *Biochemistry* **2006**, *45*, 7998–8008.
9. Rouhier, N.; Couturier, J.; Jacquot, J.P. Genome-wide analysis of plant glutaredoxin systems. *J. Exp. Bot.* **2006**, *57*, 1685–1696.
10. Xing, S.; Lauri, A.; Zachgo, S. Redox regulation and flower development: A novel function for glutaredoxins. *Plant Biol.* **2006**, *8*, 547–555.
11. Xing, S.; Rosso, M.G.; Zachgo, S. ROXY1, a member of the plant glutaredoxin family, is required for petal development in *Arabidopsis thaliana*. *Development* **2005**, *132*, 1555–1565.
12. Ndamukong, I.; Abdallat, A.A.; Thurow, C.; Fode, B.; Zander, M.; Weigel, R.; Gatz, C. SA-inducible *Arabidopsis* glutaredoxin interacts with TGA factors and suppresses JA-responsive *PDF1.2* transcription. *Plant J.* **2007**, *50*, 128–139.
13. Li, S.; Lauri, A.; Ziemann, M.; Busch, A.; Bhave, M.; Zachgo, S. Nuclear activity of ROXY1, a glutaredoxin interacting with TGA factors, is required for petal development in *Arabidopsis thaliana*. *Plant Cell* **2009**, *21*, 429–441.
14. Wang, Z.; Xing, S.; Birkenbihl, R.P.; Zachgo, S. Conserved functions of *Arabidopsis* and rice CC-type glutaredoxins in flower development and pathogen response. *Mol. Plant* **2009**, *2*, 323–335.
15. Kelliher, T.; Walbot, V. Hypoxia triggers meiotic fate acquisition in maize. *Science* **2012**, *337*, 345–348.
16. Hong, L.; Tang, D.; Zhu, K.; Wang, K.; Li, M.; Cheng, Z. Somatic and reproductive cell development in rice anther is regulated by a putative glutaredoxin. *Plant Cell* **2012**, *24*, 577–588.
17. Bandyopadhyay, S.; Gama, F.; Molina-Navarro, M.M.; Gualberto, J.M.; Claxton, R.; Natik, S.G.; Huynh, B.H.; Herrero, E.; Jacquot, J.P.; *et al.* Chloroplast monothiol glutaredoxins as scaffold proteins for the assembly and delivery of [2Fe-2S] clusters. *EMBO J.* **2008**, *27*, 1122–1133.

18. Cheng, N.H.; Liu, J.Z.; Brock, A.; Nelson, R.S.; Hirschi, K.D. AtGRXcp, an *Arabidopsis* chloroplastic glutaredoxin, is critical for protection against protein oxidative damage. *J. Biol. Chem.* **2006**, *281*, 26280–26288.
19. Rouhier, N.; Unno, H.; Bandyopadhyay, S.; Masip, L.; Kim, S.K.; Hirasawa, M.; Gualberto, J.M.; Lattard, V.; Knaff, D.B.; Georgiou, G.; Hase, T.; Johnson, M.K.; *et al*. Functional, structural, and spectroscopic characterization of a glutathione-ligated [2Fe–2S] cluster in poplar glutaredoxin C1. *Proc. Natl. Acad. Sci. USA* **2007**, *104*, 7379–7384.
20. Sundaram, S.; Rathinasabapathi, B.; Ma, L.Q.; Rosen, B.P. An arsenate-activated glutaredoxin from the arsenic hyperaccumulator fern *Pteris vittata* L. regulates intracellular arsenite. *J. Biol. Chem.* **2008**, *283*, 6095–6101.
21. Cobbett, C.S.; May, M.J.; Howden, R.; Rolls, B. The glutathione-deficient, cadmium-sensitive mutant, *cad2-1*, of *Arabidopsis thaliana* is deficient in gamma-glutamylcysteine synthetase. *Plant J.* **1998**, *16*, 73–78.
22. Vernoux, T.; Wilson, R.C.; Seeley, K.A.; Reichheld, J.P.; Muroy, S.; Brown, S.; Maughan, S.C.; Cobbett, C.S.; Montagu, M.V.; Inze, D.; *et al*. The *ROOT MERISTEMLESS1/CADMIUM SENSITIVE2* gene defines a glutathione-dependent pathway involved in initiation and maintenance of cell division during postembryonic root development. *Plant Cell* **2000**, *12*, 97–109.
23. Ogawa, K.; Tasaka, Y.; Mino, M.; Tanaka, Y.; Iwabuchi, M. Association of glutathione with flowering in *Arabidopsis thaliana*. *Plant Cell Physiol.* **2001**, *42*, 524–530.
24. Ogawa, K.; Hatano-Iwasaki, A.; Yanagida, M.; Iwabuchi, M. Level of glutathione is regulated by ATP-dependent ligation of glutamateand cysteine through photosynthesis in *Arabidopsis thaliana*: Mechanism of strong interaction of light intensity with flowering. *Plant Cell Physiol.* **2004**, *45*, 1–8.
25. Reichheld, J.P.; Khafif, M.; Riondet, C.; Droux, M.; Bonnard, G.; Meyer, Y. Inactivation of thioredoxin reductases reveals a complex interplay between thioredoxin and glutathione pathways in *Arabidopsis* development. *Plant Cell* **2007**, *19*, 1851–1865.
26. Bashandy, T.; Guilleminot, J.; Vernoux, T.; Caparros-Ruiz, D.; Ljung, K.; Meyer, Y.; Reichheld, J.P. Interplay between the NADP-Linked thioredoxin and glutathione systems in *Arabidopsis* auxin signalling. *Plant Cell* **2010**, *22*, 376–391.
27. Beinert, H.; Kiley, P.J. Fe-S proteins in sensing and regulatory functions. *Curr. Opin. Chem. Bio.* **1999**, *3*, 152–157.
28. Beinert, H.; Holm, R.H.; Münck, E. Iron-sulfur clusters: Nature's modular, multipurpose structures. *Science* **1997**, *277*, 653–659.
29. Lill, R.; Kispal, G. Maturation of cellular Fe-S proteins: An essential function of mitochondria. *Trends Biochem. Sci.* **2000**, *25*, 352–356.
30. Johnson, M.K.; Smith, A.D. Iron-sulfur proteins. In *Encyclopedia of Inorganic Chemistry*; King, R.B., Ed.; John Wiley & Sons: Chichester, UK, 2005; pp. 2589–2619.
31. Balk, J.; Lobreaux, S. Biogenesis of iron-sulfur proteins in plants. *Trends Plant Sci.* **2005**, *10*, 324–331.
32. Ye, H.; Abdel-Ghany, S.E.; Anderson, T.D.; Pilon-Smits, E.A.; Pilon, M. CpSufE activates the cysteine desulfurase CpNifS for chloroplastic Fe-S cluster formation. *J. Biol. Chem.* **2006**, *281*, 8958–8969.

33. Mühlenhoff, U.; Gerber, J.; Richhardt, N.; Lill, R. Components involved in assembly and dislocation of iron-sulfur clusters on the scaffold protein Isu1p. *EMBO J.* **2003**, *22*, 4815–4825.
34. Riondet, C.; Desouris, J.P.; Montoya, J.G.; Chartier, Y.; Meyer, Y.; Reichheld, J. A dicotyledon-specific glutaredoxin GRXC1 family with dimer-dependent redox regulation is functionally redundant with GRXC2. *Plant Cell Environ.* **2012**, *35*, 360–373.
35. Picciocchi, A.; Saguez, C.; Boussac, A.; Cassier-Chauvat, C.; Chauvat, F. CGFS type monothiol glutaredoxins from the *cyanobacterium Synechocystis PCC6803* and other evolutionary distant model organisms possess a glutathione-ligated [2Fe-2S] cluster. *Biochemistry* **2007**, *46*, 15018–15026.
36. Rodriguez-Manzaneque, M.T.; Tamarit, J.; Belli, G.; Ros, J.; Herrero, E. Grx5 is a mitochondrial glutaredoxin required for the activity of iron/sulfur enzymes. *Mol. Biol. Cell.* **2002**, *13*, 1109–1121.
37. Couturier, J.; Didierjean, C.; Jacqot, J.P.; Rouhier, N. Engineered mutated glutaredoxins mimicking peculiar plant class III glutaredoxins bind iron-sulfur centers and possess reductase activity. *Biochem. Biophys. Res. Commun.* **2010**, *403*, 435–441.
38. Coen, E.S.; Meyerowitz, E.M. The war of the whorls: Genetic interactions controlling flower development. *Nature* **1991**, *353*, 31–37.
39. Jakoby, M.; Weisshaar, B.; Droge-Laser, W.; Vicente-Carbajosa, J.; Tiedemann, J.; Kroj, T.; Parcy, F. bZIP transcription factors in *Arabidopsis*. *Trends Plant Sci.* **2002**, *1*, 106–111.
40. Després, C.; Chubak, C.; Rochon, A.; Clark, R.; Bethune, T.; Desveaux, D.; Fobert, P.R. The *Arabidopsis* NPR1 disease resistance protein is a novel cofactor that confers redox regulation of DNA binding activity to the basic domain/leucine zipper transcription factor TGA1. *Plant Cell* **2003**, *15*, 2181–2191.
41. Zhang, Y.L.; Tessaro, M.J.; Lassner, M.; Li, X. Knockout analysis of *Arabidopsis* transcription factors TGA2, TGA5, and TGA6 reveals their redundant and essential roles in systemic acquired resistance. *Plant Cell* **2003**, *15*, 2647–2653.
42. Kesarwani, M.; Yoo, J.; Dong, X. Genetic interactions of TGA transcription factors in the regulation of pathogenesis-related genes and disease resistance in *Arabidopsis*. *Plant Physiol.* **2007**, *144*, 336–346.
43. Gatz, C. From pioneers to team players: TGA transcription factors provide a molecular link between different stress pathways. *Mol. Plant Microbe Interact.* **2013**, *26*, 151–159.
44. Running, M.P.; Meyerowitz, E.M. Mutations in the *PERIANTHIA* gene of *Arabidopsis* specifically alter floral organ number and initiation pattern. *Development* **1996**, *122*, 1261–1269.
45. Chuang, C.F.; Running, M.P.; Williams, R.W.; Meyerowitz, E.M. The *PERIANTHIA* gene encodes a bZIP protein involved in the determination of floral organ number in *Arabidopsis thaliana*. *Genes Dev.* **1999**, *13*, 334–344.
46. Murmu, J.; Bush, M.J.; DeLong, C.; Li, S.; Xu, M.; Khan, M.; Malcolmson, C.; Fobert, P.R.; Zachgo, S.; Hepworth, S.R. *Arabidopsis* basic leucine-zipper transcription factors TGA9 and TGA10 interact with floral glutaredoxins ROXY1 and ROXY2 and are redundantly required for anther development. *Plant Physiol.* **2010**, *154*, 1492–1504.
47. Li, S.; Gutsche, N.; Zachgo, S. The ROXY1 C-terminal L ** LL motif is essential for the interaction with TGA transcription factors. *Plant Physiol.* **2011**, *157*, 2056–2068.

48. Li, S.; Zachgo, S. Glutaredoxins in development and stress responses of plants. *Adv. Bot. Res.* **2009**, *52*, 333–361.
49. Schenk, P.M.; Kazan, K.; Wilson, I.; Anderson, J.P.; Richmond, T.; Somerville, S.C.; Manners, J.M. Coordinated plant defense responses in *Arabidopsis* revealed by microarray analysis. *Proc. Natl. Acad. Sci. USA* **2000**, *97*, 11655–11660.
50. De Vos, M.; van Oosten, V.R.; van Poecke, R.M.P.; van Pelt, J.A.; Pozo, M.J.; Mueller, M.J.; Buchala, A.J.; Metraux, J.P.; van Loon, L.C.; *et al.* Signal signature and transcriptome changes of *Arabidopsis* during pathogen and insect attack. *Mol. Plant-Microbe Interact.* **2005**, *18*, 923–937.
51. Glazebrook, J. Contrasting mechanisms of defense against biotrophic and necrotrophic pathogens. *Annu. Rev. Phytopathol.* **2005**, *43*, 205–227.
52. Spoel, S.H.; Johnson, J.S.; Dond, X. Regulation of tradeoffs between plant defenses against pathogens with different lifestyles. *Proc. Natl. Acad. Sci. USA* **2007**, *104*, 18842–18847.
53. Sticher, L.; Mauch-Mani, B.; Metraux, J.P. Systemic acquired resistance. *Ann. Rev. Phytopathol.* **1997**, *35*, 235–270.
54. Koornneef, A.; Pieterse, C.M.J. Cross talk in defense signaling. *Plant Physiol.* **2008**, *146*, 839–844.
55. La Camera, S.; L'haridon, F.; Astier, J.; Zander, M.; Abou-Mansour, E.; Page, G.; Thurow, C.; Wendehenne, D.; Gatz, C.; Métraux, J.P.; *et al.* The glutaredoxin ATGRXS13 is required to facilitate *Botrytis cinerea* infection of *Arabidopsis thaliana* plants. *Plant J.* **2011**, *68*, 507–519.
56. Blanco, F.; Salinas, P.; Cecchini, N.M.; Jordana, X.; Van Hummelen, P.; Alvarez, M.E.; Holuigue, L. Early genomic responses to salicylic acid in *Arabidopsis*. *Plant Mol. Biol.* **2009**, *70*, 79–102.
57. Zarei, A.; Körbes, A.P.; Younessi, P.; Montiel, G.; Champion, A.; Memelink, J. Two GCC boxes and AP2/ERF-domain transcription factor ORA59 in jasmonate/ethylene-mediated activation of the *PDF1.2* promoter in *Arabidopsis*. *Plant Mol. Biol.* **2011**, *75*, 321–331.
58. Pre, M.; Atallah, M.; Champion, A.; de Vos, M.; Pieterse, C.M.J.; Memelink, J. The AP2/ERF domain transcription factor ORA59 integrates jasmonic acid and ethylene signals in plant defense. *Plant Physiol.* **2008**, *147*, 1347–1357.
59. Leon-Reyes, A.; Du, Y.; Koornneef, A.; Proietti, S.; Körbes, A.P.; Memelink, J.; Pieterse, C.M.J.; Ritsema, T. Ethylene signaling renders the jasmonate response of *Arabidopsis* insensitive to future suppression by salicylic acid. *Mol. Plant Microbe Interact.* **2010**, *23*, 187–197.
60. Zander, M.; Chen, S.; Imkampe, J.; Thurow, C.; Gatz, C. Repression of the *Arabidopsis thaliana* jasmonic acid/ethylene induced defense pathway by TGA-interacting glutaredoxins depends on their C-terminal ALWL motif. *Mol. Plant* **2012**, *5*, 831–840.
61. Qiu, D.; Xiao, J.; Xie, W.; Liu, H.; Li, X.; Xiong, L.; Wang, S. Rice gene network inferred from expression profiling of plants overexpressing *OsWRKY13*, a positive regulator of disease resistance. *Mol. Plant* **2008**, *1*, 538–551.
62. Qiu, D.; Xiao, J.; Ding, B.; Xiong, M.; Cai, M.; Cao, Y.; Li, X.; Xu, C.; Wang, S. OsWRKY13 mediates rice disease resistance by regulating defense-related genes in salicylate- and jasmoate-dependent signalling. *Mol. Plant Microbe Interact.* **2007**, *20*, 492–499.

63. Zander, M.; La Camera, S.; Lamotte, O.; Métraux, J.P.; Gatz, C. *Arabidopsis thaliana* class-II TGA transcription factors are essential activators of jasmonic acid/ethylene-induced defense responses. *Plant J.* **2010**, *61*, 200–210.
64. Hou, X.L.; Hu, W.W.; Shen, L.S.; Lee, L.Y.C.; Tao, Z.; Han, J.H.; Yu, H. Global identification of DELLA target genes during *Arabidopsis* flower development. *Plant Physiol.* **2008**, *147*, 1126–1142.
65. Ziemann, M.; Bhave, M.; Zachgo, S. Origin and diversification of land plant CC-type glutaredoxins. *Genome Biol. Evol.* **2009**, *1*, 265–277.
66. Michelet, L.; Zaffagnini, M.; Massot, V.; Keryer, E.; Vanacker, H.; Miginiac-Maslow, M.; Issakidis-Bourguet, E.; Lemaire, S.D. Thioredoxins, glutaredoxins, and glutathionylation: New crosstalks to explore. *Photosynth. Res.* **2006**, *89*, 225–245.
67. Foyer, C.H.; Noctor, G. Redox homeostasis and antioxidant signaling: A metabolic interface between stress perception and physiological responses. *Plant Cell* **2005**, *17*, 1866–1875.
68. Mittler, R.; Vanderauwera, S.; Gollery, M.; Breusegem, F.V. Reactive oxygen gene network of plants. *Trends Plant Sci.* **2004**, *10*, 1360–1385.
69. Cheng, N.H. AtGRX4, an *Arabidopsis* chloroplastic monothiol glutaredoxin, is able to suppress yeast *grx5* mutant phenotypes and respond to oxidative stress. *FEBS Lett.* **2008**, *582*, 848–854.
70. Cheng, N.H.; Liu, J.Z.; Liu, X.; Wu, Q.; Thompson, S.M.; Lin, J.; Chang, J.; Whitham, S.A.; Park, S.; Cohen, J.D.; *et al. Arabidopsis* monothiol glutaredoxin, AtGRXS17, is critical for temperature-dependent postembryonic growth and development via modulating auxin response. *J. Biol. Chem.* **2011**, *286*, 20398–20406.
71. Laporte, D.; Olate, E.; Salinas, P.; Salazar, M.; Jordana, X.; Holuigue, L. Glutaredoxin GRXS13 plays a key role in protection against photooxidative stress in *Arabidopsis*. *J. Exp. Bot.* **2012**, *63*, 503–515.
72. Wu, Q.; Lin, J.; Liu, J.Z.; Wang, X.; Lim, W.; Oh, M.; Park, J.; Rajashekar, C.B.; Whitham, S.A.; Cheng, N.H.; *et al.* Ecotopic expression of *Arabidopsis* glutaredoxin AtGRXS17 enhances thermotolerance in tomato. *Plant Biotechnol. J.* **2012**, *10*, 945–955.
73. Guo, Y.; Huang, C.; Xie, Y.; Song, F.; Zhou, Y. A tomato glutaredoxin gene *SlGRX1* regulates plant responses to oxidative, drought and salt stresses. *Planta* **2010**, *232*, 1499–1509.
74. Ma, L.Q.; Komar, K.M.; Tu, C.; Zhang, W.; Cai, Y.; Kennelley, E.D. A fern that hyperaccumulates arsenic. *Nature* **2001**, *409*, doi:10.1038/35054664.
75. Liu, Z.; Shen, J.; Carbrey, J.M.; Mukhopadhyay, R.; Agre, P.; Rosen, B.P. Arsenite transport by mammalian aquaglyceroporins AQP7 and AQP9. *Proc. Natl. Acad. Sci. USA* **2002**, *99*, 6053–6058.
76. Sundaram, S.; Wu, S.; Ma, L.Q.; Rathinasabapathi, B. Expression of a *Pteris vittata* glutaredoxin PvGRX5 in transgenic *Arabidopsis thaliana* increases plant arsenic tolerance and decreases arsenic accumulation in the leaves. *Plant Cell Environ.* **2009**, *32*, 851–858.
77. Sundaram, S.; Rathinasabapathi, B. Transgenic expression of fern *Pteris vittata* glutaredoxin PvGrx5 in *Arabidopsis thaliana* increases plant tolerance to high temperature stress and reduces oxidative damage to proteins. *Planta* **2010**, *231*, 361–369.
78. Rouhier, N.; Villarejo, A.; Srivastava, M.; Gelhaye, E.; Keech, O.; Droux, M.; Finkemeier, I.; Samuelsson, G.; Dietz, K.J.; Jacquot, J.P.; *et al.* Identification of plant glutaredoxin targets. *Antioxid. Redox Signal.* **2005**, *7*, 919–929.

79. Meyer, Y.; Belin, C.; Delorme-Hinoux, V.; Reichheld, J.P.; Riondet, C. Thioredoxin and glutaredoxin systems in plants: Molecular mechanisms, crosstalks, and functional significance. *Antioxid. Redox Signal.* **2012**, *17*, 1124–1160.
80. Arsova, B.; Hoja, U.; Wimmelbacher, M.; Greiner, E.; Ustun, S.; Melzer, M.; Petersen, K.; Lein, W.; Bornke, F. Plastidial thioredoxin z interacts with two fructokinase-like proteins in a thiol-dependent manner: Evidence for an essential role in chloroplast development in *Arabidopsis* and *Nicotiana benthamiana*. *Plant Cell* **2010**, *22*, 1498–1515.
81. Dai, S.; Schwendtmayer, C.; Schurmann, P.; Ramaswamy, S.; Eklund, H. Redox signalling in chloroplasts: Cleavage of disulfides by an iron-sulfur cluster. *Science* **2000**, *287*, 655–658.
82. Marty, L.; Siala, W.; Schwarzlaender, M.; Fricker, M.D.; Wirtz, M.; Sweetlove, L.J.; Meyer, Y.; Meyer, A.J.; Reichheld, J.P.; Hell, R. The NADPH-dependent thioredoxin system constitutes a functional backup for cytosolic glutathione reductase in *Arabidopsis*. *Proc. Natl. Acad. Sci. USA* **2009**, *106*, 9109–9114.
83. Cairns, N.G.; Pasternak, M.; Wachter, A.; Cobbett, C.S.; Meyer, A.J. Maturation of *Arabidopsis* seeds is dependent on glutathione biosynthesis within the embryo. *Plant Physiol.* **2006**, *141*, 446–455.
84. Howden, R.; Andersen, C.R.; Goldsbrough, P.B.; Cobbett, C.S. A cadmium-sensitive, glutathione-deficient mutant of *Arabidopsis thaliana*. *Plant Physiol.* **1995**, *107*, 1067–1073.
85. Sanchez-Fernandez, R.; Fricker, M.; Corben, L.B.; White, N.S.; Sheard, N.; Leaver, C.J.; Montagu, M.V.; Inze, D.; May, M.J. Cell proliferation an hair tip growth in the *Arabidopsis* root are under mechanistically different forms of redox control. *Proc. Natl. Acad. Sci. USA* **1997**, *94*, 2745–2750.
86. Noctor, G.; Foyer, C.H. Ascorbate and glutathione: Keeping active oxygen under control. *Ann. Rev. Plant Physiol. Plant Mol. Biol.* **1998**, *49*, 249–279.
87. Cobbert, C.S. Phytochelatin biosynthesis and function in heavy-metal detoxification. *Curr. Opin. Plant Biol.* **2000**, *3*, 211–216.
88. Cheng, J.C.; Seeley, K.; Sung, Z.R. *RML1* and *RML2* *Arabidopsis* genes required for cell proliferation at the root tip. *Plant Physiol*. **1995**, *107*, 365–376.
89. Parisy, V.; Poinssot, B.; Owsianowski, L.; Buchala, A.; Glazebrook, J.; Mauch, F. Identification of PAD2 as a gamma-glutamylcysteine synthetase highlights the importance of glutathione in disease resistance of *Arabidopsis*. *Plant J.* **2007**, *49*, 159–172.
90. Reichheld, J.P.; Meyer, E.; Khafif, M.; Bonnard, G.; Meyer, Y. AtNTRB is the major mitochondrial thioredoxin reductase in *Arabidopsis thaliana*. *FEBS Lett.* **2005**, *579*, 337–342.
91. Gelhaye, E.; Rounier, N.; Jacquot, J.P. Evidence for a subgroup of thioredoxin h that requires GSH/GRX for its reduction. *FEBS Lett.* **2003**, *555*, 443–448.
92. Bennett, R.S.M.; Alvarez, J.; Bossinger, G.; Smyth, D.R. Morphogenesis in pinoid mutants of *Arabidopsis thaliana*. *Plant J.* **1995**, *8*, 505–520.
93. Christensen, S.K.; Dagenais, N.; Chory, J.; Weigel, D. Regulation of auxin response by the protein kinase PINOID. *Cell* **2000**, *100*, 469–478.
94. Gälweiler, L.; Guan, C.; Müller, A.; Wisman, E.; Mendgen, K.; Yephremov, A.; Palme, K. Regulation of polar auxin transport by AtPIN1 in *Arabidopsis* vascular tissue. *Science* **1998**, *282*, 2226–2230.

95. Hardtke, C.S.; Ckurshumova, W.; Vidaurre, D.P.; Singh, S.A.; Stamatiou, G.; Tiwari, S.B.; Hagen, G.; Guilfoyle, T.J.; Berleth, T. Overlapping and non-redundant functions of the *Arabidopsis* auxin response factors MONOPTEROS and NONPHOTOTROPIC HYPOCOTYL 4. *Development* **2004**, *131*, 1089–1100.
96. Cheng, Y.; Qin, G.; Dai, X.; Zhao, Y. NPY1, a BTB-NPH3-like protein, plays a critical role in auxin-regulated organogenesis in *Arabidopsis*. *Proc. Natl. Acad. Sci. USA* **2007**, *104*, 18825–18829.
97. Li, S.; Zachgo, S. TCP3 interacts with R2R3-MYB proteins, promotes flavonoid biosynthesis and negatively regulates the auxin response in *Arabidopsis thaliana*. *Plant J.* **2013**, *76*, 901–913.
98. Li, S. Transcriptional control of flavonoid biosynthesis: Fine-tuning of the MYB-bHLH-WD40 (MBW) complex. *Plant Signal. Behav.* **2014**, *9*, e27522.
99. Backstrom, S.; Elfving, N.; Nilsson, R.; Wingsle, G.; Bjorklund, S. Purification of a plant Mediator from *Arabidopsis thaliana* identifies PFT1 as the Med25 subunit. *Mol. Cell* **2007**, *26*, 717–729.
100. Lai, Z.; Schluttenhofer, C.M.; Bhide, K.; Shreve, J.; Thimmapuram, J.; Lee, S.Y.; Yun, J.D.; Mengiste, T. MED18 interaction with distinct transcription factors regulates multiple plant functions. *Nat. Commun.* **2014**, *5*, 3064.
101. Lorang, J.M.; Carkaci-Salli, N.; Wolpert, T.J. Identification and characterization of victorin sensitivity in *Arabidopsis thaliana*. *Mol. Plant Microbe Interact.* **2004**, *17*, 577–582.
102. Jiang, K.; Schwarzer, C.; Lally, E.; Zhang, S.; Ruzin, S.; Machen, T.; Remington, S.J.; Feldman, L. Expression and characterization of a redox-sensing green fluorescent protein (reduction-oxidation-sensitive green fluorescent protein) in *Arabidopsis*. *Plant Physiol.* **2006**, *141*, 397–403.
103. Meyer, A.J.; Brach, T.; Marty, L.; Kreye, S.; Rouhier, N.; Jacquot, J.P.; Hell, R. Redox-sensitive GFP in *Arabidopsis thaliana* is a quantitative biosensor for the redox potential of the cellular glutathione redox buffer. *Plant J.* **2007**, *52*, 973–986.
104. Meyer, A.J.; Dick, T.P. Fluorescent protein-based redox probes. *Antioxid. Redox Signal.* **2010**, *13*, 621–650.
105. Garg, R.; Jhanwar, S.; Tyagi, A.K.; Jain, M. Genome-wide survey and expression analysis suggests diverse roles of glutaredoxin gene family members during development and response to various stimuli in rice. *DNA Res.* **2010**, *17*, 353–367.
106. Belin, C.; Bashandy, T.; Cela, J.; Delorme-Hinoux, V.; Riondet, C.; Reichheld, J.P. A comprehensive study of thiol reduction gene expression under stress conditions in *Arabidopsis thaliana*. *Plant Cell Environ.* **2014**, doi:10.1111/pce.12276.
107. Meyer, Y.; Siala, W.; Bashandy, T.; Riondet, C.; Vignols, F.; Reichheld, J.P. Glutaredoxins and thioredoxins in plants. *Biochim. Biophys. Acta* **2008**, *1783*, 589–600.
108. Ito, H.; Iwabuchi, M.; Ogawa, K. The sugar-metabolic enzymes aldolase and triose-phosphate isomerase are targets of glutathionylation in *Arabidopsis thaliana*: Detection using biotinylated glutathione. *Plant Cell Physiol.* **2003**, *44*, 655–660.
109. Go, Y.M.; Jones, D.P. The redox proteome. *J. Biochem. Chem.* **2013**, *288*, 26512–26520.
110. Lee, K.; Lee, J.; Kim, Y.; Bae, D.; Kang, K.Y.; Yoon, S.C.; Lim, D.B. Defining the plant disulfide proteome. *Electrophoresis* **2004**, *25*, 532–541.

111. Dixon, D.P.; Skipsey, M.; Grundy, N.M.; Edwards, R. Stress-induced protein S-glutathionylation in *Arabidopsis*. *Plant Physiol.* **2005**, *138*, 2233–2244.
112. Parker, J.; Zhu, N.; Zhu, M.; Chen, S. Profiling thiol redox proteome using isotope tagging mass spectrometry. *J. Visual. Exp.* **2012**, *61*, e376661.
113. Zaffagnini, M.; Michelet, L.; Massot, V.; Trost, P.; Lemaire, S.D. Biochemical characterization of glutaredoxins from *Chlamydomonas reinhardtii* reveals the unique properties of a chloroplastic CGFS-type glutaredoxin. *J. Biol. Chem.* **2008**, *283*, 8868–8876.
114. Rouhier, N. Plant glutaredoxins: Pivotal players in redox biology and iron-sulphur centre assembly. *New Phytol.* **2010**, *186*, 362–372.

Permissions

All chapters in this book were first published in Plants, by MDPI; hereby published with permission under the Creative Commons Attribution License or equivalent. Every chapter published in this book has been scrutinized by our experts. Their significance has been extensively debated. The topics covered herein carry significant findings which will fuel the growth of the discipline. They may even be implemented as practical applications or may be referred to as a beginning point for another development.

The contributors of this book come from diverse backgrounds, making this book a truly international effort. This book will bring forth new frontiers with its revolutionizing research information and detailed analysis of the nascent developments around the world.

We would like to thank all the contributing authors for lending their expertise to make the book truly unique. They have played a crucial role in the development of this book. Without their invaluable contributions this book wouldn't have been possible. They have made vital efforts to compile up to date information on the varied aspects of this subject to make this book a valuable addition to the collection of many professionals and students.

This book was conceptualized with the vision of imparting up-to-date information and advanced data in this field. To ensure the same, a matchless editorial board was set up. Every individual on the board went through rigorous rounds of assessment to prove their worth. After which they invested a large part of their time researching and compiling the most relevant data for our readers.

The editorial board has been involved in producing this book since its inception. They have spent rigorous hours researching and exploring the diverse topics which have resulted in the successful publishing of this book. They have passed on their knowledge of decades through this book. To expedite this challenging task, the publisher supported the team at every step. A small team of assistant editors was also appointed to further simplify the editing procedure and attain best results for the readers.

Apart from the editorial board, the designing team has also invested a significant amount of their time in understanding the subject and creating the most relevant covers. They scrutinized every image to scout for the most suitable representation of the subject and create an appropriate cover for the book.

The publishing team has been an ardent support to the editorial, designing and production team. Their endless efforts to recruit the best for this project, has resulted in the accomplishment of this book. They are a veteran in the field of academics and their pool of knowledge is as vast as their experience in printing. Their expertise and guidance has proved useful at every step. Their uncompromising quality standards have made this book an exceptional effort. Their encouragement from time to time has been an inspiration for everyone.

The publisher and the editorial board hope that this book will prove to be a valuable piece of knowledge for researchers, students, practitioners and scholars across the globe.

List of Contributors

Hisahiro Kai
Department of Pharmaceutical Health Sciences, School of Pharmaceutical Sciences, Kyushu University of Health and Welfare, 1714-1 Yoshino-machi, Nobeoka, Miyazaki 882-8508, Japan
Research Promotion Bureau for Collaboration of Regional Entities, Miyazaki Prefectural Industrial Support Foundation, 16500-2 Higashi-Kaminaka, Sadowara-cho, Miyazaki, Miyazaki 880-0303, Japan

Ena Akamatsu
Research Promotion Bureau for Collaboration of Regional Entities, Miyazaki Prefectural Industrial Support Foundation, 16500-2 Higashi-Kaminaka, Sadowara-cho, Miyazaki, Miyazaki 880-0303, Japan

Eri Torii
Research Promotion Bureau for Collaboration of Regional Entities, Miyazaki Prefectural Industrial Support Foundation, 16500-2 Higashi-Kaminaka, Sadowara-cho, Miyazaki, Miyazaki 880-0303, Japan

Hiroko Kodama
Research Promotion Bureau for Collaboration of Regional Entities, Miyazaki Prefectural Industrial Support Foundation, 16500-2 Higashi-Kaminaka, Sadowara-cho, Miyazaki, Miyazaki 880-0303, Japan
Miyazaki Prefectural Food Research and Development Center, 16500-2 Higashi-Kaminaka, Sadowara-cho, Miyazaki, Miyazaki 880-0303, Japan

Chizuko Yukizaki
Miyazaki Prefectural Food Research and Development Center, 16500-2 Higashi-Kaminaka, Sadowara-cho, Miyazaki, Miyazaki 880-0303, Japan

Isao Akagi
Research Promotion Bureau for Collaboration of Regional Entities, Miyazaki Prefectural Industrial Support Foundation, 16500-2 Higashi-Kaminaka, Sadowara-cho, Miyazaki, Miyazaki 880-0303, Japan
Department of Biochemical Science and Technology, Faculty of Agriculture, Kagoshima University, 1-21-24 Korimoto, Kagoshima, Kagoshima 890-0065, Japan

Hisatoshi Ino
Miyazaki Agricultural Experiment Station, 5805 Shimonaka, Sadowara-cho, Miyazaki, Miyazaki 880-0212, Japan

Yoichi Sakakibara
Department of Biochemistry and Applied Biosciences, Faculty of Agriculture, University of Miyazaki, 1-1 Gakuenkibanadai-nishi, Miyazaki, Miyazaki 889-2192, Japan

Masahito Suiko
Department of Biochemistry and Applied Biosciences, Faculty of Agriculture, University of Miyazaki, 1-1 Gakuenkibanadai-nishi, Miyazaki, Miyazaki 889-2192, Japan

Ikuo Yamamoto
Department of Rheumatology, Infectious Diseases and Laboratory Medicine, Faculty of Medicine, University of Miyazaki, 5200 Kihara Kiyotake, Miyazaki, Miyazaki 889-1692, Japan

Akihiko Okayama
Department of Rheumatology, Infectious Diseases and Laboratory Medicine, Faculty of Medicine, University of Miyazaki, 5200 Kihara Kiyotake, Miyazaki, Miyazaki 889-1692, Japan

Kazuhiro Morishita
Division of Tumor and Cellular Biochemistry, Department of Medical Sciences, Faculty of Medicine, University of Miyazaki, 5200 Kihara Kiyotake, Miyazaki, Miyazaki 889-1692, Japan

Hiroaki Kataoka
Section of Oncopathology and Regenerative Biology, Department of Pathology, Faculty of Medicine, University of Miyazaki, 5200 Kihara Kiyotake, Miyazaki, Miyazaki 889-1692, Japan

Koji Matsuno
Department of Pharmaceutical Health Sciences, School of Pharmaceutical Sciences, Kyushu University of Health and Welfare, 1714-1 Yoshino-machi, Nobeoka, Miyazaki 882-8508, Japan

Julia M. Davies
Department of Plant Sciences, University of Cambridge, Cambridge CB2 3EA, UK

Yumi Uenishi
Graduate School of Humanities and Sciences, Nara Women's University, Nara 630-8506, Japan
NIPPON Funmatsu Yakuhin CO., Ltd, Osaka 541-0045, Japan

Yukari Nakabayashi
Faculty of Science, Nara Women's University, Nara 630-8506, Japan
Graduate School of Information Science, Nagoya University, Nagoya 464-8601, Japan

Ayako Tsuchihira
Graduate School of Bioagricultural Sciences, Nagoya University, Nagoya 464-8601, Japan

Mari Takusagawa
Graduate School of Humanities and Sciences, Nara Women's University, Nara 630-8506, Japan
Graduate School of Science, Kyoto University, Kyoto 606-8502, Japan

Kayo Hashimoto
Graduate School of Humanities and Sciences, Nara Women's University, Nara 630-8506, Japan

Masayoshi Maeshima
Graduate School of Bioagricultural Sciences, Nagoya University, Nagoya 464-8601, Japan

Kumi Sato-Nara
Faculty of Science, Nara Women's University, Nara 630-8506, Japan

Sarah Russell French
Molecular and Cellular Biology, University of New England, Armidale, New South Wales 2351, Australia

Yousef Abu-Zaitoon
Department of Biology, Faculty of Science, Islamic University, Madinah 170, Saudi Arabia

Md. Myn Uddin
Department of Microbiology, University of Chittagong, Chittagong 4331, Bangladesh

Karina Bennett
Molecular and Cellular Biology, University of New England, Armidale, New South Wales 2351, Australia

Heather M. Nonhebel
Molecular and Cellular Biology, University of New England, Armidale, New South Wales 2351, Australia

Ines Ben Rejeb
Faculty of Sciences, Institute of Biology, University of Neuchâtel, Rue Emile Argand 11, 2000 Neuchâtel, Switzerland

Victoria Pastor
Faculty of Sciences, Institute of Biology, University of Neuchâtel, Rue Emile Argand 11, 2000 Neuchâtel, Switzerland

Brigitte Mauch-Mani
Faculty of Sciences, Institute of Biology, University of Neuchâtel, Rue Emile Argand 11, 2000 Neuchâtel, Switzerland

Beatriz Sánchez-Parra
Center for Plant Biotechnology and Genomics (U.P.M.-I.N.I.A.), Technical University Madrid, Montegancedo Campus, Crta. M-40, km 38, 28223 Pozuelo de Alarcón (Madrid), Spain

Henning Frerigmann
Department of Plant Physiology, Faculty of Biology and Biotechnology, Ruhr-University Bochum, Universitätsstraße 150, 44801 Bochum, Germany
Institute of Botany Chair II, University of Cologne, Zülpicher Straße 47b, 50674 Cologne, Germany

Marta-Marina Pérez Alonso
Center for Plant Biotechnology and Genomics (U.P.M.-I.N.I.A.), Technical University Madrid, Montegancedo Campus, Crta. M-40, km 38, 28223 Pozuelo de Alarcón (Madrid), Spain

Víctor Carrasco Loba
Center for Plant Biotechnology and Genomics (U.P.M.-I.N.I.A.), Technical University Madrid, Montegancedo Campus, Crta. M-40, km 38, 28223 Pozuelo de Alarcón (Madrid), Spain

Ricarda Jost
School of Plant Biology, University of Western Australia, 35 Stirling Highway, Crawley, WA 6009, Australia

Mathias Hentrich
Department of Plant Physiology, Faculty of Biology and Biotechnology, Ruhr-University Bochum, Universitätsstraße 150, 44801 Bochum, Germany

Stephan Pollmann
Center for Plant Biotechnology and Genomics (U.P.M.-I.N.I.A.), Technical University Madrid, Montegancedo Campus, Crta. M-40, km 38, 28223 Pozuelo de Alarcón (Madrid), Spain

Nobutoshi Yamaguchi
Department of Biology, University of Pennsylvania, Philadelphia, PA 19104, USA

Miin-Feng Wu
Department of Biology, University of Pennsylvania, Philadelphia, PA 19104, USA

Cara M. Winter
Department of Biology, University of Pennsylvania, Philadelphia, PA 19104, USA
Department of Biology, Duke University, Box 90338, Durham, NC 27708, USA

Doris Wagner
Department of Biology, University of Pennsylvania, Philadelphia, PA 19104, USA

Anastasia A. Kirpichnikova
Department of Plant Physiology and Biochemistry, St. Petersburg State University, Universitetskaya emb. 7/9, St. Petersburg 199034, Russia

Elena L. Rudashevskaya
Department of Plant Physiology and Biochemistry, St. Petersburg State University, Universitetskaya emb. 7/9, St. Petersburg 199034, Russia
Institute of Medical Chemistry, Medical University of Vienna; Waehringer strasse 10, Vienna 1090, Austria

Vladislav V. Yemelyanov
Department of Plant Physiology and Biochemistry, St. Petersburg State University, Universitetskaya emb. 7/9, St. Petersburg 199034, Russia
Department of Genetics and Biotechnology, St. Petersburg State University, Universitetskaya emb. 7/9, St. Petersburg 199034, Russia

Maria F. Shishova
Department of Plant Physiology and Biochemistry, St. Petersburg State University, Universitetskaya emb. 7/9, St. Petersburg 199034, Russia

Alicja Banasiak
Institute of Experimental Biology, University of Wroclaw, 50-328 Wroclaw, Poland

Farid M. Ibatullin
Division of Glycoscience, School of Biotechnology, Royal Institute of Technology (KTH), AlbaNova University Centre, 106 91 Stockholm, Sweden
Biophysics Division, Petersburg Nuclear Physics Institute, National Research Center Kurchatov Institute, Gatchina 188300, Russia

Harry Brumer
Division of Glycoscience, School of Biotechnology, Royal Institute of Technology (KTH), AlbaNova University
Michael Smith Laboratories, University of British Columbia, 2185 East Mall, Vancouver, BC V6T 1Z1, Canada
Department of Chemistry, University of British Columbia, 2185 East Mall, Vancouver, BC V6T 1Z1, Canada

Ewa J. Mellerowicz
Department of Forest Genetics and Plant Physiology, Swedish University of Agricultural Sciences, Umea Plant Science Centre, 90183 Umea, Sweden

Marina Mellenthin
Plant Sciences, Heinrich-Heine-Universität, Universitätsstraße 1, Düsseldorf 40225, Germany

Ulrike Ellersiek
Plant Sciences, Heinrich-Heine-Universität, Universitätsstraße 1, Düsseldorf 40225, Germany

Anna Börger
Plant Physiology, Freie Universität Berlin, Königin-Luise-Straße 12-16, Berlin 14195, Germany

Margarete Baier
Plant Physiology, Freie Universität Berlin, Königin-Luise-Straße 12-16, Berlin 14195, Germany

Xiquan Gao
State Key Laboratory of Crop Genetics and Germplasm Enhancement, College of Agriculture, Nanjing Agricultural University, Nanjing 210095, China

Kevin L. Cox Jr.
Department of Plant Pathology and Microbiology, Institute for Plant Genomics and Biotechnology, Texas A&M University, College Station, TX 77843, USA

Ping He
Department of Biochemistry and Biophysics, Institute for Plant Genomics and Biotechnology, Texas A&M University, College Station, TX 77843, USA

Yuning Chen
Department of Chemistry and Biochemistry, Ohio University, Athens, OH 45701, USA

Dening Ye
Department of Chemistry and Biochemistry, Ohio University, Athens, OH 45701, USA

Michael A. Held
Department of Chemistry and Biochemistry, Ohio University, Athens, OH 45701, USA

Maura C. Cannon
Department of Biochemistry and Molecular Biology, University of Massachusetts, Amherst, MA 01003, USA

Tui Ray
Department of Biochemistry and Molecular Biology, University of Massachusetts, Amherst, MA 01003, USA

Prasenjit Saha
Department of Biochemistry and Molecular Biology, University of Massachusetts, Amherst, MA 01003, USA

Alexandra N. Frye
Department of Chemistry and Biochemistry, Ohio University, Athens, OH 45701, USA

Andrew J. Mort
Biochemistry and Molecular Biology, Oklahoma State University, Stillwater, OK 74078, USA

Marcia J. Kieliszewski
Department of Chemistry and Biochemistry, Ohio University, Athens, OH 45701, USA

Ramesh R. Chavan
School of Biological Sciences, The University of Auckland, Private Bag 92019, Auckland 1142, New Zealand

Leona M. Fahey
School of Biological Sciences, The University of Auckland, Private Bag 92019, Auckland 1142, New Zealand

Philip J. Harris
School of Biological Sciences, The University of Auckland, Private Bag 92019, Auckland 1142, New Zealand

Shutian Li
Department of Botany, Osnabrück University, 49076 Osnabrück, Germany